Introduction to the
Principles and Practice of
Soil Science

Introduction to the
Principles and Practice of
Soil Science

R. E. WHITE
Professor of Soil Science, Massey University
Palmerston North, New Zealand

SECOND EDITION

BLACKWELL SCIENTIFIC PUBLICATIONS

OXFORD LONDON EDINBURGH

BOSTON PALO ALTO MELBOURNE

© 1979, 1987 by
Blackwell Scientific Publications
Editorial offices:
Osney Mead, Oxford OX2 0EL
8 John Street, London WC1N 2ES
23 Ainslie Place, Edinburgh EH3 6AJ
52 Beacon Street, Boston
    Massachusetts 02108, USA
667 Lytton Avenue, Palo Alto
    California 94301, USA
107 Barry Street, Carlton
    Victoria 3053, Australia

First published 1979
Reprinted 1981, 1983
Second edition 1987

Photoset by Enset (Photosetting),
Midsomer Norton, Bath, Avon and
printed and bound in Great Britain by
Butler & Tanner Ltd, Frome and London

DISTRIBUTORS

USA and Canada
    Blackwell Scientific Publications Inc
    PO Box 50009, Palo Alto
    California 94303

Australia
    Blackwell Scientific Publications
    (Australia) Pty Ltd
    107 Barry Street
    Carlton, Victoria 3053

British Library
Cataloguing in Publication Data

White, R. E.
    Introduction to the principles and
    practice of soil science. —2nd ed.
    1. Soils
    I. Title
    631.4   S591

ISBN 0-632-01606-X
ISBN 0-632-00052-X (pbk.)

Library of Congress
Cataloging-in-Publication Data

White, R. E.
    Introduction to the principles and
    practice of soil science.

    Bibliography: p.
    Includes index.
    1. Soil science.   I. Title.
    S591.W48 1986   631.4   86-23210

ISBN 0-632-01606-X
ISBN 0-632-00052-X (pbk.)

Contents

PART 3 · UTILIZATION OF SOIL

'And earth is so surely the food of all plants that with the proper share of the elements, which each species requires, I do not find but that any common earth will nourish any plant.'
Jethro Tull (1733) in
Horse-hoeing Husbandry,
republished by William Cobbett in 1829

Preface to the Second Edition

There is a growing need for information about soils, their behaviour and their influence on land use. The demand for food and other natural products is rising as the world's population continues to grow; and an increasing proportion of that population has an awakening expectation of an improved quality of life—better food, shelter and recreational areas in which to enjoy their leisure.

In the preface to the first edition of this book, I drew attention to the fundamental contribution that soil science has made to many of the applied sciences associated with management of the environment—either natural or man-made. This contribution is even more important today. Commodity surpluses in many developed Western countries have depressed world market prices to the extent that producers in the developing countries have had to adopt more exploitative farming techniques—reducing fertilizer inputs and deferring improvements such as irrigation, drainage and soil conservation measures—to reduce their costs of production. Soil scientists therefore need to find ways of producing food and fibre crops more efficiently, in terms of both yield per hectare and value of product relative to cost of inputs, while preserving the intrinsic value of the soil resource for posterity. Similarly, scientists whose brief is to maintain a healthy environment for people living in a highly technological society, need a better quantitative understanding of the fate of toxic chemicals (pesticides, industrial residues) and waste materials (domestic refuse, sewage sludge) in the soil, and their possible dissemination into the atmosphere or to surface and groundwaters.

A final point concerns the advent of a new 'buzzword'—biotechnology. Although biotechnology can mean different things to different people, there is a tendency to restrict the term to the use of 'microbes, or cells obtained from plants and animals' (Prentis, 1984), which can be cultured under controlled conditions in the laboratory or in relatively aseptic industrial plants. The phenomenal technology of the dairy cow producing milk from humble grass, or of photosynthesis in waving fields of wheat, appears to be excluded by this definition. But as Magnus Pyke comments in his foreword to Prentis' book—'biotechnology is . . . as old as the hills', implying that biotechnology has its origins in the advent of agriculture some 8,000 to 9,000 years ago. Certainly, the soil is a flourishing centre for microbiological activity where a prodigious number of diverse organisms play an essential part in many physicochemical and biological reactions. Thus, while soil science will continue to underpin the traditional land-based industries, it can also make an important contribution to any future biotechnological revolution.

R.E. White

Palmerston North, May 1986

REFERENCE

PRENTIS S. (1984) *Biotechnology—a New Industrial Revolution.* Orbis, London.

A note on using this book

All chapters of the book have been revised and extensive use is made of cross-referencing from one section to another. The reader is encouraged to make use of the Table of Contents and the Index for tracing information on specific items of interest.

Acknowledgements

I am greatly indebted to many people who helped in the preparation of this book: to Philip Beckett without whose inspiration and constant encouragement I would neither have begun nor completed the task; and to my colleagues Peter Nye, Frank Thompson, Roger Hall, Bill Handley and others farther afield—David Rowell, Grant Thomas, Doug Lathwell, Duncan Greenwood and T.E. Tomlinson, who criticized various chapters. Jeni Hughes, H.F. Woodward, John Baker, John Habgood, Olive Godwin, Rosetta Plummer, Cathy McIntosh and my wife Annette gave invaluable help with the illustrations, literature research and typing the manuscript, and Robert Campbell of Blackwell Scientific Publications can justly claim to represent a humane publishing house. Finally, I appreciate the patience and forbearance that my family have shown during the writing of this book.

I should like to thank those scientists, mentioned individually beneath the appropriate figures, who either sent me photographs to be used as illustrations or gave permission for their originals to be reproduced. Acknowledgement is also due to the publishers who gave permission for original material to be reproduced from the following publications:

Figure 1.2—from *Journal of the Australian Institute of Agricultural Science* (1976) **42**, 75–93.

Figure 2.4—from *Methods for Analysis of Irrigated Soils* (Ed.) J. Loveday, Tech. Comm. No. 54, Commonwealth Agricultural Bureaux, Farnham Royal.

Figures 2.6, 2.7—from *Environmental Chemistry* by J.W. and E.A. Moore, Academic Press, London.

Figure 3.4—from *Soil Microbiology* (Ed.) N. Walker, Butterworth, London.

Figure 3.6—from *Nature New Biology* (1973) **246**, 269–271, Macmillan (Journals), London.

Figure 3.7—from *Soil Conditions and Plant Growth* by E.W. Russell, Longman, London; with permission of the Director, Rothamsted Research Station.

Figure 3.8—from *Biology of Earthworms* by C.A. Edwards and J.R. Lofty, Chapman and Hall, London.

Figure 4.8—from *Fabric and Mineral Analysis of Soils* by R. Brewer, Wiley and Sons, London and Krieger Publishing Co., New York.

Figure 5.5—from *Agriculture and Water Quality* MAFF Tech. Bull. No. 32, HMSO, London.

Figures 11.7, 11.15—from *Journal of Soil Science* (1972) **23**, 363–380, Oxford University Press, Oxford.

Figure 12.4—from *Chemistry and Technology of Fertilizers* by A.V. Slack, Wiley and Sons, London.

Figure 12.6—from *Soil Fertility and Fertilizers* by S.L. Tisdale and W.L. Nelson, Macmillan Publishing, New York.

Figure 12.9—from *Pesticides and Human Welfare* (Eds.) D.L. Gunn and J.G.R. Stevens, Oxford University Press, Oxford.

Units of Measurement and Abbreviations used in this Book

SI units

Basic unit	Symbol
metre	m
kilogram	kg
second	s
ampere	A
kelvin	K
mole	mol

Derived and SI related units

Unit	Symbol	Value
newton	N	$kg\,m\,s^{-2}$
joule	J	$N\,m$
pascal	Pa	$N\,m^{-2}$
volt	V	$J\,A^{-1}\,s^{-1}$
siemen	S	$A\,V^{-1}$
coulomb	C	$A\,s$
hectare	ha	$10^4\,m^2$
litre	l or dm^3	$10^{-3}\,m^3$
tonne	t	$10^3\,kg$
bar	bar	$10^5\,Pa$
Faraday's constant	F	$96{,}500\,J\,mol^{-1}\,V^{-1}$

Non-SI units used in soil science

Physical quantity	Unit	Symbol	Value
length	Ångstrom	Å	$10^{-10}\,m$
temperature	degree Celsius	°C	$K{-}273$
concentration	moles per litre	M	$mol\,l^{-1}$
cation exchange capacity	milli-equivalent per 100 gram	me per 100 g	cmol charge (+) kg^{-1}

radioactivity	Curie	Ci	3.7×10^{10} disintegrations s^{-1}
electrical conductivity	millimho per cm	mmho cm^{-1}	dS m^{-1}

Prefixes to SI and derived SI units

kilo-	k	10^3
deca-	da	10^1
deci-	d	10^{-1}
centi-	c	10^{-2}
milli-	m	10^{-3}
micro-	μ	10^{-6}
nano-	n	10^{-9}
pico-	p	10^{-12}

ABBREVIATIONS USED IN THIS BOOK

AEC	Anion exchange capacity
API	Air photo interpretation
AR	Activity ratio
AWC	Available water capacity
BD	Bulk density
BP	Before present
c.	about, approximately
CEC	cation exchange capacity
concn.	concentration
CSIRO	Commonwealth Scientific and Industrial Research Organization
DDL	Diffuse double layer
DM	Dry matter
EC	Electrical conductivity
EI	Soil erodibility index

Abbreviations used in this book *continued*

ESP	Exchangeable sodium percentage
ET	Evapotranspiration
FAO	Food and Agriculture Organization
FC	Field capacity
h	hour
IAEA	International Atomic Energy Agency
IEP	Isoelectric point
IHP	Inner Helmholtz plane
IR	Infiltration rate
KE	Kinetic energy
LF	Leaching fraction
LHS	Left-hand-side
LR	Leaching requirement
MAFF	Ministry of Agriculture, Fisheries and Food
MW	Molecular (formula) weight
o.d.	oven-dry
OHP	Outer Helmholtz plane
ppm	parts per million
PSR	Pore space ratio
PWP	Permanent wilting point
PZC	Point of zero charge
RHS	Right-hand-side
RQ	Respiratory quotient
SAR	Sodium adsorption ratio
SG	Specific gravity
SI	Saturation Index
SMD	Soil moisture deficit
soln	solution
SPAC	Soil–plant–atmosphere continuum
sp., spp.	species (singular; plural)
SSEW	Soil Survey of England and Wales
TDS	Total dissolved solids
UK	United Kingdom
UNESCO	United Nations Education, Scientific and Cultural Organization
USA	United States of America
USDA	United States Department of Agriculture
USLE	Universal soil loss equation
USSR	Union of Soviet Socialist Republics
y	year

Some chemical and miscellaneous symbols

2,4-D	2,4-dichlorophenoxyacetic acid
2,4,5-T	2,4,5-trichlorophenoxy acetic acid
DAP	Diammonium phosphate
DCP	Dicalcium phosphate
DCPD	Dicalcium phosphate dihydrate
DDT	Dichlorodiphenyltrichloroethane
GRP	Ground rock phosphate
HA	Hydroxyapatite
IAA	Indolacetic acid
KCP	Potassium calcium pyrophosphate
KMP	Potassium metaphosphate
MAP	Monoammonium phosphate
MCP	Monocalcium phosphate
MCPA	(4-chloro-*o*-tolyl) oxyacetic acid
OCP	Octacalcium phosphate
PAM	Polyacrylamide
PEG	Polyethylene glycol
PVA	Polyvinyl alcohol
PVAc	Polyvinylacetate
TSP	Triple superphosphate
( )	denotes activity
[ ]	denotes concentration
$\simeq$	approximately equal to
$\sim$	of the order of
$<$	less than
$>$	greater than
$\leq$	less than or equal to
ln	$\log_e$

Part 1
The Soil Habitat

'The soil is teeming with life. It is a world of darkness, of caverns, tunnels and crevices, inhabited by a bizarre assortment of living creatures . . .'
J.A. Wallwork (1975) in
The Distribution and Diversity of Soil Fauna

Chapter 1
Introduction to
the Soil

1.1 SOIL IN THE MAKING

With the exposure of rock to a new environment—by the extrusion and solidification of lava, the uplift of sediments, the recession of water or the retreat of a glacier—a soil begins to form. Decomposition proceeds inexorably towards decreased free energy and increased entropy. The *free energy* of a closed system, such as a rock fragment, is that portion of its total energy that is available for work, other than work done in expanding its volume. Part of the energy released in a spontaneous reaction, such as rock weathering, appears as entropic energy, and entropy is a measure of the degree of disorder of the system. For example, as the rock weathers, crystalline minerals are broken down into simple molecules and ions, some of which are leached out by water or escape as gases.

Weathering is hastened by the appearance of primitive forms of plant life on the rock surfaces. These plants—lichens, mosses and liverworts—can store radiant energy from the sun as chemical energy in the products of photosynthesis. Lichens, which are a symbiotic association of an alga and fungus, are also able to 'fix' atmospheric nitrogen and incorporate it into plant protein. On the death of each generation of these primitive plants, some of the rock elements and a variety of complex organic molecules are returned to the weathering surface where they nourish the

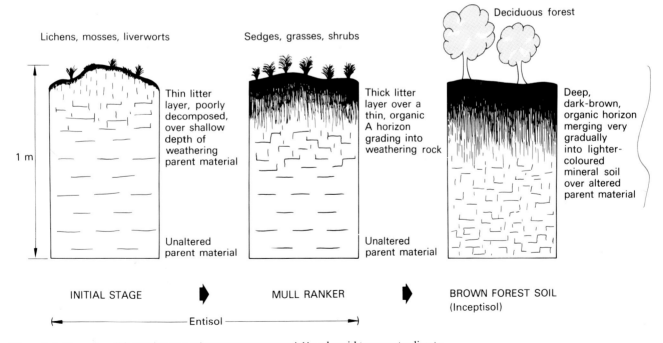

Figure 1.1. Stages in soil formation on a calcareous parent material in a humid temperate climate.

succession of organisms gradually colonizing the embryonic soil.

A very simple example is that of soil formation under the extensive deciduous forests of the cool humid areas of Europe, Asia and North America, on calcareous deposits exposed by the retreat of the Pleistocene ice cap. The profile development is summarized in Figure 1.1. The initial state is little more than a thin layer of weathered material stabilized by primitive plants. Within one to two centuries, as the organo-mineral material accumulates, more advanced species of sedge and grass, which are adapted to the harsh habitat, appear, and the developing soil is described as a *mull ranker* (Entisol).* The dead plant remains foster the growth of pioneering microorganisms and animals, which gradually increase in abundance and variety. The litter deposited on the surface is mixed into the soil by burrowing animals and insects, where its decomposition is hastened. The eventual appearance of larger plants—shrubs and trees—with their deeper rooting habit, pushes the zone of rock weathering farther below the soil surface. After a thousand years and more, a *brown forest soil* (Inceptisol) emerges. We shall return to the topic of soil formation, and the processes involved in Chapters 5 and 9.

1.2   CONCEPTS OF SOIL

The soil is at the interface between the *atmosphere* and *lithosphere* (the mantle of rocks making up the Earth's crust). It also has an interface with bodies of fresh and salt water (collectively called the *hydrosphere*). The soil sustains the growth of many plants and animals, and so forms part of the *biosphere*.

There is little merit in attempting to give a rigorous

---

*Because this book is not primarily concerned with soil classification, descriptive Great Soil Group names (see section 5.3) will be used instead of the more abstruse class names derived from the United States' Comprehensive System (Soil Survey Staff, 1975), which is discussed in Chapter 14. Nevertheless, the order or suborder in which a soil would probably be classified in the US System is given throughout the book in parentheses.

definition of soil because of the complexity of its make-up and of the physical, chemical and biological forces to which it is exposed. Nor is it necessary to do so, for the soil means different things to different users. To the geologist and engineer, the soil is no more than finely divided rock material. The hydrologist sees the soil as a storage reservoir affecting the water balance of the catchment he is studying, while the ecologist may be interested in only those soil properties that influence the growth and distribution of different plant species. The farmer is naturally concerned about the multiplicity of ways in which the soil can influence the growth of his crops and the health of his livestock, although frequently his interest will not extend below the depth of soil which is disturbed by the plough (20–25 cm).

In view of this wide spectrum of potential user need, it is appropriate when introducing the topic of soil to readers, perhaps for the first time, to review briefly the evolution of man's relationship with the soil and identify some of the past and present concepts of soil.

Soil as a medium for plant growth

Man's exploitation of the soil for food production began some two or three thousand years after the close of the last Pleistocene ice age, which occurred about 11,000 years BP (before present). Neolithic man and his primitive agriculture spread outwards from settlements in the fertile crescent embracing the ancient lands of Mesopotamia and Canaan (Figure 1.2) and reached as far as China and the Americas within a few thousand years. In China, for example, the earliest records of soil survey (4,000 BP) show how the fertility of the soil was used as a basis for levying taxes on the incumbent landholders. The study of the soil was a practical exercise of everyday life, and the knowledge of soil husbandry that had been acquired by Roman times was faithfully passed on by peasants and landlords, with little innovation, until the early 18th century.

From that time onwards, however, the rise in demand for agricultural products in Europe was dramatic. Conditions of comparative peace, and rising

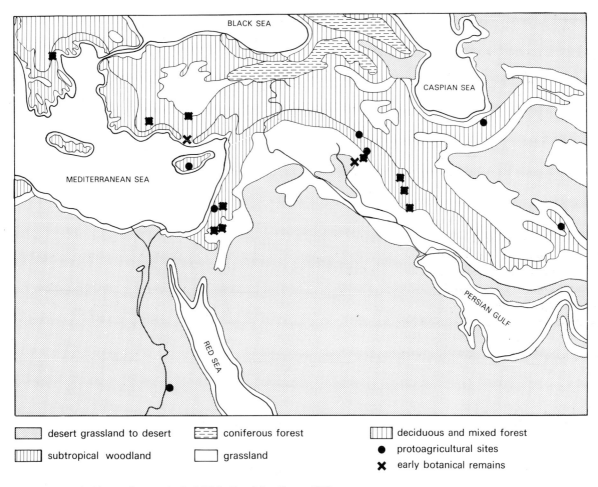

▭ desert grassland to desert	▭ coniferous forest
▭ subtropical woodland	▭ grassland

▭ deciduous and mixed forest
● protoagricultural sites
✖ early botanical remains

Figure 1.2. Sites of primitive settlements in the Middle East (after Gates, 1976).

living standards as a result of the Industrial Revolution, further stimulated this demand throughout the 19th century. The period was also one of great discoveries in the sciences of physics and chemistry, the implications of which sometimes burst with shattering effect on the conservative world of agriculture. In 1840, von Liebig firmly debunked the 'humus theory' of plant nutrition, and established that plants absorbed nutrients as inorganic compounds from the soil. In the 1850s Way discovered the process of cation exchange in soils; and during the years from 1860 to 1890 eminent

bacteriologists including Pasteur, Warington and Winogradsky elucidated the role of microorganisms in the decomposition of plant residues and the conversion of ammonia to nitrate.

Over the same period, botanists such as von Sachs and Knop, by careful experimentation in water culture, identified the major elements which were essential for healthy plant growth. Agricultural chemists drew up balance sheets of the quantities of these elements taken up by crops and, by inference, the quantities which had to be returned to the soil in fertilizers or animal manure to sustain growth. This approach, whereby the soil is regarded as a relatively inert medium providing water, mineral* ions and physical support for plant growth, has been called the 'nutrient bin' concept.

Soil and the influence of geology

The pioneering chemists who investigated the soil's ability to supply nutrients to plants tended to see the soil merely as a medium in which chemical and bio-chemical reactions occurred. They little appreciated the soil as part of a landscape, moulded by natural forces acting on the earth's crust. Great contributions were made to our knowledge of the soil *in situ* by the application of the field survey methods of the new science of geology in the late 19th century. Geologists defined the mantle of loose, weathered material on the Earth's surface as the *regolith*, of which only the upper few decimetres could be called soil, since this was the usual extent to which organic material was incor-porated (Figure 1.3). Below the soil was the subsoil which was largely devoid of organic matter. However, the mineral matter of both soil and subsoil was recognized as being derived from the weathering of underlying rock. This led to intensive investigations of

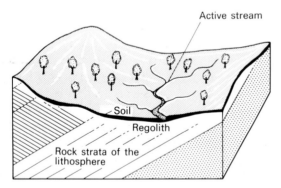

Figure 1.3. Soil development in relation to the landscape and under-lying regolith.

*The term 'mineral' is used in two contexts: first, as an adjective referring to the inorganic constituents of the soil (ions, salts and particulate matter); second, as a noun referring to specific inorganic compounds found in rocks and soil, such as quartz and feldspars.

the relationship between various soils of the landscape and the geology of the underlying rocks, with the result that soils were loosely classified in geological terms, such as loessial, glacial, alluvial, granitic or marly (derived from limestone).

The influence of Russian soil science

Russia at the end of the 19th century was a vast country covering many climatic zones in which, on the whole, crop production was limited not by infertility of the soil, but by the primitive methods of agriculture. Russian soil science was therefore concerned not so much with soil fertility, but with observing soils in the field, and studying relationships between the proper-ties of a soil and the environment in which it had formed. From 1870 onwards, Dokuchaev and his school emphasized the distinctive features of a soil that developed gradually and distinguished it from the undifferentiated weathering rock or *parent material* below. This was the beginning of the science of *pedology*.*

Following the Russian lead, scientists who had pre-viously studied only small samples of soil in the laboratory or who were interested in the soil only as a medium for plant growth, began to appreciate that factors, such as climate, parent material, vegetation, relief and time, interacted in many ways to produce an almost infinite variety of soil types. For any particular combination of these *soil-forming factors* (see Chapter 5), a unique physicochemical and biological environ-ment was established that led to the development of a distinctive soil body—the process of *pedogenesis*.

Fundamental features of the soil were identified and described by new terms. The vertical dimension, exposed by excavation from the surface to the parent material, constitutes the *soil profile*. Layers in the soil, distinguished on the basis of colour, hardness, the occurrence of included structures and other visible or tangible properties, are called *horizons*. The upper layer, from which materials are generally washed downwards, is described as *eluvial*; lower layers in which these materials can accumulate are called

*From the Greek word for ground or earth.

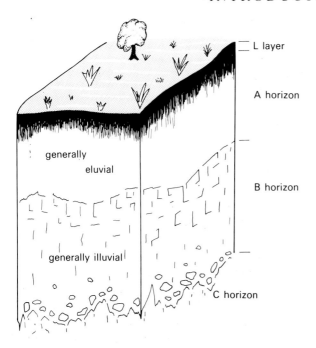

Figure 1.4. Diagram of a soil profile.

illuvial. In 1932, an international gathering of soil scientists adopted the notation of A and B for the eluvial and illuvial horizons, and C horizon for the parent material (Figure 1.4). The A and B horizons comprise the *solum*. If unweathered rock exists below the parent material it is labelled *bedrock*, R. Organic litter on the surface, which is not incorporated in the soil, is designated the L layer.

Soil genesis is now known to be much more complex than this early work suggested. It is acknowledged, for example, that many soils are *polygenetic* in origin; that is, they have undergone successive phases of development due to changes in climate and other environmental factors with time. Nevertheless, the Russian approach was a considerable advance on traditional thinking, and the recognition of the intimate relationship between soil and the environment encouraged soil scientists in their efforts to survey and map the distribution of soils. The wide range of soil morphology that was described was in turn

a stimulus to studies of pedogenesis, an understanding of which, it was believed, would enable the mass of field data on soils to be collated in a systematic manner. Thus, Russian soil science provided the inspiration for many of the early attempts at *soil classification*.

A contemporary view of soil
Between the two World Wars of the 20th century, the philosophy of the soil as a 'nutrient bin' was much in evidence, particularly in the Western World. More and more land was brought into cultivation in areas marginal for crop production because of limitations of climate, soil and topography. With the balance between crop success and failure rendered even more precarious than in favoured areas, the age-old problems of wind and water erosion, the encroachment of weeds and the accumulation of salts in irrigated lands became more serious. In the period since 1945, escalating demand for food, fibre and forest products by a world population (now in excess of 4,000 million) has led to increased use of fertilizers to improve crop yields and pesticides to control pests and diseases. Such practices inevitably resulted in a steady rise of fertilizer residues (particularly nitrates) in surface and underground waters, as well as the widespread dispersal of the more stable pesticides throughout the biosphere and their accumulation to potentially toxic levels in some species of birds and fish.

More recently, however, soil scientists, producers and planners have begun to acknowledge the need for a compromise between maximum crop production and the conservation of a valuable natural resource and man's environment. The trend now is towards preserving the soil's natural state by causing minimum disturbance when crops are grown, matching fertilizer additions more closely to crop demand in order to reduce losses, introducing legumes that can fix nitrogen from the air and returning plant residues and animal waste to the soil to supply some of the crop's nutrient requirements. In short, more emphasis is being placed on the soil as a natural body, clearly distinguished from the inert rock material by:
(1) the presence of plant and animal life;

(2) a structural organization that reflects the action of pedogenic processes;

(3) a capacity to respond to environmental change which may alter the balance between gains and losses within the profile, and predetermine the formation of a different kind of soil in equilibrium with the new set of environmental conditions.

Point (3) indicates that the soil has no fixed inheritance, because it depends on the conditions prevailing during its formation. Nor is it possible to define unambiguously the boundaries of the soil body: the soil atmosphere is continuous with the air above ground; many of the soil organisms can live as well on the surface as within the soil; the litter layer usually merges gradually with the decomposed organic matter of the soil and likewise. The boundary between soil and parent material is difficult to demarcate. It is common, therefore, to speak of the soil as a *three-dimensional body that is continuously variable in time and space.*

1.3 COMPONENTS OF THE SOIL

As we have seen, a combination of physical, chemical and biotic forces acts on organic and weathered rock fragments to produce a soil with a porous fabric that can retain water and gases. The water contains dissolved organic and inorganic solutes and is called the *soil solution.* While the soil air consists primarily of nitrogen and oxygen, it usually contains higher concentrations of carbon dioxide than the atmosphere, and traces of other gases that are byproducts of microbial metabolism. The relative proportions of the four major components—*mineral matter, organic matter, water* and *air*—may vary widely, but generally lie within the ranges indicated in Figure 1.5. These components are discussed in more detail in the subsequent chapters of Part 1.

1.4 SUMMARY

Soil forms at the interface between the atmosphere and the consolidated or loose deposits of the Earth's crust. Physical and chemical weathering, denudation and redeposition combined with the activities of a succession of colonizing plants and animals moulds a distinctive soil body from the milieu of rock minerals in the parent material. The process of soil formation, called *pedogenesis,* culminates in the remarkably variable differentiation of the soil material into a series of horizons that constitute the *soil profile.* The horizons are distinguished, in the field at least, by their visible and tangible properties such as colour, hardness and structural organization. The intimate mixing of mineral and organic materials to form a porous fabric that is permeated by water and air creates a favourable habitat for a variety of plant and animal life.

REFERENCES

Gates C.T. (1976) China in a world setting: agricultural response to climatic change. *Journal of the Australian Institute of Agricultural Science* **42**, 75–93.

FURTHER READING

Epstein E. (1972) *Mineral Nutrition of Plants: Principles and Perspectives.* Wiley, New York.
Jenny H. (1980) *The Soil Resource—Origin and Behaviour.* Springer-Verlag, New York.

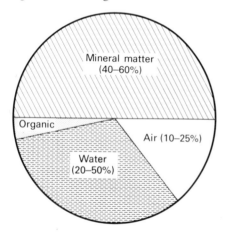

Figure 1.5. Proportions of the main soil components by volume.

KELLOGG C.E. (1950) Soil. *Scientific American* **821**, 2–11.
RUSSELL E.J. (1957) *The World of Soil.* Collins, London.
RUSSELL E.W. (1973) *Soil Conditions and Plant Growth*, 10th Ed. Longmans, London.
SIMONSON R.W. (1968) Concepts of Soil. *Advances in Agronomy* **20**, 1–47.

SOIL SURVEY STAFF (1975) *Soil Taxonomy. A basic system of soil classification for making and interpreting soil surveys.* Agriculture Handbook No. 436. United States Department of Agriculture, Washington DC.
UNITED STATES DEPARTMENT OF AGRICULTURE YEARBOOK (1938) *Soils and men.* US Government Printing Office, Washington DC.

Chapter 2
The Mineral Component
of the Soil

2.1 THE SIZE RANGE

Mineral fragments in soil vary enormously in size from boulders and stones down to sand grains and minute particles that are beyond the resolving power of an optical microscope (< 0.2 μm in length or diameter). Particles smaller than *c*. 1 μm are classed as *colloidal*. If they do not settle quickly when mixed with water, they are said to form a colloidal solution or *sol*; if they settle within a few hours they are said to form a suspension. Colloidal solutions are distinguished from true solutions (dispersions of ions and molecules) by the *Tyndall effect*. This occurs when the path of a beam of light passing through the solution can be seen from either side at right angles to the beam, indicating a scattering of the light rays.

An arbitrary division is made by size-grading soil into material that passes through a sieve with 2-mm diameter holes, the *fine earth*, and that which is retained on the sieve, the *stones* or *gravel*. Several categories of stone between 2 and 600 mm in diameter are described: fragments > 600 mm are called boulders. The separation by sieving is carried out on air-dry soil which has been gently ground by mortar and pestle, or crushed between wooden rollers, to break up the obvious *aggregates* of smaller particles. Air-dry soil is soil allowed to dry in air at ambient temperatures.

Particle-size distribution of the fine earth

The distribution of particle sizes determines the soil *texture*, which may be assessed subjectively in the field or more rigorously by particle-size analysis in the laboratory.

SIZE CLASSES

In any natural material, such as soil, one expects a

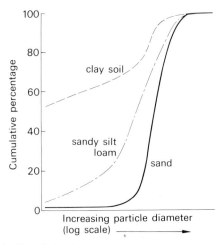

Figure 2.1. Cumulative frequency distributions of soil particle sizes in a clay, sandy silt loam and sandy soil.

continuous spectrum of particle sizes—this is called a frequency distribution, wherein the number of particles of any given size is plotted against the actual size. Some examples of cumulative frequency distributions of soil-particle sizes are given in Figure 2.1. In practice, however, it is convenient to subdivide the continuous distribution into several class intervals that define the size limits of the *sand*, *silt* and *clay* fractions. The extent of this subdivision, and the class limits chosen, vary from country to country and between institutions within countries. The major systems in use today are those adopted by the Soil Survey Staff of the USDA, the British Standards Institution and the International Society of Soil Science. These are illustrated in Figure 2.2. All three systems set the upper limit for clay at 2 μm, but differ in the upper limit chosen for silt and the way in which the sand fraction is subdivided.

FIELD TEXTURE

A soil surveyor determines the texture of a soil by moistening a sample until it glistens: it is then kneaded

International or Atterberg system

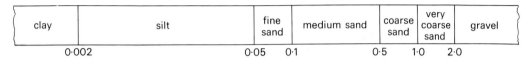

U.S.D.A. system

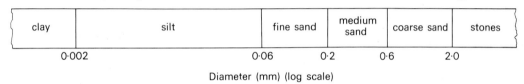

Soil Survey of England and Wales, British Standards and Mass. Institute of Technology

Diameter (mm) (log scale)

Figure 2.2. Particle-size classes most widely adopted internationally.

between fingers and thumb until the aggregates are broken down and the soil grains thoroughly wetted. The proportions of sand, silt and clay are estimated according to the following qualitative criteria.

Coarse sand grains are large enough to grate against each other and can be detected individually by sight and feel.

Fine sand grains are much less obvious, but when they comprise more than about 10 per cent of the sample they can be detected by feel.

Silt grains cannot be detected by feel, but their presence makes the soil feel smooth and soapy and only very slightly sticky.

Clay is characteristically sticky, although some dry clays require much moistening and kneading before they develop their maximum stickiness.

High organic matter contents tend to reduce the stickiness of clayey soils and to make sandy soils feel more silty. Finely divided calcium carbonate also gives a silt-like feeling to the soil.

The soil can be assigned to a *textural class* according to the triangular diagram of Figure 2.3, which is used by the Soil Survey of England and Wales and is based on the British Standards system of particle-size grading (Figure 2.2). In this system, all the possible combinations of sand, silt and clay are condensed into 11 major textural classes. The field assessment of texture can be checked by carrying out a complete analysis of the particle-size distribution by physical means.

Particle-size analysis in the laboratory

The success of the method relies upon the complete disruption of soil aggregates and the addition of chemicals which ensure the dispersion of the soil colloids in water. Full details of the methods employed are given in standard texts, for example, Black (1965)

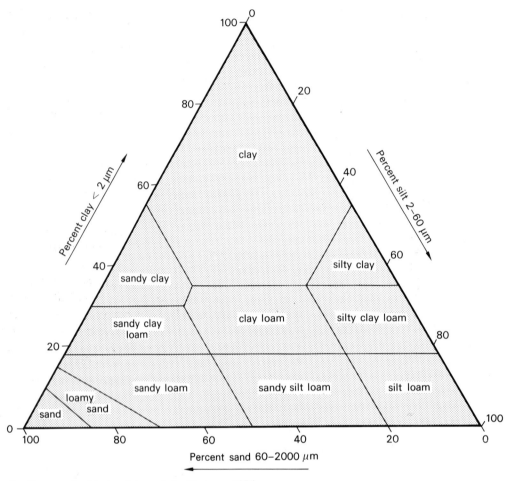

Fig. 2.3. Triangular diagram of soil textural classes (after Hodgson, 1974).

and Loveday (1974). The course sand particles are separated by sieving; fine sand, silt and clay are separated by making use of the differences in their settling velocities in the suspension. The principle of the latter technique is as follows.

A rigid particle falling freely through a liquid of a lower density will attain a constant velocity when the force opposing movement is equal and opposite to the force of gravity acting on the particle. The frictional force acting vertically upward on a spherical particle is calculated from Stoke's law. The net gravitational force acting downwards is equal to the weight of the submerged particle. At equilibrium, these expressions can be combined to give an equation for the terminal settling velocity, v, as

$$v = \frac{2}{9}\frac{g}{\eta}(\rho_p - \rho)r^2 \qquad (2.1)$$

where g = acceleration due to gravity
   η = coefficient of viscosity of the liquid (water)
   ρ_p = density of the particle
   ρ = density of water
   r = radius of the particle

When all the constants in this equation are collected

into one term A, we derive the simple relationship

$$v = \frac{h}{t} = Ar^2 \qquad (2.2)$$

where h is the depth, measured from the liquid surface, below which all particles of radius, r, will have fallen in time, t. To illustrate the use of equation 2.2, we may note that all particles $> 2\ \mu m$ in diameter settling in a suspension at 20°C will have fallen below a depth of 10 cm in 8 hours. Thus, by sampling the mass per unit volume of the suspension at this depth after 8 hours, the amount of clay present can be calculated. The clay

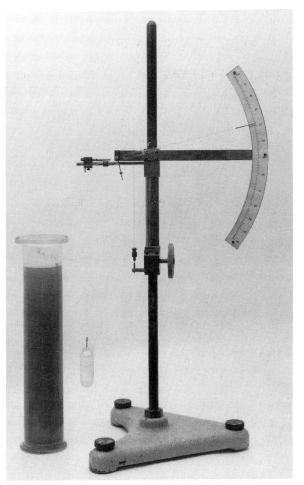

Figure 2.4. A settling soil suspension and plummet balance (courtesy of J. Loveday).

content is normally measured with a Bouyoucos hydro-meter placed in the suspension, or calculated from the loss of weight of a bulb of known volume immersed in the suspension—the *plummet balance* method, illustrated in Figure 2.4.

Great accuracy in measuring the mass per unit volume of the suspension is not warranted because of the simplifying assumptions made in the calculation of settling velocity. In particular, it should be noted that:
(a) the clay and silt particles are not smooth spheres, they have irregular plate-like shapes;
(b) the density of the particle varies with its mineralogy (see section 2.3).
In practice, we take an average value for ρ_p of 2.65, and speak of the 'equivalent spherical diameter' of the particle size being measured.

The result of the analysis is expressed as the mass of the individual fractions per 100 g of oven-dry soil (fine earth only). Oven-dry soil is soil dried to a constant weight at 105°C. By combining the coarse and fine sand fractions, the soil may be represented by one point on the triangular diagram of Figure 2.3. Alternatively, by the stepwise addition of particle-size percentages, graphs of cumulative percentage against particle diameter of the kind shown in Figure 2.1 are obtained.

2.2 THE IMPORTANCE OF SOIL TEXTURE

Soil scientists are primarily interested in the texture of the fine earth of the soil. Nevertheless, in some soils the size and abundance of the stones cannot be ignored, as they can have a marked influence on the soil's suit-ability for agriculture. As the stone content increases, the soil will hold less water than a stoneless soil of the same fine-earth texture, so that crops become more susceptible to drought. Conversely, the drainage of such soils may be better under wet conditions. Stoni-ness also determines the ease, and to some extent the cost of cultivation, as well as the abrasive effect of the soil on tillage implements.

Texture is one of the most stable of properties and

is a useful index of several other properties that
determine the soil's agricultural potential. The fine and
medium-textured soils, such as the clays, clay loams,
silty clay loams and sandy silt loams, are generally the
more desirable because of their superior retention of
nutrients and water. Conversely, where high rates of
infiltration and good drainage are required, as in
irrigation schemes, sandy or coarse-textured soils are
preferred. In farming parlance, clay soils are described
as 'heavy' and sandy soils as 'light': this does not refer
to their mass per unit volume, which may be greater for
a sandy soil than a clay, but to the power required to
draw implements through the soil.

Texture has a pronounced effect on soil *temperature*.
Clays hold more water than sandy soils, and the
presence of water considerably modifies the heat
requirements of the soil: first, because its specific heat
capacity is 3–4 times that of the soil solids, and second,
because of the quantity of latent heat either absorbed
or evolved during a change in the physical state of
water, for example, from ice to liquid. Thus it is found
that the temperature of wet clay soils responds more
slowly than that of sandy soils to changes in air
temperature in spring and autumn.

Texture should not be confused with *tilth*, of which it
has been said that a good farmer can recognize it with
his boot, but no soil scientist can describe it (Russell,
1971). Tilth refers to the condition of the surface of a
ploughed soil: how sticky it is when wet and how hard it
sets when dry. Tilth therefore depends on the soil's
consistency (section 13.4) which reflects both texture
and structure. The action of frost in cold climates can
break down the massive clods left on the surface of a
heavy clay soil after autumn ploughing, producing
a mellow 'frost tilth' of numerous small granules
(section 4.2).

2.3 MINERALOGY OF THE SAND AND SILT FRACTIONS

Simple crystalline structures
Sand and silt consist almost entirely of the resistant

residues of *primary* rock minerals, although small
amounts of secondary mineral (salts, oxides and
hydroxides) also occur. The primary rock minerals are
predominantly *silicates*, which have a crystalline
structure based upon a simple unit—the *silicon tetrahedron*, SiO_4^{4-} (Figure 2.5). An electrically neutral
crystal is formed when cations, such as Al^{3+}, Fe^{3+},
Fe^{2+}, Ca^{2+}, Mg^{2+}, K^+ and Na^+, become covalently
bonded to the oxygen atoms. The surplus valencies of
the oxygens in the SiO_4^{4-} group are then satisfied.

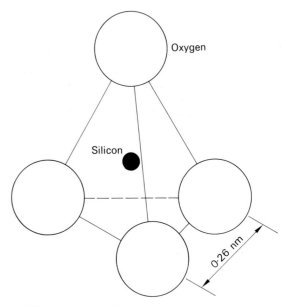

Figure 2.5. Diagram of a silicon tetrahedron (interatomic distances
not to scale).

Coordination number
The packing of the oxygen atoms, which are by far the
largest of the more abundant elements in the silicates,
determines the crystalline dimensions. In quartz, for
example, oxygen occupies 98.7 per cent and silicon
only 1.3 per cent of the mineral volume. The size ratio
of Si to O is such that four oxygens can be packed
around one silicon. Larger cations can accommodate
more oxygen atoms, the ratio of the ionic radius of the
cation to the ionic radius of oxygen being the determinant of the *coordination number* of the cation (Table

2.1). Many of the common metal cations have radius ratios between 0.4 and 0.7 which means that an octahedral arrangement of six oxygen atoms around the cation (coordination number 6) is preferred. The larger alkali and alkaline earth cations have radius ratios > 0.7 and therefore form complexes of co-ordination number 8 or greater. Aluminium, which has a cation/oxygen radius ratio close to the maximum for coordination number 4 and the minimum for coordination number 6, can exist in either fourfold (IV) or sixfold (VI) coordination.

Table 2.1. Cation/oxygen radius ratios and coordination numbers for common elements in the silicate minerals.

Radius of cation : radius of oxygen							
Coordination No. 3		Coordination No. 4		Coordination No. 6		Coordination No. 8 and greater	
B	0.15	Be	0.26	Al	0.43	Na	0.74
				Ti	0.48	Ca	0.80
		Si	0.30	Fe^{3+}	0.51		
		Mn^{4+}	0.39	Mg	0.59	Sr	0.96
				Fe^{2+}	0.63		
				Zr^{4+}	0.69	K	1.01
				Mn^{2+}	0.69	Ba	1.08

(After Marshall, 1964.)

Isomorphous substitution
Elements of the same valency and coordination number frequently substitute for one another in a silicate structure—a process called *isomorphous substitution*. The structure remains electrically neutral. However, when elements of the same coordination number but different valency are exchanged, there is an imbalance of charge. The most common substitutions are Mg^{2+}, Fe^{2+} or Fe^{3+} for Al^{3+} in octahedral coordination, and Al^{3+} for Si^{4+} in tetrahedral coordination. The excess negative charge is neutralized by the incorporation of additional cations, such as K^+, Na^+, Mg^{2+} or Ca^{2+} into the crystal lattice, or by structural arrangements that allow an internal compensation of charge (see chlorites and smectites).

More complex crystalline structures

CHAIN STRUCTURES
These are represented by the *pyroxene* and *amphibole* groups of minerals, which collectively make up the ferromagnesian minerals. In the pyroxenes, each silicon tetrahedron is linked to adjacent tetrahedra by the sharing of two out of three basal oxygen atoms to form a single extended chain (Figure 2.6). In the

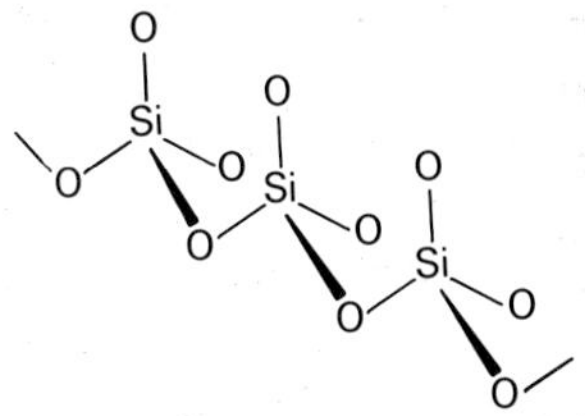

Figure 2.6. Single chain structure of silicon tetrahedra (after Moore and Moore, 1976).

amphiboles, a double parallel chain structure is formed when all three of the basal oxygen atoms of a silicon tetrahedron are shared with adjacent tetrahedra (Figure 2.7).

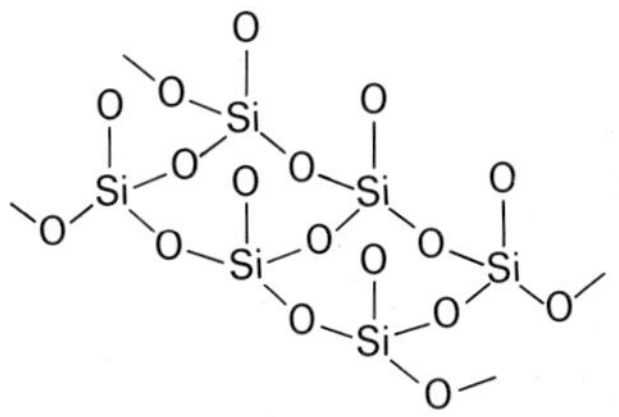

Figure 2.7. Double-chain structure of silicon tetrahedra (after Moore and Moore, 1976).

SHEET STRUCTURES
It is easy to visualize an essentially one-dimensional chain structure being extended in two dimensions to form a sheet of silicon tetrahedra linked by sharing the basal oxygen atoms. When seen in plan view (projection along the crystallographic c axis), the bases of the

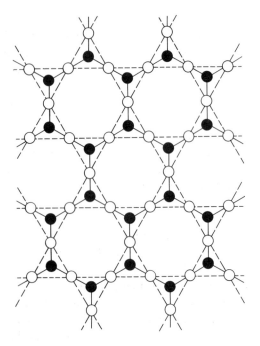

Figure 2.8. A silica sheet in plan view showing the pattern of hexagonal holes (after Fitzpatrick, 1971).

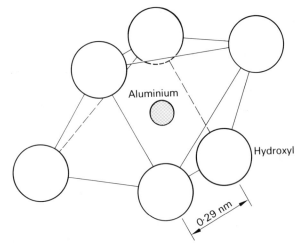

Figure 2.9. Diagram of an aluminium octahedron.

linked tetrahedra form a network of hexagonal holes (Figure 2.8). The apical O atoms (shown as the black spheres in Figure 2.8) can form ionic–covalent bonds with other metal cations by, for example, displacing OH groups from their coordination positions around a trivalent aluminium ion. (Because the unhydrated proton is so small, OH occupies virtually the same space as O.) When aluminium octahedral units (shown diagrammatically in Figure 2.9) are linked by the sharing of edges, they form an *alumina* sheet. Aluminium atoms normally occupy only two-thirds of the available octahedral positions—this is a *dioctahedral* structure, characteristic of the mineral *gibbsite*, $[Al_2OH_6]_n$. If Mg is present instead of Al, however, all the available octahedral positions are filled—this is a *trioctahedral* structure, characteristic of the mineral *brucite* $[Mg_3(OH)_6]_n$.

The bonding together of silica and alumina sheets through the apical O atoms of the silicon tetrahedra causes the bases of the tetrahedra to twist slightly so

that the cavities in the sheet become trigonal rather than hexagonal in shape. When *two* silica sheets sandwich *one* alumina sheet, the result is a 2:1 layer lattice (or phyllosilicate) structure characteristic of the *micas*, *chlorites* and many soil *clay minerals* (section 2.4). Each 2:1 layer or *lamella* is a recognizable crystalline

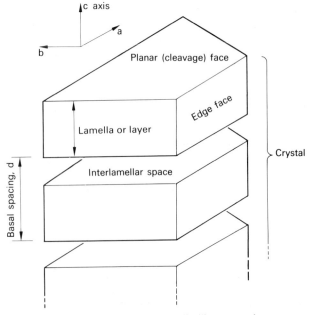

Figure 2.10. General structure of a phyllosilicate crystal.

entity which may be stacked regularly in the *c* direction to form *crystals* (Figure 2.10).

Additional structural complexity is introduced by isomorphous substitution. In the dioctahedral mica *muscovite*, for example, one-quarter of the tetrahedral Si is replaced by Al resulting in a net 2 moles of negative charge* per unit cell, which has the composition:

$$[(OH)_4(Al_2Si_6)^{IV} Al_4^{VI} O_{20}]^{2-} 2K^+$$

Note that the negative charge is neutralized by K^+ ions held in the spaces formed by the juxtaposition of the trigonal cavities of adjacent silica sheets (Figure 2.11). Because the isomorphous substitution in muscovite occurs in the tetrahedral sheet, the negative charge is distributed over only three surface O atoms. The magnitude of the layer charge and its localization are sufficient to cause cations of relatively low ionic potential†, such as K^+, to lose their water of hydration. The unhydrated K^+ ions have a diameter comparable to that of the ditrigonal cavity formed between opposing siloxane surfaces and their presence provides very strong bonding between the layers. Such complexes where an unhydrated ion forms an ionic–covalent bond with atoms of the mineral surface are called *inner-sphere* complexes.

Biotite is a trioctahedral mica which has Al substituted for Si in the tetrahedral sheet and Fe^{2+} for Mg^{2+} in the octahedral sheet. Again, because of the localization of charge in the tetrahedral sheet, unhydrated K^+ ions are retained in the interlayer spaces to give a unit cell formula of

$$[(OH)_4(Al_2Si_6)^{IV} (MgFe)_6^{VI} O_{20}]^{2-} 2K^+$$

The most complex of the two-dimensional structures belongs to the *chlorites*, which have a brucite layer sandwiched between two mica layers. In the type-mineral *chlorite*, the negative charge of the two biotite layers is neutralized by a positive charge in the brucite, developed due to the replacement of two-thirds of the Mg atoms by Al (Figure 2.12). Chlorite is an example of a regular mixed layer mineral. Predictably, the bonding between lamellae is strong, but the high content of Mg renders this mineral susceptible to dissolution in acidic solutions. For the same reason, and also because of its ferrous iron content, biotite is much less stable than muscovite.

Three-dimensional structures

The most important silicates in this group are *silica* and the *feldspars*. Silica minerals consist entirely of polymerized silicon tetrahedra of general composition $(SiO_2)_n$. Silica occurs as the residual mineral *quartz*, which is very inert, and as a secondary mineral precipitated after the hydrolysis of more complex silicates, which when metamorphosed appears as *flint* or *chert*. It also occurs as amorphous or microcrystalline silica of biological origin, the fossilized remains of the diatom, a minute aquatic organism, being almost pure silica. Silica absorbed by terrestial plants (grasses and hardwood trees in particular) is also returned to the soil as microscopic, opaline structures called *phytoliths*, or as amorphous material in the ash.

Quartz or flint fragments of greater than colloidal size are very insoluble, and hence are abundant in the sand and silt fractions of many soils. The *feldspars*, on the other hand, are chemically more reactive and are rarely of significance in the sand fraction of mature soils. Their structure consists of a three-dimensional framework of polymerized silicon tetrahedra in which some of the Si is replaced by Al. The cations balancing the excess negative charge are all of high coordination number, such as K^+, Na^+ and Ca^{2+}, and less commonly, Ba^{2+} and Sr^{2+}. The range of composition encountered is shown in Figure 2.13. Details of the structure, composition and chemical stability of the feldspars are given in specialist texts by Marshall (1964) and Loughnan (1969).

---

*One mole (mol) of charge is the charge of an electron (or proton) × Avogadro's constant (6.0×10^{23}); it is the same as an equivalent of charge which is the charge carried by 1 g of H^+ ions.

†Determined by the charge to radius ratio.

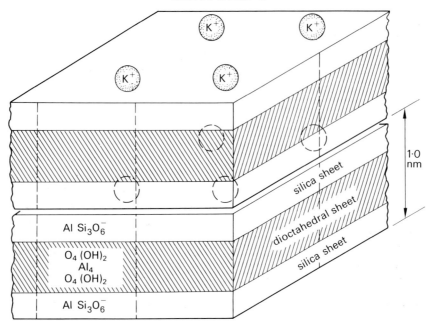

Figure 2.11. Structure of muscovite mica—a 2:1 non-expanding, layer-lattice mineral.

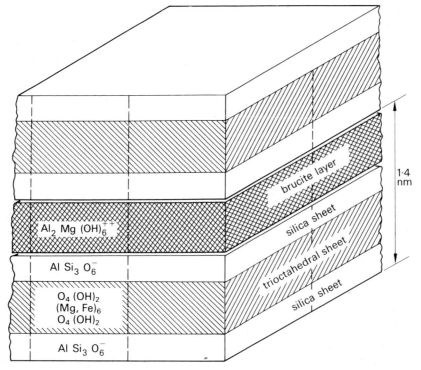

Figure 2.12. Structure of chlorite—2:2 or mixed layer mineral.

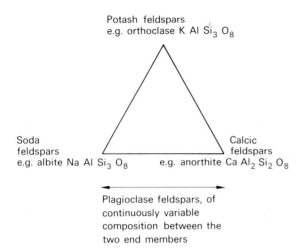

Figure 2.13. Type minerals of the feldspar group.

2.4 MINERALOGY OF THE CLAY FRACTION

For many years the small size of colloidal particles prevented scientists from elucidating their mineral structure. It was thought that the clay fraction consisted of inert mineral fragments enveloped in an amorphous gel of hydrated *sesquioxides* ($Fe_2O_3.n$ H_2O and $Al_2O_3.n$ H_2O) and *silicic acid* ($Si(OH)_4$). The surface gel was *amphoteric*, the balance between acidity and basicity being dependent on the pH of the soil. Between pH 5 and 8, the surface was usually negatively charged (proton-deficient), which could account for the observed cation exchange properties of the soil (section 2.5).

During the 1930s, however, the crystalline nature of the clay minerals was established unequivocally by the technique of X-ray diffraction. Most of the minerals were found to have a plate-like structure similar to that of the micas and chlorite. The various mineral groups were identified from their characteristic *basal spacings* (Figure 2.10), as measured by X-ray diffraction, and their unit cell compositions deduced from elemental analyses. Subsequent studies using the scanning and transmission electron microscopes have fully confirmed the conclusions of the early work. However, it is

known that *accessory minerals*, the weathered residues of resistant primary minerals that have been comminuted to colloidal size, also occur in the clay fraction, as well as variable amounts (~ 20 per cent by weight) of *amorphous material*. This is of considerable importance because it may occur as a thin covering on the surface of the clay minerals.

The crystalline clay minerals

The clay minerals are aluminosilicates that are important sites of physical and chemical reactions in soil because of their high specific surface area and electrical charge. Most have a layer–lattice structure but a small group similar to the amphiboles (the sepiolite–attapulgite series) has chain structures and another group, the *allophanes*, although of layer–lattice structure, forms hollow spherical crystals 3.5–5 nm in diameter.

Under mild (generally physical) weathering conditions, they may be *inherited* as colloidal fragments of primary layer silicates, such as the micas and chlorites. Under more intense weathering, the primary minerals may be *transformed* to secondary clay minerals, as when soil illites and vermiculites are formed by the leaching of interlayer K from primary micas. *Neoformation* of clay minerals is a feature of intense weathering, or of the metamorphosis of sedimentary deposits (section 5.2), when minerals completely different from the original primary minerals are formed. When the soluble silica concentration in the weathering environment is high, 2:1 layer minerals like the smectites are likely to form. Subsequent leaching and removal of silica, however, can produce kaolinite and aluminium hydroxide. Increased lattice negative charge due to isomorphous substitution of Al for Si in the smectites leads to K^+ being the favoured interlayer cation with the resultant formation of illite.

MINERALS WITH A Si/Al MOL RATIO ≤ 1
Three groups of clay minerals—*imogolite*, the *allophanes* and the *kaolinites*—have Si/Al ratios ≤ 1.

Imogolite and allophane are most commonly found in relatively young soils (< 10,000 years) formed on volcanic ash and pumice (the Andepts of the order Inceptisol). Imogolite has also been identified in the B horizon of certain podzols (Spodosols). By X-ray diffraction these minerals appear amorphous, but their crystallinity can be demonstrated by high-resolution electron diffraction. Imogolite has the structural formula:

$$6OH \ 4Al^{VI} \ 6O \ 2Si \ 2OH$$

and occurs in threads 10–30 nm in diameter and several μm long. The threads are curved to permit a reduced number of silicon tetrahedra to bond to the Al octahedral sheet, as shown in Figure 2.14.

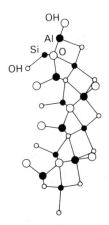

Figure 2.14. Projection along the imogolite c axis showing the curvature produced as the Al octahedral sheet distorts to accommodate the Si tetrahedra (after Wada, 1980).

One kind of allophane, which has a Si/Al ratio of 0.5 and the structural formula

$$2H_2O, \ 4OH \ 3 \ Al^{VI} \ 2O,$$
$$4OH(2Si, Al)^{IV} \ 3O, \ 2OH, \ H_2O$$

contains most of its Al in sixfold coordination and has charge properties very similar to imogolite, i.e., very little permanent negative charge due to isomorphous substitution, but variable positive and negative charge due to proton association or dissociation at surface OH groups (section 7.1). At the other extreme, allophane

with a Si/Al ratio of 1 and the structural formula

$$H_2O, \ 2OH \ Al^{VI} \ O, \ 2OH,$$
$$H_2O(2Si, Al)^{IV} \ 3O, \ 2OH, \ H_2O$$

contains half its Al in the tetrahedral sheet and half in the octahedral sheet. Although a large layer charge arises because of the substitution of Al for Si, it is neutralized to a variable extent, depending on the ambient pH, by the association of protons at surface OH groups. Thus, the allophanes and imogolite have pH-dependent surface charges at pH > 4.5.

Minerals of the kaolinite group have a well-defined 1:1 layer–lattice structure formed by the sharing of O atoms between one silicon tetrahedral sheet and one dioctahedral gibbsite sheet. The unit cell composition is $Si_2^{IV} \ O_5 \ Al_2^{VI} \ OH_4$. The common member of the group is *kaolinite*, the dominant clay mineral in many weathered tropical soils. *Halloysite* is found in weathered soils formed on volcanic ash parent material.

Kaolinite has a basal spacing fixed at 0.72 nm due to hydrogen-bonding between the hydrogen and oxygen atoms of adjacent lamellae (Figure 2.15). The lamellae are stacked fairly regularly in the c direction to form crystals from 0.05 to 2 μm thick, the larger crystals occurring in relatively pure deposits of China Clay. The crystals are hexagonal in plan view and usually larger than 0.2 μm in diameter, as shown in the electron-micrograph of Figure 2.16. Halloysite has the same structure as kaolinite, with the addition of two layers of water molecules between the crystal lamellae which increases the basal spacing of the mineral to 1 nm. The presence of this hydrogen-bonded water so alters the distribution of stresses within the crystal that the layers curve to form a tubular structure.

There is some isomorphous substitution of Al for Si in kaolinite which produces < 0.005 mols of negative charge per unit cell. In addition, a variable charge can develop at the crystal edge faces (Figure 2.17) due to the association or dissociation of protons at exposed O and OH groups. At a pH > 9, this can contribute as much as 8×10^{-4} centimols (cmol) of charge (−) per m² of edge area (*cf.* Table 2.3).

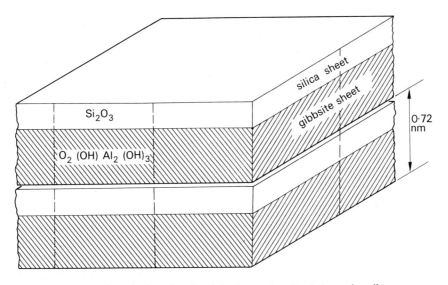

Figure 2.15. Structure of kaolinite—a 1:1 layer-lattice mineral with hydrogen-bonding between lamellae.

MINERALS WITH A Si/Al MOL RATIO OF 2:1

This category comprises the *mica, vermiculite* and *smectite* groups of clay minerals, which all have 2:1 layer–lattice structures. The minerals differ mainly in the extent and location of isomorphous substitution, and hence in the type of interlayer cation that predominates.

Within the mica group, the dioctahedral mineral *illite* has substitution of Al for SI in the tetrahedral sheets and some substitution of Mg and Fe^{2+} for Al in the octahedral sheet. An average of 1.6 mols of negative charge per unit cell of formula

$$[(OH)_4(Si_{7.1}Al_{0.9})^{IV} (Al_{3.3}Mg_{0.7})^{VI} O_{20}]^{-1.6} 1.6 K^+$$

and its location partly in the tetrahedral sheets are sufficient, however, to favour the formation of inner-sphere complexes with cations, primarily K^+, which fits into the ditrigonal cavities between opposing siloxane surfaces. The basal spacing is characteristically 0.96–1.01 nm. As illite weathers and the K^+ is gradually replaced by cations of higher ionic potential such as Ca^{2+} and Mg^{2+}, which remain partially hydrated in the interlayer regions, the mineral expands to a basal spacing of 1.4–1.5 nm. The additional spacing is equivalent

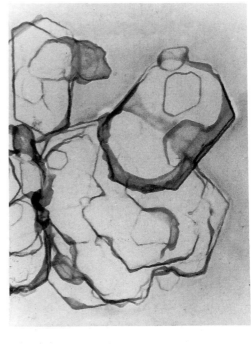

Figure 2.16. Electron micrograph of kaolinite crystals.

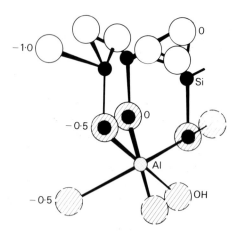

Figure 2.17. Charges on the edge face of kaolinite at high pH (after Hendricks, 1945).

to a bimolecular layer of water molecules between the mineral layers. Consequently, the interlayer bonding is weaker than in the primary micas and the stacking of the layers to form crystals is much less regular.

Vermiculites are trioctahedral minerals with both di- and trivalent cations occupying all the available sites in the octahedral sheet, thereby generating a net positive charge which in part neutralizes the negative charge developed through substitution of Al for Si in the tetrahedral sheet. The net layer charge of the vermiculites is therefore slightly lower than that of the soil micas at 1.2–1.9 moles of negative charge per unit cell. As a result, ions of lower ionic potential are not likely to form inner-sphere complexes and partially hydrated Ca^{2+} and Mg^{2+} are the dominant interlayer cations. The basal spacing of a Mg–vermiculite, shown in Figure 2.18, is typically 1.43 nm, but collapses to

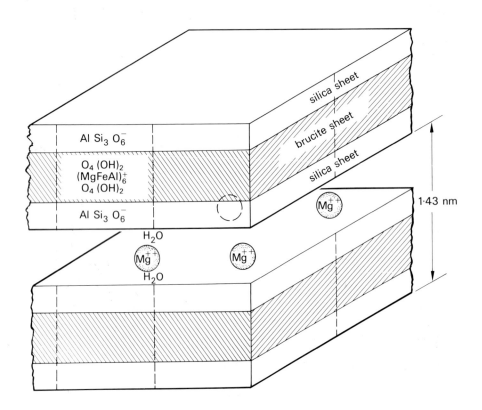

Figure 2.18. Structure of vermiculite—basal spacing fixed in the presence of partially hydrated Mg^{2+} ions.

1 nm on heating. Thus, the vermiculites show limited reversible swelling. In a moderately acid environment, hydrated Al^{3+} ions replace the Ca^{2+} and Mg^{2+} in the interlayers and 'islands' of $Al(OH)_3$ may form. This gives rise to an aluminous chlorite type of clay mineral.

Smectite minerals exhibit a wide range of composition with isomorphous substitution occurring in both the tetrahedral and octahedral sheets. The layer charge is lower (0.4–1.2 moles of negative charge per unit cell) than in either the soil micas or vermiculites. Two smectite minerals—*nontronite* and *beidellite*—have isomorphous substitution in the tetrahedral sheet, but the commonest smectite in soil, *montmorillonite*, is a dioctahedral mineral with substitution in the octahedral sheet only. It has the unit cell formula

$$[(OH)_4 Si_8^{IV} (Al_{3.3} Mg_{0.7})^{VI} O_{20}]^{-0.7} 0.35\, Ca^{2+}$$

When isomorphous substitution occurs in the octahedral sheet, the excess negative charge is distributed over 10 surface O atoms and the field strength is weak. Consequently, there is no tendency for cations of low ionic potential to dehydrate and to draw adjacent mineral layers close together, as occurs in the micas. Rather, the cations neutralizing the layer charge

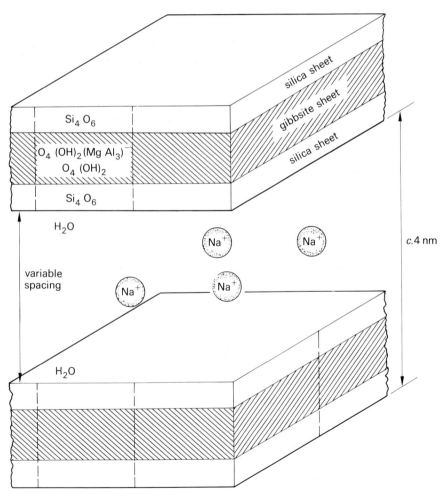

Figure 2.19. Structure of montmorillonite—basal spacing variable depending on the dominant exchangeable cation.

Figure 2.20. A transmission electron micrograph showing clusters of many small acicular goethite crystals (courtesy of A. Suddhiprakarn and R.J. Gilkes).

remain hydrated and may form *outer-sphere* complexes with the surface, as in the case of Ca-montmorillonite. The interlayer cations are freely exchangeable with other cations in solution, and depending on the nature of the exchangeable cations, the basal spacing may vary from *c*. 1.5–4 nm, as illustrated in Figure 2.19. Ca-montmorillonite, a form common in soil, has a basal spacing of 1.9 nm at full hydration (relative humidity > 98 per cent), when three molecular layers of water exist between the mineral layers.

Because of the relatively weak interlayer bonding and the free movement of water and cations into and out of this region, the smectites have been called expanding-lattice clays. The stacking of the layers is very irregular and the average crystal size is therefore much smaller than in the micas and kaolinites. The planar interlamellar surfaces provide a large internal area which augments the external cleavage face area so that the total surface area of a smectite clay is very large indeed (see Table 2.2).

MIXED LAYER MINERALS

Regular and irregular interstratification of clay minerals produces mixed layer minerals. The chlorites, for example, are formed by the *regular* interstratification of brucite or gibbsite layers and biotite mica. The basal spacing is a consistent 1.4 nm. With the improvement of X-ray diffraction techniques, *irregular* interstratification has been found to occur much more frequently in clay minerals separated from soil than was previously believed. Mixed layers of the 2:1 minerals are most common, especially intergrades of vermiculite and illite, and smectite and illite. However, interstratification of montmorillonite (2:1) and kaolinite (1:1) occurs in acid environments, and *anauxite* is an example of a 1:1 mixed layer mineral in which a double silica sheet is inserted at random between kaolinite layers.

Accessory minerals and amorphous materials

The only salt of importance found in the clay fraction is

calcite ($CaCO_3$), which can accumulate in soils formed on chalk and limestone and in high pH soils of arid regions. The accessory minerals are predominantly *free oxides*—compounds in which a single cation species is coordinated with oxygen, which include crystalline and amorphous forms of silica, iron oxides, aluminium oxides, manganese oxides and anatase (TiO_2).

SILICA

The various forms of silica have been described (section 2.3). Silicic acid in solution has a marked tendency to polymerize as the pH rises so that colloidal silica gels can occur as cementing agents in the lower B and C horizons of leached soils. Amorphous silica also occurs in soils formed on recent deposits of volcanic ash, as has been observed in Japan, New Zealand and Hawaii.

MANGANESE OXIDES

There is a variety of oxides and hydroxides of manganese in soil because Mn can occur in oxidation states of 2, 3 and 4. A continuous range in composition from manganous oxide, MnO, to manganese dioxide, MnO_2, is possible, with several stable and metastable minerals having been identified. For example, the most stable form of MnO_2 is *pyrolusite*, but the commonest Mn oxides found in soil are the *birnessite* group in which Mn^{2+}, Mn^{3+} and Mn^{4+} ions are bonded to O^{2-} and OH^-. Small black deposits (1–2 mm diameter) of manganese oxides are common in soils that experience alternating aerobic and anaerobic conditions (section 8.4). The surface of these minerals has a high affinity for heavy metals, especially Co and Pb, and the availability of Co to plants is often controlled by the solubility of these Mn oxides.

IRON OXIDES

Iron oxides tend to accumulate in soils that are highly weathered, especially those derived from the more basic rocks (section 5.2). They have strong colours ranging from yellow to reddish-brown to black, even when disseminated through the soil profile. The most common mineral is the oxyhydroxide *goethite* (α-FeOOH), which can precipitate directly from the soil solution as small acicular crystals, generally in clusters (Figure 2.20). As it is the most thermo-dynamically stable oxide of iron, it also forms by the slow transformation of ferrihydrite and lepidocrocite (see below). *Ferrihydrite* ($Fe_2O_3 . 2FeOOH . nH_2O$), which has a rusty red colour, readily precipitates in drainage ditches and in the B horizon of podzols (Spodosols): it is a necessary precursor of *hematite* (α-Fe_2O_3) formation by precipitation from solution. Because ferrihydrite formation is inhibited by the presence of organic ligands in solution, hematite (and hence red-coloured soils) are more common in regions where organic matter is rapidly oxidized than in regions where it is not.

In poorly drained soils, where Fe^{2+} and Fe^{3+} ions can co-exist in solution (section 8.4), rapid oxidation of Fe^{2+} leads to the precipitation of *lepidocrocite* (γ-FeOOH), which through crystal rearrangement slowly reverts to goethite. If the oxidation is slow, however, partial dehydration of the oxide can occur and the ferrimagnetic mineral *maghemite* (γ-Fe_2O_3) forms. The latter is therefore more common in soils of the tropics and subtropics.

With the exception of maghemite and hematite, which can be formed by direct alteration during weathering of the primary mineral *magnetite* (Fe_3O_4), all the iron oxides form by precipitation from solution and so inclusions of other elements can occur, especially Al which has been found to replace as much as 30 per cent of the Fe in goethite on a molar basis. Al is also found in the crystal structure of hematite and maghemite.

ALUMINIUM OXIDES

The aluminium oxides have a non-distinctive greyish-white colour which is easily masked in soils except when large concentrations occur as in bauxite ores. In acid soils, deposits of amorphous aluminium hydroxide form in the interlayers of expanding lattice clays such as the vermiculites and smectites, and occur as surface coatings on clay minerals generally. This poorly ordered material slowly crystallizes to *gibbsite* (γ-$Al(OH)_3$), the principal aluminium hydroxide mineral in soil. At temperatures only slightly higher than ambient, amorphous $Al(OH)_3$ undergoes dehydration to form *boehmite* (α-$AlOOH$), which is common in bauxite deposits but less common in soil.

Aluminium and iron oxides accumulate in old soils that are in an advanced stage of weathering and in younger soils forming under the present-day intense weathering conditions of the tropics (section 9.2). Soils of the order Oxisol are characterized by an *oxic* horizon which is greatly enriched in Fe and Al oxides. The presence of these oxides has a profound effect on the physical and chemical properties of the soils.

Table 2.2. Specific surface areas according to mineral type and particle size.

Mineral or size class	Specific surface $(m^2\,g^{-1})$	Method of measurement
Coarse sand	0.01	Low temperature N_2
Fine sand	0.1	adsorption (BET)
Silt	1.0	
Kaolinites	5–100	
Hydrous micas (illites)	100–200	BET
Vermiculites and mixed layer minerals	300–500	Ethylene glycol adsorption
Montmorillonite (Na-saturated)	700–800	
Allophanes	200–500	
Amorphous and hydrated iron and aluminium oxides	100–300	BET

(After Fripiatt, 1965.)

2.5 SURFACE AREA AND SURFACE CHARGE

Specific surface area

We can demonstrate that the smaller the size of an object, imagined as an ideal sphere, the greater the ratio of its surface to volume. Thus, when the radius $r = 1$ mm:

$$\frac{\text{Surface area}}{\text{volume}} = \frac{4\pi r^2}{4/3\pi r^3} = 3 \text{ mm}^{-1}$$

and when $r = 0.001$ mm (or 1 micron):

$$\frac{\text{Surface area}}{\text{volume}} = \frac{4\pi \times 10^{-6}}{4/3\pi \times 10^{-9}} = 3 \times 10^3 \text{ mm}^{-1}$$

The ratio of surface area to volume defines the specific surface: in practice, the *specific surface area* of a soil particle is measured as the surface area per unit mass, which implies a constant particle density. Representative values for the specific surfaces areas of sand, silt and clay-size minerals are given in Table 2.2.

ADSORPTION

When a boulder is reduced to small rock fragments by weathering, part of the energy put in is conserved as free energy associated with the new particle surfaces. The higher the specific surface of a substance, therefore, the greater is its surface free energy. Because free energy tends to a minimum (Second Law of Thermodynamics), the surface energy is reduced by the work done in attracting substances to the surface, which gives rise to the phenomenon of surface *adsorption*. The adsorption of nitrogen gas at very low temperatures on thoroughly dry clay is used to measure the specific surface of the clay (the Branauer, Emmet and Teller or BET method) since the adsorbed N_2 forms a monomolecular layer with an area of 0.16 nm^2 per molecule. Where the mineral has internal surfaces, as in montmorillonite, an additional measurement is required because the non-polar N_2 molecules do not penetrate into the interlamellar space. However, small polar molecules like water or ethylene glycol do penetrate so that monolayer adsorption of these compounds gives a measure of the *internal* and *external* surfaces. In the absence of an internal surface, as in kaolinite, the BET and ethylene glycol methods should give identical results. As shown in Table 2.2, the total specific surface of Na-montmorillonite approaches 800

$m^2\ g^{-1}$, which is about the maximum for a completely dispersed clay of 1-nm lamellar thickness.

SURFACE CHARGES

In addition to the adsorption of neutral molecules, cations and anions are adsorbed at clay and oxide surfaces as a result of:
(a) the permanent negative charges of the clays due to isomorphous substitution;
(b) the amphoteric properties of the edge faces of kaolinite and the surfaces of imogolite, allophane and the oxides and hydroxides of Fe and Al.

The permanent charge due to isomorphous substitution can be expressed as a surface density of charge by dividing the layer charge per unit cell by the surface area per unit cell. In practice, the charge on the mineral surface is measured as the difference between the moles of charge contributed per unit mass of mineral by the cations and anions adsorbed from an electrolyte solution of known pH. The cation and anion charge adsorbed gives rise respectively to the *cation exchange capacity (CEC)* and anion exchange capacity *(AEC)* of the mineral, which may be expressed in milliequivalents (me) per 100 g, or as cmols charge per kg. For the crystalline 1:1 and 2:1 phyllosilicates, permanent negative charge and hence *CEC* is the more important, and representative values of *CEC* for the main mineral groups are given in Table 2.3. Note that the *CEC* is quite variable within and between mineral groups, and not always the value expected from the

number of moles of unbalanced negative charge per unit cell. Of the 2:1 clays, for example, the *CEC* of the illites is low because much of the charge is neutralized by interlamellar K^+ which is *non-exchangeable*; on the other hand, the interlamellar Ca^{2+} and Mg^{2+} of the soil vermiculites are exchangeable and the *CEC* is high.

The functional surface charge density (Γ) of the clays is obtained from the ratio *CEC*/specific surface area (cmols (+) m^{-2}). As shown in Table 2.3, despite the large range in charge per unit mass (the *CEC*), the values of Γ are reasonably similar, suggesting that the clay crystals cannot acquire an indefinite number of negative charges and remain stable. Kaolinite, which has very little isomorphous substitution, exists as large crystals and so has a much lower specific surface than montmorillonite; on the other hand, the large kaolinite crystals have a relatively greater edge area at which additional negative charges can develop and so augment the total surface charge density of the clay.

For minerals bearing predominantly pH-dependent charges (the sesquioxides, imogolite and allophane), both the *CEC* and *AEC* can be significant, depending on the prevailing pH. Imogolite and allophane have roughly comparable positive and negative charges in the pH range 6–7. The sesquioxides are positively charged up to pH 8 so in most soils they make no contribution to the *CEC*. Maximum *AEC* values (measured at pH 3.5) range from 30 to *c.* 50 cmols (−) per kg for gibbsite and goethite, respectively. The development of charges on the surfaces of clays and oxides is discussed more fully in Chapter 7.

Table 2.3. Cation exchange capacities and surface charge densities of the clay mineral groups.

Clay mineral group	CEC^* (cmols(+) kg^{-1})	Surface charge density (Γ) (cmols(+) m^{-2})
Kaolinites	3–20	$2-6 \times 10^{-4}$
Illites	10–40	$1-2 \times 10^{-4}$
Smectites	80–120	$1-1.5 \times 10^{-4}$
Vermiculites	100–150	3×10^{-4}

*Although the charge on the surface is negative, it is measured by the number of mols of cation (+) adsorbed.

2.6 SUMMARY

There is a continuous distribution of particle sizes in soil from boulders and stones down to clay minerals less than 2 μm in equivalent diameter. The material that passes through a 2-mm sieve, the *fine earth*, is divided into *sand*, *silt* and *clay*, the relative proportions of which determine the soil *texture*. Texture is a property that changes only slowly with time, and is an important determinant of the soil's response to water—its stickiness, mouldability and permeability; its capacity to retain cations, and its rate of adjustment to ambient temperature changes.

The adsorption of water and solutes by clay fraction particles ($< 2~\mu m$ size) depends not only on their large *specific surface area*, but also on the nature of the minerals present. The basic crystal structures consist of sheets of Si in tetrahedral coordination with $O[SiO_2]_n$ and sheets of Al in octahedral coordination with $OH[Al_2(OH)_6]_n$. Crystal layers are formed by the sharing of O atoms between contiguous silica and alumina sheets, giving rise to minerals with Si/Al mol ratios ≤ 1 (*imogolite* and the *allophanes*), the 1:1 layer lattice minerals (*kaolinites*) and the 2:1 layer lattice minerals (*illites*, *vermiculites* and *smectites*). Isomorphous substitution ($Al^{3+} \rightarrow Si^{4+}$ and Fe^{2+}, $Mg^{2+} \rightarrow Al^{3+}$) in the lattice results in an overall net layer charge ranging from < 0.005 to 2 moles of negative charge per unit cell for kaolinite to muscovite, respectively. The formation of *inner-sphere* complexes between the surface and unhydrated K^+ ions in the interlayer spaces of the micas effectively neutralizes the layer charge and permits large crystals to form. On weathering, K^+ is replaced by cations of higher ionic potential, such as Ca^{2+} and Mg^{2+} and layers of water molecules intrude into the interlayer spaces—the *basal spacing* of the mineral increases. The Ca^{2+} and Mg^{2+}, which form *outer-sphere* complexes with the surfaces, are freely exchangeable with other cations in the soil solution.

The edge faces of the clay crystals, especially kaolinite, and the surfaces of iron and aluminium oxides (e.g. *goethite* and *gibbsite*) bear variable charges depending on the association or dissociation of protons at exposed O and OH groups. The surfaces are positively charged at low pH and negatively charged at high pH. The cation and anion charges adsorbed (in mols of charge per unit mass) measure the *CEC* and *AEC* respectively of the mineral. *CEC* ranges from 3 to 150 cmols(+) per kg for the clay minerals and *AEC* from 30 to 50 cmols(−) per kg for the sesquioxides. Imogolite and allophane have roughly comparable *CEC* and *AEC* at pH 6–7.

Clays in which the interlayer surfaces are freely accessible to water and solutes are called expanding lattice clays. The total surface area is then the sum of the internal and external areas, and approaches $800~m^2$ g^{-1} for a fully dispersed Na-montmorillonite. However, when the crystal layers are strongly bonded to one another, as in kaolinite, the surface area comprises only the external surface and may vary from 5 to $100~m^2$ g^{-1} depending on the crystal size.

REFERENCES

BLACK C.A. (1965) (Ed.) *Methods of soil analysis. Part I. Physical and mineralogical properties, including statistics of measurement and sampling.* American Society of Agronomy Monograph No. 9.

FRIPIAT J.J. (1965) Surface chemistry and soil science, in *Experimental Pedology* (Eds. E.G. Hallsworth and D.V. Crawford). Butterworth, London.

HENDRICKS S.B. (1945) Base exchange of crystalline silicates. *Industrial and Engineering Chemistry* **37**, 625–630.

HODGSON J.M. (1974) (Ed.) *Soil survey field handbook.* Soil Survey of England and Wales, Technical Monograph No. 5, Harpenden.

LOUGHNAN F.C. (1969) *Chemical Weathering of the Silicate Minerals.* American Elsevier, New York.

LOVEDAY J. (1974) (Ed.) *Methods for analysis of irrigated soils.* Commonwealth Bureau of Soils, Technical Communication No. 54.

MARSHALL C.E. (1964) *The Physical Chemistry and Mineralogy of Soils.* **I**. *Soil Materials.* Wiley, New York.

MOORE J.W. & MOORE E.A. (1976) *Environmental Chemistry.* Academic Press, New York.

RUSSELL E.W. (1971) Soil structure: its maintenance and improvement. *Journal of Soil Science* **22**, 137–151.

WADA K. (1980) Mineralogical characteristics of Andisols, in *Soils with Variable Charge.* New Zealand Society of Soil Science, Lower Hutt.

FURTHER READING

BOLT G.H. & BRUGGENWERT M.G.M. (1978) (Eds.) *Soil Chemistry A. Basic Elements,* 2nd Ed. Elsevier, Amsterdam.

BROWN G. (1984) Crystal structures of clay minerals and related phyllosilicates. *Philosophical Transactions of the Royal Society, London A* **311**, 221–240.

DIXON J.B. & WEED S.B. (1977) (Eds.) *Minerals in Soil Environments.* Soil Science Society of America, Madison.

GRIM R.E. (1968) *Clay Mineralogy,* 2nd Ed. McGraw-Hill, New York.

SPOSITO G. (1984) *The Surface Chemistry of Soils.* Oxford University Press, Oxford.

THENG B.K.G. (1980) (Ed.) *Soils with Variable Charge.* New Zealand Society of Soil Science, Lower Hutt.

Chapter 3
Soil Organic Matter

3.1 ORIGIN OF SOIL ORGANIC MATTER

The carbon cycle

The organic matter of the soil arises from the debris of green plants, animal residues and excreta that are deposited on the surface and admixed to a variable extent with the mineral component. The dead organic matter is colonized by a variety of *soil organisms*, most importantly microorganisms (section 3.2), which derive energy for growth from the oxidative decomposition of complex organic molecules. During decomposition, essential elements are converted from organic combination to simple inorganic forms, a process called *mineralization*. For example, organically combined N, P and S appear as NH_4^+, $H_2PO_4^-$ and SO_4^{2-} ions, and about half the C is released as CO_2. Mineralization, especially the release of CO_2, is vital for the growth of succeeding generations of green plants. The remainder of the substrate C used by the microorganisms is incorporated into their cell substance or *biomass*, as is a variable proportion of the other essential elements N, P, S, etc. This incorporation renders these elements unavailable for plant growth until the organisms die and decay, so the process is called *immobilization*. The residues of the organisms, together with the more recalcitrant parts of the original substrate, accumulate in the soil. The various interlocking processes of synthesis and decomposition by which carbon is circulated through soil, plants, animals and air—collectively the *biosphere*—comprise the *carbon cycle* (Figure 3.1).

Inputs of plant and animal residues

The annual return of plant and animal residues to the soil varies greatly with the climatic region and the type of vegetation or land use. The amount of above-ground

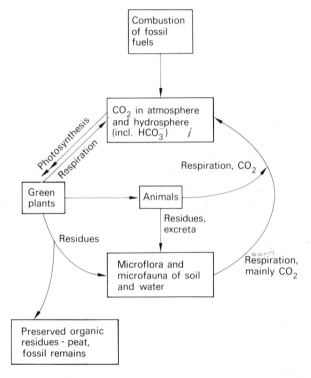

Figure 3.1. The carbon cycle.

material—the annual litter fall—is easily measured and some average values, in terms of organic C, are given in Table 3.1. To this must be added timber fall in forests (although this does not provide a readily decomposable substrate) and the contribution from plant roots, which comprises dead root material and C compounds released from living roots into the surrounding soil (the rhizosphere C, section 8.1). Root contributions are difficult to measure, but experiments in which the plant shoot has been placed in an atmosphere of CO_2 labelled with ^{14}C indicate that some 10–20 per cent of the total plant C may be released into the rhizosphere, and the same proportion may remain as root residues. Thus, the below-ground return of C to the soil may range from 20 to 40 per cent of the total photosynthate produced during a season of growth. The return of C in animal excreta is obviously highly variable, but the figure for dung produced by dairy cows grazing highly productive pastures is $3.5-4.5$ t ha^{-1} y^{-1}.

Table 3.1. Annual rate of litter return to the soil.

Land use or vegetation type	Organic C* (t ha^{-1})
Alpine and arctic forest	0.1–0.4
Arable farming (cereals)	1–2
Temperate grassland	2–4
Coniferous forest	1.5–3
Deciduous forest	1.5–4
Tropical rainforest (Colombia)	4–5
Tropical rainforest (West Africa)	10

*To convert organic C to organic matter, divide these figures by 0.6.

Composition of plant litter

Plant cells are made up primarily of carbohydrates, proteins and fats, plus smaller amounts of organic acids, lignin, pigments, waxes and resins. The bulk of the material is carbohydrate of which sugars and starch are rapidly decomposed, while hemi-cellulose, a pentose sugar polymer, and cellulose, a $\beta(1-4)$ glucose polymer, are less readily decomposed. These and other decomposition reactions are catalysed by enzymes produced by specialized groups of microorganisms, as identified in section 3.2. Proteins are rapidly metabolized and *proteolysis* may begin in the senescing leaf before it has reached the soil. The ultimate products are amino-acids (see Figure 3.2) some or all of which may be used in protein synthesis by the microorganisms. Whether or not there is amino-N surplus to the needs of the microorganisms, so that *net* mineralization can occur, depends on the C:N ratio of the

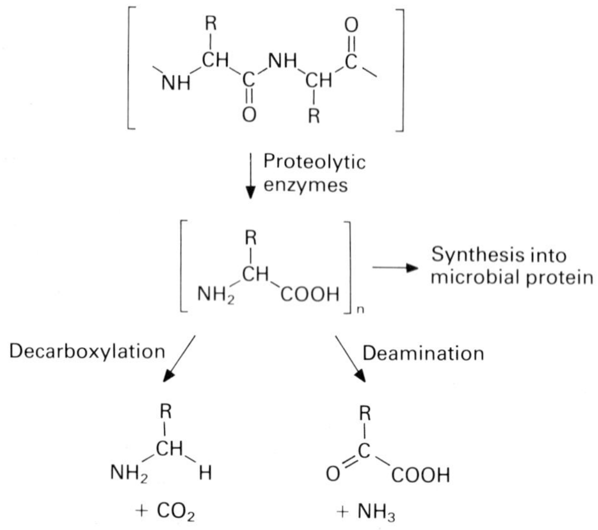

Figure 3.2. Steps in proteolysis.

substrate and on the properties of the decomposing organisms (section 3.3). Generally, when the C:N ratio is > 25, net immobilization occurs, whereas at ratios < 25 net mineralization is likely.

As can be seen from Table 3.2, the C:N ratio of fresh litter is highly variable among plant species, but as the organic matter passes through successive cycles of decomposition in the soil the C:N ratio gradually

Table 3.2. C:N ratios of freshly-fallen litter and of soils.

	Range	Mean
Litter		
Herbaceous legumes	—	20
Cereal straw	40–120	80
Tropical forest species (Colombia)	27–32	30
Temperate hardwoods (elm, ash, lime, oak, birch)	25–44	35
Scots pine	—	91
Soil		
Soils from 63 sites in the USA	7–26	12.8
Rothamsted: old woodland	—	9.5
Old pastures, fertilized and unfertilized	11–12	11.8
Old arable, manured and unmanured	8–10	9.4

narrows, and we find the C:N ratio of well-drained soils of pH ~ 7 to be very close to 10. Exceptions occur with poorly drained soils or those on which *mor humus* forms (section 3.3).

N in complex heterocyclic ring compounds, such as chlorophyll, is not easily mineralized, but the chitin of insect cuticles and of fungal cell walls, a glucosamine polymer, is eventually hydrolysed to glucose and NH_4^+. Most plant organic acids are readily decomposed: not so the fats, waxes and resins which can persist in the soil for some time. Similarly lignin, which is a complex phenolic polymer, is more stable than the carbohydrates and accumulates relative to these constituents during the decomposition of fresh residues. Lignin is a significant proportion of the dry matter of cereal straw (10–20 per cent) and wood (20–30 per cent). Its hydrolytic oxidation, catalysed by extracellular enzymes from various actinomycetes and fungi, produces monocyclic phenols as shown in Figure 3.3. These degradative products probably serve as precursors for the synthesis of humic macromolecules by the soil organisms.

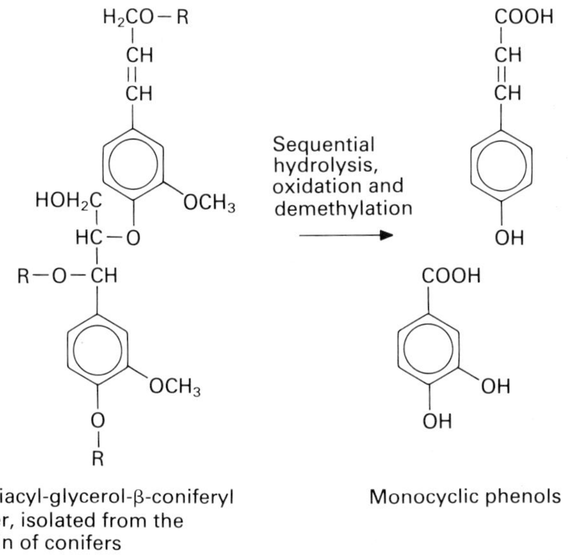

Figure 3.3. Hydrolytic oxidation of a lignin-type compound (after Andreux, 1982).

3.2 THE SOIL ORGANISMS

The soil organisms may be grouped, according to size, into:

the *macrofauna*—vertebrate animals mainly of the burrowing type, such as moles and rabbits, which live wholly or partly underground;

the *mesofauna*—small invertebrate animals representative of the phyla Arthropoda, Annelida, Nematoda and Mollusca;

the *microorganisms*, comprising the *microfauna* (soil animals < 0.2 mm in length) and the *microflora* (members of the plant kingdom).

In respect of their evolutionary development, the microorganisms may be subdivided into:

Prokaryotes (organisms without a true nucleus), which include the bacteria, actinomycetes and Cyanophyceae, and

Eukaryotes (organisms with a membrane-bound nucleus), which include the fungi, algae and protozoa.

Soil organisms may also be classified on their mode of nutrition. Broadly, the *heterotrophs*, which include many species of bacteria and all the fungi, require C in the form of organic molecules for growth, whereas the *autotrophs*, which include the remaining bacteria and most algae, can synthesize their cell substance from the C of CO_2, harnessing the energy of sunlight (in the case of the photosynthetic bacteria and algae) or chemical energy from the oxidation of inorganic compounds (the chemoautotrophs). Another way of subdividing the microorganisms is on the basis of their requirement for molecular O_2; that is into:

Aerobes—those requiring O_2 as the terminal acceptor of electrons in respiration;

Facultative anaerobes—those normally requiring O_2 but able to adapt to oxygen-free conditions by using NO_3^- and other inorganic compounds as electron acceptors in respiration;

Obligate anaerobes—those which grow only in the absence of O_2 because O_2 is toxic to them.

BIOMASS
Soil *biomass* refers to the mass of organisms in the soil.

Because the macro- and mesofauna can be physically separated from the soil, their mass can be measured directly and is usually expressed as kg (liveweight) ha^{-1} to a certain depth. The macro- and mesofaunal biomass ranges from 2 to 5 t ha^{-1}, with earthworms making the largest single contribution (see Table 3.3). However, the microorganisms are intimately mixed with dead organic matter in soil and, being very small, are difficult to isolate for counting or weighing. An indirect method now widely used to estimate the *microbial* biomass is based on the flush of CO_2 evolved following the inoculation and incubation of soil fumigated with $CHCl_3$ (section 3.5). After a standard period of incubation (10 days), the microbial biomass is calculated using the equation:

$$\text{Biomass C} = \frac{CO_2\text{-C evolved}}{K_c} \qquad (3.1)$$

where K_c is the empirically derived fraction of killed biomass C that is evolved as CO_2 (Jenkinson and Powlson, 1976). This is related to the growth yield, F, the fraction of substrate C that is converted into cell substance, by the expression:

$$K_c = 1 - F \qquad (3.2)$$

K_c values between 0.4 and 0.5 have been used in (3.1). This is consistent with a growth yield of *c.* 0.5 for soil bacteria and a value between 0.3 and 0.4 for actinomycetes and fungi.

Most of the organisms are concentrated in the top 15–25 cm of soil because C substrates are more plentiful there. Estimates of microbial biomass C range from 500 to 2,000 kg ha^{-1} to 15-cm depth (section 4.5), with higher values being recorded on a per ha basis when the soil is sampled to greater depths. Alternatively, biomass C can be expressed as a percentage of total soil C when it ranges from about 2 per cent in arable soils to 3 or 4 per cent in grassland or woodland soils.

Types of microorganisms

BACTERIA, INCLUDING ACTINOMYCETES

These organisms can be as small as 1 μm in length and 0.2 μm in breadth and therefore live in water films around soil particles in all but the finest pores. They can be motile or non-motile, coccoid (round) or rod-shaped (Figure 3.4), and can reproduce very rapidly in the soil under favourable conditions—as little as 8–24 h for the production of two daughter organisms from a single parent by fission. Accordingly, their number in the soil is enormous provided that living conditions, especially the food supply, are suitable. Bacterial numbers are estimated by observing the growth of colonies on special nutrient media which have been inoculated with drops of a very dilute soil suspension (the *dilution plate method*), or by direct microscopic observation of suitably stained thin soil sections, recording all the microorganisms—living, dead and those which have formed spores. Estimates of between 1 and 4×10^9 organisms g^{-1} of soil are obtained by the latter method.

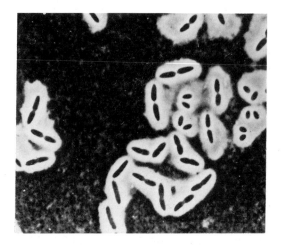

Figure 3.4. Soil bacteria—*Azotobacter* sp. (after Hepper, 1975).

Bacteria exhibit almost limitless variety in their metabolism and ability to decompose diverse substrates. Three major subgroups are recognized:

(a) *unicellular Eubacteria*—the most numerous group including both heterotrophs and autotrophs;

(b) *branched Eubacteria* or *Actinomycetes*—these are heterotrophs that form mycelial growths more delicate than the fungi (see below). They include such genera as *Streptomyces* and *Nocardia* which produce antibiotics and can degrade the more recalcitrant C compounds such as lignin. They are aerobic spore-formers;

(c) *Myxobacteria* (slime bacteria)—these unicellular organisms differ from the Eubacteria in the flexibility of their cell walls and their mode of locomotion. Many are specialized in degrading cellulose and chitin.

The activities of some of the specialized groups of bacteria are discussed in Chapters 8 and 10.

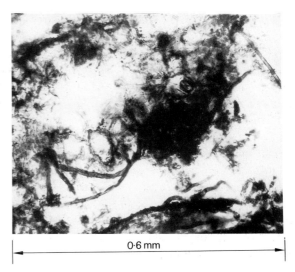

0·6 mm

Figure 3.5. Fungal mycelium in litter layer (courtesy of P. Bullock).

FUNGI

Some fungi, such as the yeasts, are unicellular, but the majority produce long filamentous *hyphae*, 1–10 μm in diameter, which may be segmented and/or branched. The network of hyphae or *mycelium* develops fruiting bodies on which the spores formed are often highly coloured. Thus, members of the Basidiomycetes (white-rot fungi) and the abundant spore-formers of the Fungi Imperfecti, such as *Penicillium* and *Aspergillus*, are often conspicuous on decaying wood and leaf litter in moist situations.

Fungi occur as *soil inhabitants*, those residing in the soil and feeding on dead organic matter (the saprophytes), and as *soil invaders*, those whose spores are normally deposited in the soil but which only germinate and grow when living tissue of a suitable host plant appears close by. Many of the latter fungi are pathogenic, growing parasitically on the host. Of special benefit, however, are the *mycorrhizal fungi* which live symbiotically in the host tissue, deriving C compounds from the host and in turn supplying it with mineral nutrients, especially P (section 10.3).

Although soil fungal colonies are much less numerous than bacteria ($1–4\times10^5$ organisms g^{-1}),

because of their filamentous growth their biomass can be comparable to that of the bacterial biomass. The fungal population is also considerably less variable than bacterial numbers, for although fungi, excluding yeasts, are intolerant of anaerobic conditions, they grown better in acid soils (particularly at pH < 5.5) and tolerate variations in soil moisture better than bacteria. The fungi are all heterotrophic and are most abundant in the litter layer (Figure 3.5) and the organically rich surface horizons of the soil, where their superior ability to decompose lignin confers a competitive advantage. Because of their slow growth rate relative to the bacteria, fungi cannot be reliably counted by the dilution plate method; methods employing the transfer of fungal hyphae from the soil to nutrient agar plates give more accurate results.

For details of methods of species identification and counting of soil microorganisms, the reader is referred to Parkinson, Gray and Williams (1971).

ALGAE

The algae are the major group of photosynthetic organisms in soil and are therefore confined to the soil surface, although many will grow heterotrophically in the absence of light if simple organic solutes are provided. The two major subgroups are the prokaryotic Cyanophyceae or blue-green algae and the eukaryotic green algae. The blue-greens, as exemplified by the common genera *Nostoc* and *Anabaena*, are important because they can reduce atmospheric nitrogen and incorporate it into amino acids, thereby making a substantial contribution to the N status of the soil (section 10.2). The 'blue-greens' prefer neutral to alkaline soils, whereas the green algae are more common in acid soils. Algae in soil are much smaller than the aquatic or marine species and may number from 100,000 to 3×10^6 organisms g^{-1}.

The prolific growth of aquatic algae during past eras in the Earth's history has produced large areas of sedimentary rocks such as the English Chalk (made up of calcified algal bodies—see Figure 5.3) and deposits of diatomaceous earth (silicified algae).

PROTOZOA

The smallest of the soil animals, ranging from 5 to 40 μm in their longer dimension, the protozoa live in water films and move by means of cytoplasmic streaming, cilia or flagellae in the case of *Euglena*. *Euglena* also contains chlorophyll and can live autotrophically, but nearly all of the protozoa prey upon other small organisms such as bacteria, algae, fungi and even nematodes. Protozoan biomass can be comparable to that of the earthworms so they are important in controlling bacterial and fungal numbers in the soil.

SOIL ENZYMES

Many extracellular enzymes exist in soil independently of living organisms. They avoid denaturation and degradation to some extent by being adsorbed onto soil mineral or organic matter. The major types present are hydrolases, transferases, oxidases, reductases and decarboxylases. A very common and stable soil enzyme is *urease* which is important in the hydrolysis of urea (section 10.2).

The mesofauna

Next to the protozoa, the threadlike *nematodes* are among the smallest of the soil fauna, ranging from a few tenths of a mm to a few mm in size. They are plentiful in soil and litter and may be carnivores or saprophages.

Other quite primitive soil animals are the *molluscs* (slugs and snails), many of which feed on living plants and are therefore pests, although some species feed on fungi and the faeces of other animals. As the mollusc biomass is only 200–300 kg ha^{-1} in most soils, their contribution to the turnover of organic matter is limited.

By far the most important groups involved in the turnover of organic matter are the *arthropods* and *annelids*.

ARTHROPODS

This group includes the isopods (wood lice), arachnids (mainly mites), insects (winged and wingless) and myriapods (centipedes and millipedes). The more numerous are the mites and many of the insects, present as adults and larvae.

Mites and *springtails* (wingless Collembola) feed on plant remains and fungi in the litter, especially where thick mats build up under forest and undisturbed grassland. Their droppings appear as characteristic pellets in the litter (see Figure 3.9). They may also be found at depth in the soil, living in the larger pores (0.3–5 mm). Some species of mite prey on springtails.

Figure 3.6. Dung beetle (*Sisyphus rubripes*) rolling a 'dung ball' (after Waterhouse, 1973).

Many *beetles* and *insect larvae* live in the soil, some feeding saprophytically, whereas others feed on living tissues and can be serious pests of agricultural crops. Of particular value, however, are the coprophagous beetles of Africa and other tropical regions that feed upon the dung of large herbivores and greatly increase the rate at which such residues are comminuted and physically mixed with the soil (Figure 3.6).

The *termite* (a member of the Isoptera) has been called the tropical analogue of the earthworm and indeed they are important comminuters of all forms of litter—tree trunks, branches and leaves—in the forest and especially the seasonal rainfall regions (savanna) of the tropics. Most species are surface feeders and build nests called termitaria by packing and cementing together soil particles with organic secretions and excrement. The small mounds only 30 cm or so high are the homes of *Cubitermes* species which feed under the cover of recent leaf fall; the less frequent large mounds, some 4–6 m high, are the homes of wood-feeders and foraging termites, such as *Macrotermes*

and *Odontotermes*, the latter usually colonizing the mounds of *Macrotermes* when they are abandoned. These species cultivate 'fungus gardens' within the termitaria to provide food for their larvae.

Some termite species or 'harvester ants' destroy living vegetation in African grasslands. In cool temperate regions, species of ant, such as *Formica fusca*, are of some importance in comminuting litter under pine trees, whereas *Lasius flavus* is active in grassland soils on chalk.

Centipedes are carnivorous whereas *millipedes* feed on vegetation, much of which is in the form of living roots, bulbs and tubers. Size comparisons between these and other soil invertebrates may be made by reference to the specimens in Figure 3.7.

ANNELIDS
These include the enchytraeid worms and the lumbricids or earthworms.

The *enchytraeids* are small (a few mm to 2–3 cm long) and threadlike. Their biomass is rarely > 100 kg ha^{-1}. However the *earthworms*, because of their size (several cm long) and physical activity are generally more important in the consumption of leaf litter than all the other invertebrates together. Exceptions occur in the drier savanna-region soils and soils of temperate regions on which mor humus forms (section 3.3). An indication of the earthworm biomass under different systems of land use is given in Table 3.3.

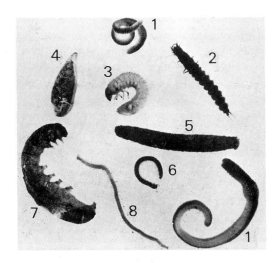

Figure 3.7. The main litter-feeding soil animals (after Russell, 1973). 1. Earthworm, *Lumbricus* sp. (Oligochaeta); 2. Beetle larva, Carabidae (Coleoptera); 3. Chafer grub, *Phyllopertha* sp. (Coleoptera); 4. Slug, *Agriolimax* sp. (Mollusca); 5. Leatherjacket, Tipulidae (Diptera); 6. Millipede (Diplopoda); 7. Cutworm, Agrotidae (Lepidoptera); 8. Centipede (Chilopoda).

Table 3.3. Estimates of earthworm biomass in soils under different land use.

Earthworm biomass	(kg ha^{-1})
Hardwood and mixed woodland	370–680
Coniferous forest	50–170
Orchards (grassed)	640–287
Pasture	500–1500
Arable land	16–760

(After Edwards and Lofty, 1977.)

Earthworms feed exclusively on dead organic matter. A few species, such as *Lumbricus rubellus*, live

Figure 3.8. (a) Litter accumulation on soil devoid of earthworms. **(b)** Uniform organic matter distribution in a soil with earthworms (after Edwards and Lofty, 1977).

only in the litter layer. The majority migrate between the litter layer and the mineral soil, and in the course of feeding ingest large quantities of clay and silt-size particles. The large populations of earthworms in soils long down to grass may consume up to 90 t of soil ha^{-1} annually. As a result, the organic and mineral matter is more homogeneously mixed when deposited in the worm faeces, which may appear as *casts* on the soil surface, as for example with *Allolobophora longa* and *A. nocturna*. Other species, such as *A. caliginosa* and *Lumbricus terrestris*, are non-casters and deposit their faeces in burrows. Organic matter from the surface rapidly becomes incorporated throughout the upper part of the soil profile, in contrast to the sharp boundary between the litter layer and mineral soil proper in the absence of earthworms (Figures 3.8a and b).

The consumption of dung and vegetable matter by earthworms can be prodigious. It has been estimated from feeding experiments that a population of 120,000 adult worms ha^{-1} is capable of consuming 25–30 t of cow dung annually. The quantities of dung and leaf litter available in the field are normally much less than this, so that the size of the earthworm population is regulated by the amount of suitable organic matter available, as well as by soil temperature, moisture and pH. For example, earthworms are rarely found in soils of pH < 4.5, and most species prefer neutral to calcareous soils. In hot dry weather they burrow deeper into the soil and aestivate.

Grass and most herbaceous leaf litter is readily acceptable to earthworms, but there is great variability in the palatability of litter from temperate forest species. Litter from elm, lime and birch is more palatable than pine needles which must age considerably before being acceptable.

3.3 CHANGES IN PLANT REMAINS DUE TO THE ACTIVITIES OF SOIL ORGANISMS

Types of humus

Decomposition begins with the invasion of ageing

plant tissue by surface saprophytes, and proceeds in parallel with biochemical changes in the senescing tissue—the synthesis of protease enzymes, the rupture of cell membranes with consequent mixing of cellular constituents, and the auto-oxidation and polymerization of phenolic-type compounds. Decomposition accelerates, as evidenced by the rise in the rate of CO_2 production, when the plant material falls to the ground and is invaded by a host of soil organisms. The changes that occur in the plant residues lead not only to mineralization and immobilization of nutrient elements, but also to the synthesis of new compounds, less susceptible to decomposition, which collectively form a dark-brown to black, amorphous material called *humus*.

MOR AND MULL HUMUS

Polyphenolic compounds formed in the senescing leaf, which are not leached out by rainwater once it has fallen to the ground, have a considerable effect on the rate of decomposition of the leaf residue. This is well illustrated by the contrasting superficial layers and A horizons of soils under temperate forests.

Typically under deciduous species, there is a loose litter layer 2–5 cm deep under which the soil is well aggregated and porous, dark-brown in colour and changes only gradually with depth to the lighter colour of the mineral matrix. This deep (30–50 cm) organic A horizon of C:N ratio 10–15 and rich in animals, especially earthworms, is characteristic of *mull humus*. Alternatively, under coniferous species, the surface litter is thick (5–20 cm) and often ramified by plant roots and a tough fungal mycelium; it is sharply differentiated from the mineral soil which is capped by a thin band of blackish humus and usually compact, poorly drained and devoid of earthworms. This superficial organic horizon, of C:N ratio > 30, is called *mor humus* and generally consists of three layers:

    L layer—undecomposed litter, primarily remains deposited during the previous year;
    F layer—the fermentation layer, partly decomposed, with abundant fungal growth and the faecal pellets of mites and springtails;

    H layer—the humified layer, comprising completely altered plant residues and faecal pellets.

The contrasting microstructure of mor and mull humus is illustrated by the thin sections of Figure 3.9.

Mor and mull humus, and intermediate types called *mor moder* and *mull moder*, occur under a variety of vegetation type: as well as the main deciduous trees (oak, elm, ash and beech), mull usually occurs under a well-drained grassland, whereas mor is found under

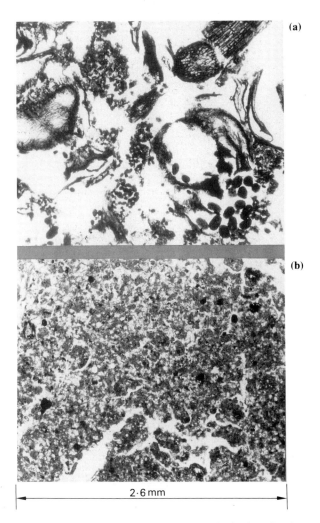

Figure 3.9. (a) Loose fabric of plant residues and mite droppings in mor humus. **(b)** Intimately mixed matrix of organic and mineral matter in mull humus (courtesy of P. Bullock).

heath vegetation (*Erica* and *Calluna* spp.) as well as under the major conifers (pine, spruce, larch and fir). However, the tendency towards mor or mull formation is determined not only by plant species, but also by the mineral status of the soil, particularly of Ca, N and P, so that the same species of deciduous tree can form mull on a fertile calcareous soil and mor on an infertile acid sand.

Colonization by soil microorganisms

The formation of more humus is predisposed by high concentrations of phenols in the senescing leaf which precipitate cytoplasmic proteins onto the mesophyll cell walls, thereby rendering both the protein and cellulose more resistant to microbial decomposition. The process has been likened to the 'tanning' of leather, whereby plant tannins (polyphenols) are used to preserve the protein of animal skins. Where plant proteins and cellulose are not protected by tanning, these constituents and the sugars and storage carbohydrates are rapidly metabolized by the soil microorganisms that colonize the litter in a fairly definite sequence:

> sugar fungi and non-spore-forming bacteria →
> spore formers → cellulolytic myxobacteria →
> actinomycetes.

As indicated in section 3.1, whether or not net mineralization occurs during decomposition depends on the C:N ratio of the organic substrate being below or above a critical value of *c*. 25. This critical value is set by the C:N ratio of the decomposing organisms and their growth yield, *F*, which may be demonstrated as follows:

For the N requirements of the decomposers to be satisfied, the C:N ratio of the substrate must be:

$$\leq \frac{C_a}{N_m} \qquad (3.3)$$

where C_a, the C assimilated, is equal to C_m, the C incorporated into microbial cells plus C_r, the C

respired as CO_2; N_m = microbial N. Equation (3.3) may be written as:

$$\frac{C_a}{C_m} \times \frac{C_m}{N_m} \qquad (3.4)$$

from which it is obvious that the critical C:N ratio of the substrate is given by:

$$\frac{C:N \text{ ratio of microbial biomass}}{\text{growth yield, } F} \qquad (3.5)$$

C:N ratios range from about 3.3 for *Escherichia coli* to 12.9 for *Penicillium*, and the mean for soil microorganisms is about 10. Taking *F* as 0.4 (section 3.2), it can be seen that the critical substrate C:N ratio for net mineralization is *c*. 10/0.4 = 25.

The role of soil animals

There is a great deal of interdependence between the activities of microorganisms and the soil mesofauna in the decomposition of litter and formation of humus. The digestive processes of soil animals are normally very inefficient: termites are an exception because they can digest cellulose and lignin by means of their intestinal microflora or the fungus gardens they cultivate in their termitaria. With the earthworms some microbial decomposition of the litter must occur before it is readily acceptable to these animals. Once ingested, the main effect of an earthworm's digestive activity is the comminution of the plant residues which vastly increases the surface area accessible to microbial attack. There is some chemical breakdown, usually less than 10 per cent, which could be due to the activities of bacteria in the earthworm gut or to the effect of chitinase and cellulase which are secreted by the earthworms. The combined result of these processes is that earthworm casts have a higher pH, exchangeable Ca^{2+}, available phosphate and mineral nitrogen content than the surrounding soil.

Furthermore, whereas the partially humified residues in the L and F layers of mor humus retain a recognizable plant structure for many years, similar residues which have passed through the earthworm's digestive tract several times are reduced to a black, amorphous humus which is indissociable from the mineral particles.

Biochemical changes and humus formation

The biochemical changes, which in total comprise *humification*, are complex because both degradative and synthetic processes occur simultaneously. Certainly plant carbohydrates and proteins are decomposed and their microbial analogues synthesized, although the latter also become substrates to be decomposed in time. Important polymerization reactions involving aromatic* compounds also occur. For example, simple *o*- and *p*-hydroxy phenols occurring naturally in plants, or produced from the degradation of lignin and the polyphenolic pigments (Figure 3.3), are oxidatively polymerized to form humic precursors.

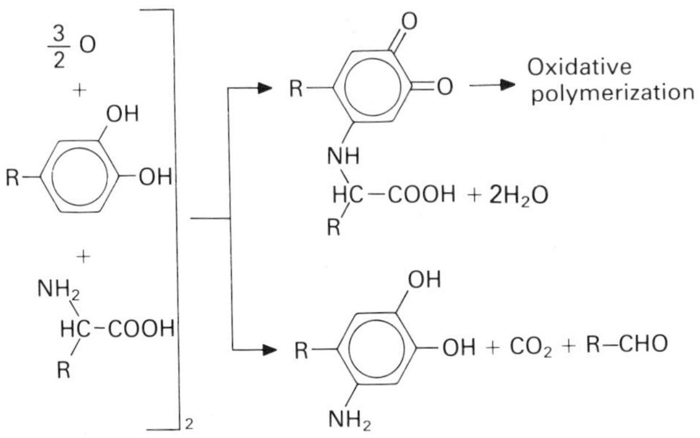

Figure 3.10. Reaction between amino acids and simple phenols under oxidizing conditions (after Andreux, 1982).

The oxidation can be auto-catalytic or catalysed by polyphenol oxidases. Nitrogen is incorporated into the polymers if amino acids condense with the phenols before polymerization occurs, as shown in Figure 3.10.

*Unsaturated ring structures of C and H. of which the simplest is benzene C_6H_6.

This is the basis of Waksman's (1938) ligno-protein theory for the formation of humic compounds. Some of the quinone rings formed during polymerization may be broken and additional carboxyl groups formed, as shown in Figure 3.11, which augment the acidity of the carboxyl groups in the non-aromatic (i.e. aliphatic) side chains. Carboxyls and the remaining reactive phenolic —OH groups form salts or chelate complexes (section 3.4) with metal ions which increase the stability of the macromolecule. Their longevity in soil also depends on physical stabilization due to adsorption on to mineral surfaces.

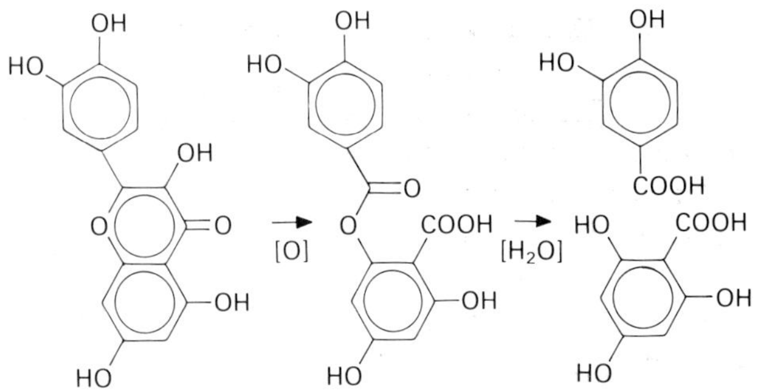

Figure 3.11. Enzymatic decomposition of a quinone polymer, with increase in carboxylic acid groups (after Andreux, 1982).

A typical elemental analysis of humic compounds on a mineral ash-free basis is 44–53 per cent C, 3.6–5.4 per cent H, 1.8–3.6 per cent N and 40.2–47 per cent O. In view of the diversity of complex molecules that can be formed, it is not surprising that the identification of the chemical structure of these compounds has proved difficult. The C:H ratio is an index of aromaticity, the minimum value being 1, as for benzene. Values > 1 reflect the degree of condensation of the rings and the substitution of other elements for H in the structure. An accurate knowledge of the structure of humic compounds is necessary when assessing the validity of the various hypotheses on the mode of humus formation and for explaining the properties of this material (see section 3.4).

3.4 PROPERTIES OF SOIL ORGANIC MATTER

Fractionation of organic matter

Physically, soil organic matter may be subdivided into those components that are readily dispersed and separated from the mineral particles, and that which can be dispersed only by drastic chemical treatment. A suggested subdivision is as follows:

(a) MACRO-ORGANIC MATTER

This consists of the most recently added plant and animal debris, which, due to the entrapment of air, has a density near 1 g cm^{-3} and can be separated from the heavier mineral particles by flotation in water or aqueous salt solutions. It is then collected on a fine sieve (0.06-mm aperture).

(b) LIGHT FRACTION

This consists of partially humidified and comminuted plant and faunal remains which can comprise up to 25 per cent of the total organic matter in grassland soils. Separation from the mineral particles is achieved after ultrasonic vibration of the dry soil, to disrupt the soil aggregates, and flotation of the light fraction in a heavy organic solvent (density 2 g cm^{-3}) with the aid of a surfactant.

(c) HUMIFIED FRACTION

Much of the soil organic matter adheres strongly to the mineral particles, particularly the clay, to form a *clay-humus complex*. This organic matter cannot be separated by density flotation, and is incompletely dissolved in metal-chelating solvents such as acetyl-acetone. It is most effectively dispersed by strong alkalis, such as sodium hydroxide, which act both by hydrolysing and depolymerizing the large organic molecules that are tightly adsorbed on the mineral surfaces.

Traditionally, the humus is subdivided further according to its solubility in alkaline and acidic solutions, as summarized in Figure 3.12. Central to the chemical characterization of humus is the problem that, because a strong solvent such as 0.1 M NaOH is necessary to recover most of the humus from the soil, there is some doubt that the compounds in the extract are the same as those originally in the soil. Further treatment with strong oxidizing agents such as KMnO$_4$ for the quantitative estimation of functional groups (—COOH, C=O etc.) produces other changes in the compounds present. Recently, however, progress has been made in the structural analysis of *fulvic acid* (FA) and *humic acid* (HA) fractions from soil by the use of cross-polarization, magic-angle spinning ^{13}C and ^{1}H nuclear magnetic resonance. This work indicates that the C aromaticity of HA is much less than previously thought ($< 50\%$), with unsaturated aliphatic chains containing —COOH groups making up the bulk of the remainder. Acidity attributable to phenolic-C groups is minor and carboxyls in the aromatic and aliphatic components contribute most of the titratable acidity (section 7.2).

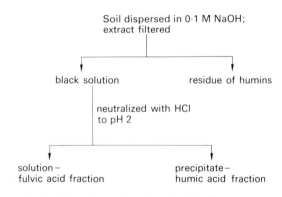

Figure 3.12. Scheme for the fractionation of soil humus.

The FA fraction generally consists of lower molecular weight molecules than the HA fraction which has MWs between 20,000 and 100,000. The polysaccharide content (mainly microbial) of FA can be as much as 10 per cent but the general nature of the FA-humic compounds is similar to that of the HA compounds. Indeed, FA may contain both the precursors and the degradation products of the HA fraction. The non-extractable *humins* are thought to be

HA-type compounds strongly adsorbed or precipitated on mineral surfaces as metal salts or chelates.

CATION EXCHANGE CAPACITY

Humification produces an organic colloid of high specific surface and high cation exchange capacity. Of the several functional groups containing oxygen, those which dissociate H^+ ions—the carboxylic and to a lesser extent phenolic groups—are the most important. The former have pK values between 3 and 5 while the latter only begin to dissociate protons at pH values > 7. A few basic $-NH_3^+$ groups may exist at pH values < 3. Thus, the *CEC* of soil organic matter is completely pH-dependent and buffered over a wide range of H^+ ion concentration; its value ranges between 150 and 300 cmols $(+)$ per kg dry matter. Organic matter can make a substantial contribution to the *CEC* of the whole soil, and hence to the retention of exchangeable cations, especially in soils of low clay content.

CHELATION

Organic compounds with a suitable spatial arrangement of reactive groups ($-OH$, $\odot$-OH and $-COOH$) can form coordination complexes with metallic cations by displacement of some of the water molecules from the cation's hydration shell. The resultant *inner-sphere* complex (section 2.3) acquires additional stability through the 'pincer effect' of the coordinating groups, giving rise to a *chelate compound* (Figure 3.13). Chelates formed with certain di- and polyvalent cations are the most stable, the stability constants falling in the order $Cu > Fe \simeq Al > Mn \simeq Co > Zn$. Metal complexes formed with the humic-acid fraction are largely

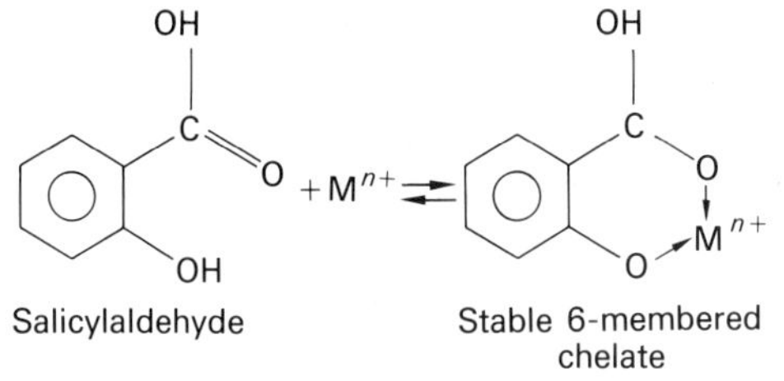

Salicylaldehyde

Stable 6-membered chelate

Figure 3.13. Organo-metallic chelate formation.

immobile, but the more ephemeral Fe^{2+}-polyphenolic complexes are soluble and their downward movement contributes to profile differentiation in *podzols* (section 9.2).

Organic matter and soil physical properties

The presence of organic matter is of great importance in the formation and stabilization of soil structure. Polymers of the FA and HA fractions are adsorbed on to mineral surfaces by a variety of mechanisms, primarily involving the functional groups—carboxyls, carbonyl ($-C=O$), alcoholic-OH, phenolic-OH, amino ($=NH$) and amine ($-NH_2$). These reactions are discussed in more detail in sections 4.4 and 7.5. Uncharged non-polar groups on the humic polymers, and large uncharged polymers, such as the polysaccharide and polyuronide gums synthesized by many bacteria, can be adsorbed at mineral surfaces by hydrogen-bonding and by van der Waals' forces (see Table 4.2), and also function as bonding agents between mineral particles.

3.5 FACTORS AFFECTING THE RATE OF ORGANIC MATTER DECOMPOSITION

Turnover

Gains of organic matter through litter decomposition and root death are offset by losses through decomposition and leaching. This is called *turnover*, which is defined as the flux of organic C through a unit volume of soil. Given a constant environment, a soil will eventually attain a steady-state equilibrium in which there is no measurable change in the organic matter content with time. The concept of turnover may be expressed mathematically by the equation:

$$\frac{dC}{dt} = A - kC \qquad (3.6)$$

where C = the organic C content of the soil (t ha^{-1}); A = the annual addition of residues, assumed to be constant; and k is a constant (y^{-1}), assuming that all the

C in the soil decomposes at the same rate according to simple first-order kinetics. At equilibrium $dC/dt = 0$ and:

$$A = kC \qquad (3.7)$$

The *turnover time* is then given by:

$$\frac{C}{A} = \frac{1}{k} \qquad (3.8)$$

Annual inputs of C for old arable soils at Rothamsted in England have been estimated at 5–6 per cent of the soil organic C, giving turnover times of 16–22 years. Theoretically, this is the time taken for all the organic C in the soil to be replaced, and in a sense is inversely related to the 'biological activity' of the soil. This picture is grossly over-simplified because the half-life of C in these soils has been determined from radioactive carbon dating as 1,450–3,700 years, depending on depth. It is obvious that there must be a great range of decomposability of C compounds in the soil, with some of the C much younger than 16–22 years and the bulk of it very much older.

This is confirmed by measurements of the rate of disappearance of ^{14}C from field soils to which labelled residues have been added. For various residues (ryegrass shoots and roots, green maize, wheat straw) and

even for soluble compounds like glucose, in a temperate climate, a surprisingly constant fraction (*c.* two-thirds) of the material decomposes and disappears in the first year (Figure 3.14). In straw, the more resistant component is mainly lignin, but in the case of glucose the residue is mainly microbial metabolites produced during decomposition. After 5 years, approximately one-fifth of the added ^{14}C remains and its decomposition rate then still exceeds that of the unlabelled native soil C. Thus, although simple models such as equation 3.6 can represent decomposition reasonably well over longer times (10–100 y), more complex models are needed to fit data for the first few years following the addition of residues to soil, or when changes in soil management are made.

Although a whole spectrum of k values must pertain in the field, changes in organic C and N have been modelled quite successfully using four or five components of organic matter, each decomposing at a distinctive rate. Jenkinson and Rayner (1977) divided the C added into easily decomposable (DPM) and resistant residues (RPM), both of which decompose to release CO_2 and form biomass (BIO) plus physically stabilized (POM) and chemically stabilized organic matter (COM). The BIO, POM and COM fractions can in turn decompose to form more CO_2, BIO, POM and COM. For unmanured soils under wheat at Rothamsted (annual input = 1.2 t C ha^{-1}), the model which fitted best the experimental data predicted that each hectare of soil should contain

0.1 t C as DPM (turnover time = 0.2 y)
0.6 t C as RPM (turnover time = 3.3 y)
0.3 t C as BIO (turnover time = 2.4 y)
13.6 t C as POM (turnover time = 71 y)
14.6 t C as COM (turnover time = 2,900 y).

A check on the model was that the predicted radio-C age of 1,240 y was close to the soil's measured radio-C age of 1,450 y. This is only one example of a multi-component dynamic model of soil organic matter turnover. Others may identify somewhat different components with different turnover times: for

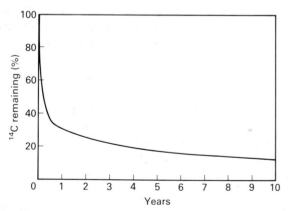

Figure 3.14. Simplified time-course for the decomposition of ^{14}C-labelled C compounds in the field (after Jenkinson, 1981 and Paul, 1984).

example, Paul's (1984) model for N turnover employs 4 components as follows:

	Per cent of total soil N	Turnover time (y)
Biomass	4–6	0.7
Active non-biomass	6–10	2.2
Stabilized N	36	32
Old N	50	866

Both of these examples demonstrate that half or more of the soil organic matter is resistant to decomposition and therefore very old.

Substrate availability

PRIMING ACTION

It was once thought that the stimulus to the resident soil microbial population provided by the addition of fresh residues was often sufficient to accelerate also the

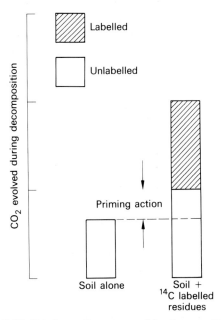

Figure 3.15. Priming action measured by the use of ^{14}C-labelled residues (after Jenkinson, 1966).

decomposition of soil humus. This effect was called *priming action*. Priming action, which can be positive or negative, has been demonstrated subsequently by following the decomposition of ^{14}C-labelled residues, as illustrated in Figure 3.15. The conclusion from this kind of experiment is that priming action generally has a negligible effect on the overall rate of decomposition. The addition of fresh residues does stimulate microbial activity, but the increase in soil biomass usually compensates for any acceleration in the rate of humus decomposition, so that the net effect is an increase in soil organic matter.

SOIL PROPERTIES AND ENVIRONMENTAL CONDITIONS

Soil organic matter levels are affected by moisture, O_2 supply, pH and temperature. The first two factors tend to counteract one another because when soil moisture is high, deficiency of O_2 may restrict decomposition, whereas when the soil is dry, moisture but not O_2 will be limiting. pH has little effect, except below 4 when the decomposition rate slows as in the case of mor humus and many upland peats (section 9.5). On the other hand, temperature has a marked effect, not only on plant growth and hence on litter return, but also on litter decomposition through an effect on microbial respiration rate (section 8.1). In a comparison of Nigerian (mean temperature 26.1°C) and English (mean temperature 8.9°C) soils, Jenkinson and Ayanaba (1977) found that the rate of disappearance of organic residues in both cases could be represented by the curve in Figure 3.14 *provided that* 10 years in southern England were equated with 2.5 years in Nigeria.

The adsorption of protein by montmorillonite, especially in the interlamellar regions, protects the protein from microbial attack. Similarly, the adsorption of various compounds by clays and sesquioxides generally serves to slow down their rate of decomposition. The organic matter held in the relatively stable pores in clay soils of diameter < 1 μm—the so-called 'sterile pores'—is also less accessible to microbial attack. Cultivation of the soil tends to break down the structure so that organic matter in sterile pores is exposed to microorganisms and its decomposition rate accelerated. Higher surface soil temperatures are also attained when the protection of the vegetative canopy and litter layer is lost.

Under English conditions, clay content has little effect on the rate of decomposition of most fresh residues. Decomposition is slightly slower in clay than in sandy soils, but part of this difference is explicable by the greater leaching losses from light-textured soils. Under tropical conditions, where the potential for decomposition is greater, the part played by clay minerals and particularly sesquioxides in protecting soil organic matter is much greater. Similarly, in soils derived from volcanic ash, the considerable accumulation of organic matter is attributed to the stabilization of the humic and fulvic acid fractions through their adsorption onto the Al sheets of imogolite and allophane clay minerals.

Activity of the soil biomass

NUMBERS OF ORGANISMS
The effect of environmental and soil factors on the size and species diversity of the populations of soil microorganisms has been referred to briefly. Soil animals are also affected by types of litter, prevailing conditions of temperature, moisture and pH, and management practices, such as cultivation and the use of agricultural chemicals.

PHYSIOLOGICAL ACTIVITY
Environmental and management factors can influence the physiological activity of the soil organisms. Striking effects occur, for example, when the soil is partially sterilized with toxic chemicals, such as chloroform or toluene. On the restoration of favourable conditions, the rapid multiplication of the few surviving organisms, feeding on the bodies of the killed organisms, produces a flush of decomposition as evidenced by a surge in the release of CO_2. Similar flushes of decomposition occur in soils subjected to extremes of wetting and drying

(section 10.2) and in soils of temperate regions on passing from a frozen to thawed state.

The dynamics of soil biomass changes in response to variations in soil and environmental factors are summarized graphically in Figure 3.16.

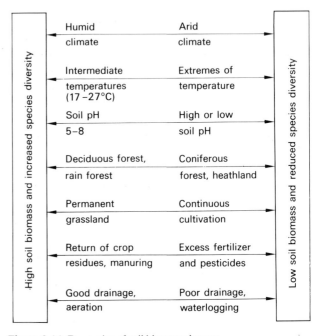

Figure 3.16. Dynamics of soil biomass changes.

3.6 SUMMARY

Plant litter, dead roots, animal remains and excreta form the *soil organic matter*, which is a repository for the essential elements required by the next generation of organisms. Release of these elements or *mineralization* depends on the decomposition rate and the demands made by the heterogeneous population of soil organisms.

The living organisms (*biomass*) vary in size from the *macrofauna*, vertebrate animals of the burrowing type, to the *mesofauna*, invertebrates such as mites, springtails, insects, earthworms and nematodes, to the *microorganisms*—broadly subdivided into the *prokaryotes* and *eukaryotes* and comprising the bacteria, actinomycetes, fungi, algae and protozoa. The most important of the mesofauna are the *earthworms* in temperate soils, especially under grassland, and *termites* in many soils of the tropics and subtropics. Earthworm biomass normally ranges from 0.5 to 1.5 t ha^{-1}, whereas the microbial biomass makes up 2–4 per cent of the total organic C in soil, or 0.5–2 t C ha^{-1}.

Annual rates of litter fall range from 0.1 t C ha^{-1} in alpine and Arctic forests to 10 t C ha^{-1} in tropical rain forests. Carbon substrates from rhizosphere deposition and the death of roots also make an important contribution that is difficult to quantify. The interaction of microorganisms and mesofauna in decomposing litter leads not only to the release of mineral nutrients but also to the synthesis of complex new organic compounds that are more resistant to attack. This is called *humification*. The chemical structure of humic compounds has proved difficult to elucidate, but analyses made of *fulvic acid* (FA) and *humic acid* (HA) fractions indicate that polysaccharides of plant and microbial origin and substituted aromatic and aliphatic polymers with a high density of —COOH groups, are important constituents. Soil humus has a high *CEC* and is active in chelating metallic cations. Humic compounds are also important in the stabilization of soil structure (Chapter 4).

The organic matter content of a soil reflects the balance between gains of C from plant residues and excreta and the loss of C through decomposition by soil organisms. Experiments with ^{14}C-labelled residues show that under temperate conditions, about one-third of the C remains after 1 year and one-fifth after 5 years. In the hotter tropics, decomposition is approximately 4 times faster than in temperate soils. Low pH retards decomposition, as does a high polyphenolic content in the senescing plant tissue, leading typically to the formation of *mor humus*. *Mull humus* forms on base-rich soils under herbaceous and deciduous forest species low in polyphenols. C:N ratios > 25 favour an increase in microbial biomass and *net immobilization* of N, whereas C:N ratios < 25 favour *net mineralization*.

REFERENCES

ANDREUX F. (1982) Genesis and properties of humic molecules, in *Constituents and Properties of Soils* (Eds M. Bonneau and B. Souchier). Academic Press, London, pp. 109–139.

EDWARDS C.A. & LOFTY J.R. (1977) *Biology of Earthworms*, 2nd Ed. Chapman and Hall, London.

HEPPER C.M. (1975) Extracellular polysaccharides of soil bacteria, in *Soil Microbiology* (Ed. N. Walker). Butterworth, London.

JENKINSON D.S. (1966) The priming action, in *The use of isotopes in soil organic matter studies*. Proceedings of FAO/IAEA Technical Meeting, Brunswick, Völkenrode. Pergamon, Oxford, pp. 199–208.

JENKINSON D.S. (1981) The fate of plant and animal residues in soil, in *The Chemistry of Soil Processes* (Eds D.J. Greenland and M.H.B. Hayes). Wiley, Chichester, pp. 505–561.

JENKINSON D.S. & AYANABA A. (1977) Decomposition of carbon-14 labeled plant material under tropical conditions. *Soil Science Society of America Journal* **41**, 912–915.

JENKINSON D.S. & POWLSON D.S. (1976) The effects of biocidal treatments on metabolism in soil. I. Fumigation with chloroform. *Soil Biology and Biochemistry* **8**, 167–177.

JENKINSON D.S. & RAYNER J.H. (1977) The turnover of soil organic matter in some of the Rothamsted classical experiments. *Soil Science* **123**, 298–305.

PARKINSON D., GRAY T.R.G. & WILLIAMS S.T. (1971) *Ecology of Soil Micro-organisms*. IBP Handbook No. 19. Blackwell Scientific Publications, Oxford.

PAUL E.A. (1984) Dynamics of organic matter in soil. *Plant and Soil* **76**, 275–285.

RUSSELL E.W. (1973) *Soil Conditions and Plant Growth*, 10th Ed. Longman, London.

WAKSMAN S.A. (1938) *Humus*, 2nd Ed. Bailiere, Tindall and Cox, London

WATERHOUSE D.F. (1973) Pest management in Australia. *Nature New Biology* **246**, 269–271.

FURTHER READING

ALEXANDER M. (1977) *Introduction to Soil Microbiology*, 2nd Ed. Wiley, New York.

FENCHEL T. & BLACKBURN T.H. (1979) *Bacteria and Mineral Cycling*. Academic Press, London.

GREENLAND D.J. (1971) Interactions between humic and fulvic acids and clays. *Soil Science* **111**, 34–41.

KONONOVA M.M. (1966) *Soil Organic Matter*, 2nd Ed. Pergamon Press, Oxford.

LADD J.N. & MARTIN J.K. (1984) Soil organic matter studies, in *Isotopes and Radiation in Agricultural Sciences, Volume 1* (Eds M.F. L'Annuziata and J.O. Legg). Academic Press, London, pp. 67–98.

LYNCH J.M. & POOLE N.J. (1979) (Eds) *Microbial Ecology. A Conceptual Approach*. Blackwell Scientific Publications, Oxford.

STEVENSON F.J. (1982) *Humus Chemistry—Genesis, Composition, Reactions*. Wiley, New York.

SWIFT M.J., HEAL O.W. & ANDERSON J.M. (1979) *Decomposition in Terrestrial Ecosystems*. Studies in Ecology, Volume 5. Blackwell Scientific Publications, Oxford.

Chapter 4
Peds and Pores

4.1 SOIL STRUCTURE

As we have seen in Chapter 3, one reason why the soil supports a diversity of plant and animal life is the abundance of energy and nutrients supplied in organic remains. An equally important reason lies in the physical protection and favourable temperature, moisture and oxygen supply afforded by the structural organization of the soil. Weathering of parent material produces the primary soil particles of various sizes—the clay, silt, sand and stones. These particles may simply 'pack', as for example in Figure 4.1a, to attain a state of minimum potential energy. However, vital forces associated with plants, animals and micro-organisms and physical forces associated with the change in state of water and its movement, arrange the soil particles into larger units of varied size called *aggregates* (Figure 4.1b). This aggregation of soil particles is the basis of *soil structure* of which several definitions have been offered: in the broadest sense, it denotes:

(a) the size, shape and arrangement of the particles and aggregates;
(b) the size, shape and arrangement of the voids or spaces separating the particles and aggregates;
(c) the combination of voids and aggregates into various types of structure.

4.2 LEVELS OF STRUCTURAL ORGANIZATION

Many soils have structural features that are readily observed in the field, and the most widely used system for describing and classifying these features—the *macrostructure*—is that laid down in the USDA Soil Survey Manual of 1951. However, much of the fine detail of aggregation and the distribution of voids—the *microstructure*—can only be determined in the laboratory using powerful tools, such as the polarizing light microscope and the scanning electron microscope which can provide magnification up to 20,000 times. Some of the macroscopic features have identifiable counterparts at the microscopic level, in which case a descriptive term common to both macro- and microstructure can be used; other features are peculiar to the microstructure and therefore require new descriptive terms, which have been devised by specialists in the study of thin soil sections—notably Kubiena (1938) and Brewer (1964, 1976).

Aggregation
Relatively permanent aggregates or *peds* are recognizable in the field because they are separated by voids and natural planes of weakness. They should persist through cycles of wetting and drying, as distinct from the less permanent aggregates formed at or near the soil surface by the mechanical disturbance of digging and ploughing (*clods*) or the rupture of the soil mass *across* natural planes of weakness (*fragments*). A *concretion* is formed when localized accumulations of an insoluble compound irreversibly cement or enclose the soil particles. *Nodules* are similarly formed but lack the symmetry and concentric internal structure of concretions.

PED TYPE
Four main types of ped have been recognized, and they are illustrated in Figure 4.2. A brief description of each type follows.

(1) *Spheroidal*. Peds that are roughly equidimensional and bounded by curved or irregular planar surfaces that are not accommodated by the faces of

(a)

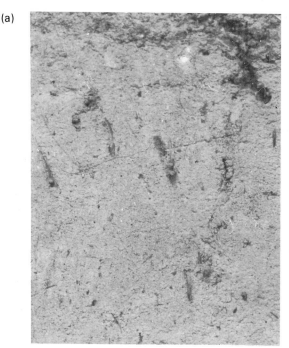

(b)

Figure 4.1. (a) Random close packing of soil particles. **(b)** Structured arrangement of soil particles.

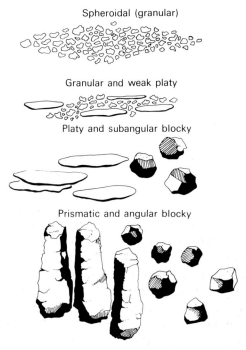

Spheroidal (granular)

Granular and weak platy

Platy and subangular blocky

Prismatic and angular blocky

Figure 4.2. Examples of common ped types (after Strutt, 1970).

adjacent peds. There are two subtypes—*granules* which are relatively non-porous, and *crumbs* which are very porous. These peds are usually evenly coloured throughout, and arise primarily from the interaction of soil organisms, roots and mineral particles. They are therefore common in the A horizons of soils under grassland and deciduous forest.

(2) *Blocky.* Peds that are bounded by curved or planar surfaces that are more or less mirror images of the faces of surrounding peds. Those with flattened faces and sharply angular vertices are called *angular*; those with a mixture of rounded and flat faces and more subdued vertices are called *subangular*. Blocky peds are features of B horizons where their natural surfaces may be identified by their smoothness and often distinctive colours. The ped faces may be coated, for example, by organic matter, clay or metal oxides (section 4.3). The voids between peds often form paths for root penetration and water flow; the ped faces may

have live or dead roots attached, or show distinct root impressions.

(3) *Platy.* The particles are arranged about a horizontal plane with limited vertical development; most of the ped faces are horizontal. When the ped is thicker in the middle than at the edges it is called *lenticular.* These peds may occur in the A horizon of a soil immediately above an impermeable B horizon, or in the surface of a soil compacted by heavy machinery.

(4) *Prismatic.* In these peds the horizontal development is limited when compared with the vertical; the

Figure 4.3. Prismatic structure of a clay subsoil prone to waterlogging.

peds are bounded by flat, vertical faces with sharp vertices. They are subdivided into prisms without caps (*prismatic*) and prisms with rounded caps (*columnar*). Prismatic structure is common in the subsoil of heavy clay soils subject to frequent waterlogging (Figure 4.3). Columnar structure is associated particularly with the B horizon of sodium-affected soils (Figure 4.4).

COMPOUND PEDS

All types of ped shown in Figure 4.2 are invariably compounded of smaller units which are clearly visible, and others which can only be seen under a microscope. For example, the large compound prismatic peds in Figure 4.5 are composed of secondary subangular blocks which in turn break down to the primary structure of angular blocky peds < 1 cm in diameter. Each of these primary peds presents an intricate internal arrangement of granules and voids which can only be elucidated by careful appraisal of the soil in thin sections (section 4.3).

CLASS AND GRADE

In addition to type, the *class* of ped is judged according to its size, and the *grade* according to the distinctness and durability of the visible peds. Grade of structure may be described as follows.

(1) *Structureless (or apedal).* No observable aggregation nor definite arrangement of natural lines of weakness. *Massive* if coherent; *single grain* if noncoherent.

(2) *Weakly developed.* Poorly formed, indistinct peds that are barely observable *in situ.* When disturbed, the soil breaks into a mixture of a few entire peds, many broken peds and much unaggregated material.

(3) *Moderately developed.* Well-formed, distinct peds that are moderately durable, but not sharply distinct in undisturbed soil. The soil, when disturbed, breaks down into a mixture of many entire peds, some broken peds and a little aggregated material.

Figure 4.4. Columnar structure in the B horizon of a *solodized solonetz* (CSIRO photograph, courtesy of G.D. Hubble).

Prismatic
compound ped

Subangular blocky
secondary ped

5 cm

Figure 4.5. Compound prismatic peds showing the breakdown into secondary structure.

(4) *Strongly developed.* Durable peds that are quite evident in undisturbed soil; they adhere weakly to one another and become separated when the soil is disturbed. The soil material consists very largely of entire peds and includes a few broken peds and little or nor unaggregated material.

Grades of structure reflect differences in the strength of *intraped* cohesion relative to *interped* adhesion. The grade will depend upon the moisture content at the time of sampling and the length of time the soil has been exposed to direct drying in air. Prolonged drying will markedly increase the structure grade, particularly of clay soils and soils of high iron oxide content.

0·6 mm

Figure 4.6. Thin section through a soil pore showing clay coatings (courtesy of P. Bullock).

4.3 SOIL MICROMORPHOLOGY

Techniques and terminology

Thin soil sections and their study come under the heading of *soil micromorphology*. Briefly, the method of thin sectioning consists of taking an undisturbed soil sample (dimensions 10×5×3.5 cm), removing the water in such a way as to minimize soil shrinkage and impregnating the voids with a resin under vacuum. Once the resin has cured, the soil block is sawn and polished to produce a section $\sim$ 25 μm in thickness which is stuck onto a glass slide for viewing. The section is examined with a petrological microscope to determine the characteristic optical properties of the minerals present in the soil. An example of a thin soil section is shown in Figure 4.6.

Micromorphologists have found it difficult to agree on an unambiguous set of terms to describe soil features seen in thin sections. A 'foundation' set of terms was introduced by Brewer in 1964 and revised in 1976; a summary of these is presented in Table 4.1. Fitzpatrick (1984) provides a valuable glossary of terms and their meanings as used by the principal proponents in the field.

Reorientation of plasma *in situ*

During the repeated cycles of wetting and drying, and sometimes freezing and thawing, that accompany soil formation, the plate-like clay particles gradually become reorganized from a randomly arranged network of small bundles, the *asepic* S-matrix, into larger bundles that show a degree of preferred orientation,

Table 4.1. Summary of descriptive terms in soil micromorphology introduced by Brewer (1964, 1976).

FABRIC—the spatial arrangement of particles, primary and compound (aggregated), and voids

Soil material with basic units or organization—the PEDS

Soil material without obvious organization—APEDAL

S-MATRIX—comprising the plasma, skeleton grains and voids of primary peds, or comprising apedal soil material that does not occur as pedological features other than plasma separations. Thus we have:

PEDOLOGICAL FEATURES—those formed due to reorganization of matrix materials by pedogenic processes (ORTHIC) and those inherited from the parent material (INHERITED). Orthic features are subdivided into:

PLASMA	SKELETON GRAINS	VOIDS	PLASMA CONCENTRATIONS	PLASMA SEPARATIONS	FOSSIL FORMATIONS
colloidal ($<2\,\mu m$) and soluble materials not bound up in skeleton grains, which can be moved, reorganized and/or concentrated during soil formation; includes mineral and organic material.	larger than colloidal size and include detrital mineral fragments, secondary crystalline and amorphous bodies which are relatively stable and not usually translocated, concentrated or reorganized during soil formation.	the spaces within the plasma and between plasma and skeleton grains.	the concentration of any component of the plasma which has moved in solution or in suspension; e.g. carbonate and oxide nodules, clay coatings.	the changed arrangement of the plasma components, as, for example, the orientation of clay minerals at the surfaces of peds, such as slickensides (see text).	preserved root channels or burrows of the soil fauna; may be filled with various materials.

Figure 4.7. Intersecting slickensides in a heavy clay subsoil (CSIRO photograph, courtesy of G.D. Hubble).

the *sepic* S-matrix. The bundles or plasma separations are called *domains*, which can be up to 5 μm long and 1–5 μm wide, depending on the dominant clay mineral present and their mode of formation.

As well as the reorganization of clay particles within peds, there is often a change in the orientation of clay particles at ped faces. In dry seasons, material from upper horizons sometimes falls into deep cracks in clay soils. On rewetting, the subsoil, which now contains more material than before, swells to a larger volume and exerts a pressure that displaces material upwards. Ped faces slide against each other to produce zones of preferred orientation called *slickensides* lying at an angle of about 45° to the vertical (Figure 4.7).

Plasma translocation and concentration

One of the major processes of soil formation is the translocation of colloidal and soluble material from upper horizons into the subsoil. In the changed physical and chemical environment of the illuviated horizon, salts and oxides precipitate, clay particles flocculate and the ped faces become coated with deposited materials. Brewer has classified these *coatings* or *cutans* according to the nature of the surface to which they adhere and their mineralogical composition. Some of the more common examples are given below.

Clay coatings (argillans). Often different in colour from and with a higher reflectance than the S-matrix of the ped, they are easily recognized in sandy and loam soils, but difficult to distinguish from slickenside surfaces in clay soils (Figure 4.6).

Sand or silt coatings (skeletans). They can be formed by the illuviation of sand or silt, or by the residual concentration of these particles after clay removal.

Sesquioxidic coatings (sesquans, mangans, ferrans). They can be formed by reduction and solution of Fe

Figure 4.8. Manganese dioxide coatings on ped faces (courtesy of R. Brewer).

Figure 4.9. A thin iron pan developed in ferruginous sand deposits.

and Mn under anaerobic conditions and their subsequent oxidation and deposition in aerobic zones (Figure 4.8). Colours range from black (mangans) through grey, red, yellow, blue and blue-green (ferrans), depending on the degree of oxidation and hydration of the iron compounds.

Salt coatings (of gypsum, calcite, sodium chloride) and *organic coatings* (organans) can also occur.

Mixtures of these materials, particularly clay, sesquioxides and organic matter, to form *compound coatings* are not uncommon.

Pans. The coatings may build up to such an extent that bridges of material form between soil particles and

the smallest peds so that a cemented layer, sometimes impenetrable to roots, may develop in the soil. Such a layer or horizon is called a *pan* and is described by its thickness (if < 1 cm it it called 'thin') and its main chemical constituent. An example of a thin iron pan is shown in Figure 4.9. Compacted horizons that are weakly cemented by silica are called *duripans*. A *fragipan* is not cemented at all, but the size distribution of the soil particles is such that they pack to a very high bulk density (section 4.5).

Voids. Cylindrical and spherical voids, usually old root channels or earthworm burrows, are called *pores* and those of roughly planar shape are called *fissures*.

Figure 4.10. Close cracking pattern in the surface of a montmorillonitic clay soil (scale in photo is 15 cm long) (CSIRO photograph, courtesy of G.D. Hubble).

Those of indeterminate shape are called *vughs*. Fissures start as cracks striking vertically downwards from the surface, especially in soils containing much of the expanding lattice-type clays (Figure 4.10). Secondary cracks develop at right angles to the vertical fissures.

4.4   THE CREATION OF SOIL STRUCTURE

Formation of aggregates

The organization of soil material to form peds and pedological features, as described previously, requires the action of physical, chemical and biotic forces. For example, swelling pressures are generated through the osmotic effects of exchangeable cations adsorbed on clay surfaces. There are also forces associated with the change in state of the soil water. Unlike any other liquid, water expands on cooling from 4 to 0°C. As the water in the largest pores freezes, liquid water is drawn from the finer pores thereby subjecting them to shrinkage forces; but the build-up of ice lenses in the large pores, and the expansion of the water on freezing, subjects these regions to intense disruptive forces. At higher temperatures tension forces develop as water evaporates from soil pores, and the prevalence of evaporation over precipitation can lead to localized accumulations of salts and the formation of concretions anywhere within the soil profile. The mechanical impact of rain, animal treading and earth-moving machinery can reorganize soil materials and the energy of running water can move them.

Biotic forces are no less important. Burrowing animals reorganize soil materials, particularly the earthworm which grinds and mixes organic matter with clay and silt particles in its gut, adding extra calcium in the process, to form a cast of considerable strength. The physical binding effect of fine roots and fungal mycelia also contributes to aggregation; and the metabolic energy expended by soil microorganisms can maintain concentration gradients for the diffusion of salts, or induce a change in valency leading to the solution and subsequent reprecipitation of manganese and ferric oxides.

INTERPARTICLE FORCES

For a ped to persist, *interparticle* forces within the ped must be sufficient to prevent the particles separating again under the continuing impact of the forces responsible for reorganization. Good structure for plant growth depends on the existence of peds between 1 and 10 mm in diameter (the primary peds mentioned in section 4.2) that are stable when wetted. A test of water stability is to place an air-dry ped in water of very low salt concentration (akin to rain water). If the ped breaks down into subunits, which may themselves be aggregates of smaller particles, it is said to *slake*. This indicates that the intraped forces are not strong enough to withstand the pressure of air entrapped as the ped wets quickly, and the pressure generated by swelling. The subunits produced by slaking may also be unstable in which case individual clay-size particles may separate—a process of *deflocculation*. Several undesirable effects follow such as the translocation of clay to form a dense clay B horizon (section 9.2), the blocking of drainage pores and, once the soil has dried, the formation of surface crusts, which impede infiltration and seedling emergence.

Flocculation, deflocculation and swelling are surface phenomena which will be discussed in more detail in section 7.4. However, it is pertinent here to consider

ways in which the primary constituents—sand, silt, clay and organic matter—are organized to form peds. Some authors distinguish between macroaggregates (> 250 μm diameter) and microaggregates (< 250 μm). This division is made by taking account of the soil's response to disruptive treatments (sonic vibration, rapid wetting, periodate oxidation) and the inferred mechanism of aggregate stabilization. Several models of *microaggregate* organization have been proposed, all of which involve some combination of clay–clay and clay–organic matter interactions.

Domains consisting of illitic or vermiculite clay crystals up to 5 nm thick, stacked in roughly parallel alignment, and extending up to 5 μm in the *a–b* direction, are thought to be the fundamental structural units of many soils. Similarly, single-layer crystals of montmorillonite (1 nm thick) can align themselves, with much overlapping, to form *quasi-crystals* several nanometres thick and exhibiting long range order up to 5 μm in the *a–b* direction. When Ca^{2+} is the dominant exchangeable cation, the *intercrystalline* spacing in the domains and the interlayer (*intracrystalline*) spacing in the quasi-crystals do not exceed 0.9 nm, even when the soil is placed in distilled water, because of the stability of the trimolecular layer hydrate of Ca^{2+} ions between opposing clay surfaces (section 2.4). Each unit—quasi-crystal or domain—acts as a stable entity in water, showing limited swelling, and is only likely to be disrupted if more than *c.* 30% of the moles of charge due to Ca is replaced by Na. In Emerson's (1959) model of a microaggregate, domains or quasi-crystals combine with organic polymers, through edge or face attractions, and also bond to the O surfaces of quartz grains (Figure 4.11). The organic polymers are most probably polysaccharides and polyuronide gums and mucilages, mainly of bacterial and fungal origin, which are sensitive to oxidation by sodium periodate, although humic polymers that complex polyvalent cations through carboxyl groups may also be involved.

Another model of microaggregate organization for soils of high-base status (Edwards and Bremner, 1967) proposes a clay (C)–polyvalent cation of (P)–organic matter (OM) complex as the fundamental structural

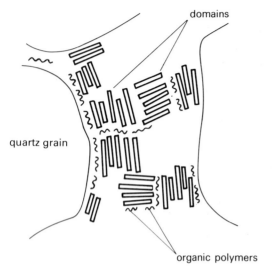

Figure 4.11. Possible arrangements of clay domains, organic polymers and quartz grains in a microaggregate (after Emerson, 1959).

Stabilization of aggregates

Flocculated clay is a prerequisite of microaggregate ($< 250\ \mu$m) stability and the ability to resist disruption on wetting and mechanical disturbance. Clay–organic matter interactions, which may involve polyvalent cations (Fe^{3+}, Al^{3+} or Ca^{2+}) and humic polymers, or polysaccharides and clay surfaces, are also important. Inorganic compounds act as interparticle cements and stabilization agents, for example, calcium carbonate in calcareous soils, sesquioxides as films between planar clay surfaces and as discrete charged particles in many acid, highly weathered soils, and silica at depth in *laterites* (section 9.4), where an indurated layer may form. The nature of microaggregate stabilization is such, therefore, that it is relatively insensitive to changes in soil management, except in so far as management induces marked changes in the dominant exchangeable cations (Na^+ replacing Ca^{2+}) or soil pH (acid $\rightarrow$ alkaline pH decreases the positive charges on kaolinite edge faces and sesquioxide surfaces).

On the other hand, macroaggregation (units > 250 μm) is much more dependent on management because it is stabilized primarily through the action of plant roots and fungal hyphae, particularly the vesicular–arbuscular *mycorrhizas* (section 10.3). Roots can

unit ($< 2\ \mu$m). C–P–OM units combine to form (C–P–OM)$_x$ and $[(C–P–OM)_x]_y$ particles, probably through cation bridging, which might also take the form C–P–C or OM–P–OM. The microaggregates formed by these units should be roughly spherical and they lack any regions of preferred clay orientation characteristic of domains and quasi-crystals. The organic matter, being tightly bonded to the cations inside the aggregate, would be largely inaccessible to microbial attack.

In more highly weathered soils, in which 1:1 clays, such as kaolinite, may predominate, individual clay crystals can flocculate edge to face to form a 'card-house' type of microstructural unit (section 7.4). Positively charged sesquioxide films on the planar surfaces also act as electrostatic bridges between crystals, or form complexes with negatively charged organic polymers. A summary of the various inter-particle forces of attraction involved in microaggregate formation is given in Table 4.2.

Macroaggregate formation depends primarily on the stabilization of microaggregate in larger structural arrangements, which is discussed below.

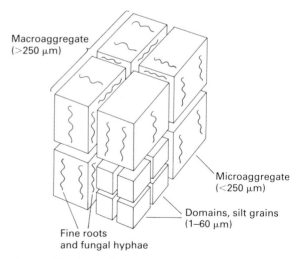

Figure 4.12. Diagrammatic representation of a soil macroaggregate (after Greenland, 1979).

Table 4.2. Summary of interparticle forces of attraction.

ELECTROSTATIC OR COULOMBIC FORCES—these involve clay–clay, clay–oxide, and clay–organic interactions and include:	VAN DER WAALS' FORCES—these involve clay–organic interactions such as:
(a) The positively charged edge of a clay mineral attracted to the negative cleavage face of an adjacent clay mineral.	(a) Specific dipole–dipole attraction between constituent groups of an uncharged organic molecule and a clay mineral surface, as in hydrogen-bonding between a polyvinyl alcohol and O or OH surface.
(b) A positively charged sesquioxide film 'sandwiched' between two negative clay mineral surfaces.	(b) Non-specific dispersion forces between large molecules when in very close proximity, which are proportional to the number of atoms in each molecule. These forces can account for the strong adsorption of poly-saccharide and polyuronide gums and mucilages by soil particles. These polymers of molecular weight 100,000–400,000 are produced in the soil by bacteria, such as *Pseudomonas* spp., which are abundant in the rhizo-sphere of many grasses. When the adsorption of such a large molecule results in the displacement of many water molecules, the increase in entropy of the system is large and the reaction is essentially irreversible. Large, flexible polymers of this kind can make contact with several clay crystals at many points and so help to hold them together in the manner of a 'coat of paint' (section 7.5).
(c) Positively charged groups (e.g. amino groups) of an organic molecule attracted to a negative clay mineral surface. Organic molecules, in displacing the more highly hydrated inorganic cations from the clay, considerably modify its swelling properties.	
(d) Clay–polyvalent cation–organic anion linkages or 'cation bridges'. If the organic anion is very large (a polyanion), it may interact with several clay particles through cation bridges, positively charged edge faces or isolated sesquioxide films. Linear polyanions of this kind (e.g. 'Krilium') are excellent flocculants of clay minerals (section 7.5).	

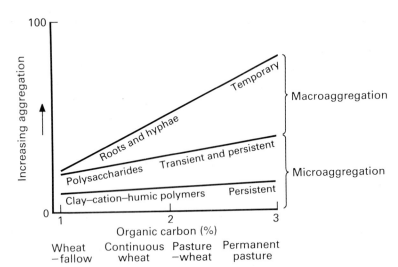

Figure 4.13. The relative contributions of various forms of organic matter to the stabilization of aggregates in soil (after Tisdall and Oades, 1982).

penetrate pores $> 10\ \mu$m diameter and fungal hyphae pores $> 1\ \mu$m, and, in so doing, they enmesh soil particles to form stable macroaggregates. A diagrammatic model of a macroaggregate is shown in Figure 4.12. Grass swards are especially effective because of the high density of roots in the surface horizon. Further to the physical binding effect, rhizosphere C deposition (section 3.1) provides substrate for the saprophytic microorganisms producing the polysaccharide gums that bind soil particles together. In all, these effects depend on maintaining a high level of biotic activity—plants, animals (mainly earthworms) and microorganisms—in the soil; they are most important in soils of pH 5.5–7, which derive little benefit through aggregate stabilization by either sesquioxides or $CaCO_3$. Substantial inputs of organic matter and minimal soil disturbance therefore favour macroaggregate stabilization, as shown in Figure 4.13.

4.5 PORE VOLUME

Porosity

Clearly, the size, shape and arrangement of the peds determines the pore space or *porosity* of the soil.

Porosity is measured as the ratio:

$$\text{Pore Space Ratio } (PSR) = \frac{\text{volume of pores}}{\text{total soil volume}} \quad (4.1)$$

The *PSR* depends on the water content of the soil, since both pore volume and the total volume of an initially dry soil may change due to swelling as clay surfaces hydrate within the peds. The soil moisture content at which the porosity is measured therefore needs to be stated.

Total pore space indicates nothing about the actual size distribution of the pores. Pores larger than about $60\ \mu$m in diameter can be seen with the naked eye, and are called *macropores*. They comprise a high proportion of the space, particularly that created by faunal activity, between the soil peds. The distribution of pore sizes in non-swelling soils can be determined reasonably accurately from the volume of water released in response to known suctions, using the relationship between suction and the radius of pores that will just retain water at that suction (section 6.1). In clay soils, however, where contraction of the soil volume as well as water release occurs as suction is applied, the use of the moisture-characteristic curve (section 6.5) to determine pore size distribution is unsatisfactory.

Instead, water should be removed with the minimum of shrinkage, by critical-point drying for instance, and intrusion of mercury used to determine the pore size distribution.

Bulk density

Porosity can be measured from the volume of a non-polar liquid, such as paraffin, that is absorbed into a dry ped under vacuum. Alternatively, porosity may be determined indirectly from the bulk density and particle density. Particle density (ρ_p) was introduced in equation 2.1; *bulk density (BD)* is defined as the mass of oven-dry soil per unit volume, and depends on the densities of the constituent soil particles (clay, organic matter, etc.) and their packing arrangement. Thus, for a sample of oven-dry (o.d.) soil, from equation 4.1:

$$PSR = \frac{\text{total soil volume} - \text{volume of solids}}{\text{total soil volume}}$$

$$= 1 - \frac{\text{mass of o.d. soil}}{\text{total soil volume}} \times \frac{\text{volume of solids}}{\text{mass of o.d. soil}}$$

$$\text{i.e. } PSR = 1 - \frac{\text{bulk density } (BD)}{\text{particle density } (\rho_p)} \quad (4.2)$$

Values of BD range from $< 1\,\text{g cm}^{-3}$ for soils high in organic matter, to 1.0–1.4 for well-aggregated loamy soils, to 1.2–2.0 for sands and compacted horizons in clay soils. Using equation 4.2, it can easily be calculated that a sandy soil of bulk density 1.50 and particle density 2.65 has a PSR value of 0.43 or 43 per cent. Soil measurements made on a unit mass basis are translated to a unit volume basis by multiplying by the BD. Where the exact BD is not known, it is common practice to assume an 'average' value of 1.33, which corresponds to a soil mass of $2.0 \times 10^6\,\text{kg ha}^{-1}$ to a depth of 15 cm.

Water-filled porosity

The pores of a soil are partly or wholly occupied by water. Water is vital to life in the soil, its importance stemming from its existence as a liquid over the temperature range most suited to living organisms. It possesses several unique physical properties, namely the greatest specific heat capacity, latent heat of vaporization, surface tension and dielectric constant of any known liquid.

Soil *wetness* can be characterized by the amount of water held in a certain mass or volume of soil. The *gravimetric water content (θ_g)* is determined by drying the soil to a constant weight at a temperature of 105°C. θ_g is then calculated from:

$$\theta_g = \frac{\text{mass of water}}{\text{mass of o.d. soil}} \quad (4.3)$$

θ_g is often expressed as a percentage by multiplying (4.3) by 100. Structural water, that is water of crystallization held within soil minerals, is not driven off at 105°C and is not measured as soil water. Soil wetness can also be expressed as a *volumetric water content (θ_v)*, which is given by:

$$\theta_v = \frac{\text{volume of water}}{\text{total soil volume}} \quad (4.4)$$

Although θ_v is a dimensionless quantity, it is important to note it has the units of cm^3 water cm^{-3} soil. Thus, the concentration of a solute in the soil solution can be expressed on a soil volume basis by multiplying by θ_v. θ_g and θ_v are related through the soil bulk density in the equation

$$\theta_v = \theta_g \times BD \quad (4.5)$$

because 1 g water occupies 1 cm^3 at normal temperatures. As measurements of θ_v are more useful for most purposes than θ_g, soil water content will be expressed in terms of volumetric water content, using the symbol θ, throughout this book. For example, θ gives a direct measure of the depth of water per unit area in the soil, which is useful for comparison with the rainfall, evaporation or depth of irrigation water applied. An 'average' value for θ of 0.25 translates into 250 mm water per metre depth of soil.

Normally, water occupies less than the total pore volume and the soil is said to be unsaturated, although during a period of prolonged rain the soil may temporarily become saturated to the surface. Typically, at a variable depth below the surface, depending on the regional hydrology and the soil's permeability, the pore space is permanently saturated with water which is called *groundwater*. The upper surface of the groundwater, which may fluctuate with the season of the year, is called the *watertable*.

Air-filled porosity

Those pores and fissures that do not contain water are filled with air. The air-filled porosity is defined by:

$$\epsilon = \frac{\text{volume of soil air}}{\text{total soil volume}} \qquad (4.6)$$

and it follows that

$$\epsilon = PSR - \theta \qquad (4.7)$$

Soils have been found to drain, after rain, to a water content in equilibrium with suctions between 50 and 100 mbars. This water content defines the field capacity, which will be discussed in Chapter 6. For many well-structured British soils, the field capacity is attained at suctions close to 50 mbars when all pores $> 60\ \mu$m diameter are drained of water. The value of ϵ for a soil in this state defines the *air capacity*, C_a (50), which, as Figure 4.14 shows, may vary from 0.1 to 0.30 for topsoil, depending on the texture. A knowledge of the air capacity is vital for the planning of agricultural under-drainage schemes (Chapter 13).

COMPOSITION OF THE SOIL AIR

The major gases in the Earth's atmosphere normally occur in the proportions 79 per cent N_2, 21 per cent O_2 and 0.03 per cent CO_2 by volume. According to Dalton's law of partial pressures, these concentrations correspond to partial pressures of 0.79, 0.21 and 0.0003 of one atmosphere, which are equivalent to pressures of 0.802, 0.213 and 0.00033 bars, respectively.

The composition of the soil air may deviate from these proportions due to soil respiration (section 8.1) which depletes the air of O_2 and raises the content of CO_2. The exchange of O_2 and CO_2 between the atmosphere and the soil air tends to establish an equilibrium that depends on the soil respiration rate and on the resistance to gas movement through the

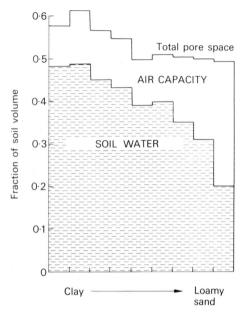

Figure 4.14. Air and water-filled porosities at field capacity in top-soils of varying texture (after Hall *et al.*, 1977).

soil. Thus, for a soil in which the gases diffuse rapidly through the macropores, O_2 partial pressures are ~ 0.20 bars and CO_2 pressures ~ 0.005 bars: the normal gas compositions of a variety of soils put to different uses are given in Table 4.3.

Table 4.3. The composition of the gas phase of well-aerated soils.

Soil and land use	Usual composition (% by volume)	
	O_2	CO_2
Arable land, fallow	20.7	0.1
unmanured	20.4	0.2
manured	20.3	0.4
Sandy soil, manured and cropped with potatoes	20.3	0.6
Serradella	20.7	0.2
Pasture land	18–20	0.5–1.5

(After Russell, 1973.)

However, if the macropores begin to fill with water and the soil becomes waterlogged, undesirable chemical and biological changes ensue as the soil becomes increasingly O_2 deficient or *anaerobic*; the O_2 partial pressure may drop to zero and that of CO_2 rises far above 0.015 bars. In the case of soil placed as capping over refuse tips (landfill), the gas phase may consist almost entirely of CO_2 and methane (CH_4) in the approximate proportions of 40:60 per cent. This is due to the proliferation of microorganisms adapted to anaerobic conditions which produce, in addition to these gases and H_2, reduced forms of N, Mn, Fe and S. These reactions are discussed more fully in section 8.3.

4.6 SUMMARY

Physical, chemical and biotic forces act upon the primary clay, silt and sand particles, in intimate combination with organic materials, to form an organized arrangement of aggregates (*peds*) and intervening voids. The magnitude of the *interparticle* forces within the structural units must be sufficient to prevent the particles separating again under the influence of the same forces that are responsible for the reorganization. On a microscopic scale, individual crystals of montmorillonite (1 nm thick) can stack in roughly parallel alignment, with much overlapping, to form *quasicrystals* which have long-range order in the *a–b* direction for up to 5 μm. The somewhat larger crystals of illite or vermiculite can orient in a similar same way to form larger units called *domains*. Both kinds of unit are stable in water when Ca^{2+} is the dominant exchangeable cation.

Domains, quasi-crystals, clay crystals, sesquioxides and organic polyanions are drawn together by the attraction of positively and negatively charged surfaces, reinforced by van der Waals forces and hydrogen-bonding to form *microaggregates* (< 250 μm) diameter). The cohesion of microaggregates to form *macroaggregates* (> 250 μm) is enhanced by the enmeshment of microbial polysaccharides and polyuronide gums, fungal hyphae and fine plant rootlets. Good structure for crop growth depends on the existence of water stable aggregates between 1 and 10 mm in diameter.

The naturally formed beds have *spheroidal, blocky, platy* or *prismatic* shapes, reflecting quite strongly the predominant pedogenic processes in the soil. The size, shape and arrangement of the peds determines the intervening void volume of cylindrical and spherical *pores* and planar cracks or *fissures*. Movement of materials in solution or suspension results in *plasma concentrations* (material < 2 μm) in the pores, on ped faces and on sand particles (skeleton grains): these coatings are variously called clay coatings (*argillans*), organic coatings (*organans*) and iron-oxide coatings (*ferrans*).

The pore volume determines the size of the reservoir of water sustaining plant and animal life in the soil. It is usually expressed as a *pore space ratio* (*PSR*) and can be calculated approximately from the *bulk density* (dry mass per unit soil volume) and the mean particle specific gravity, i.e.:

$$PSR = 1 - \frac{\text{bulk density}}{\rho_p}$$

The *PSR* ranges from 0.5 to 0.6 irrespective of soil texture. Changes in soil wetness as measured by changes in the *volumetric water content* θ not only affect life processes but also have a marked bearing on structure formation itself. The pore volume not occupied by water defines the *air-filled porosity*, ϵ, for which in a well-aerated soil, the O_2 partial pressure is approximately the same as in the atmosphere, and that of CO_2 lies between 0.001 and 0.015 bars.

REFERENCES

BREWER R. (1964) *Fabric and Mineral Analysis of Soils*. Wiley, New York.

BREWER R. (1976) *Fabric and Mineral Analysis of Soils*, 2nd Ed. Krieger, New York.

EDWARDS C.A. & BREMNER J.M. (1967) Microaggregates in soils. *Journal of Soil Science* **18**, 64–73.

EMERSON W.W. (1959) The structure of soil crumbs. *Journal of Soil Science* **10**, 235–244.

FITZPATRICK E.A. (1984) *Micromorphology of Soils*. Chapman and Hall, London.

GREENLAND D.J. (1979) Structural organization of soils and crop production, in *Soil Physical Properties and Crop Production in the Tropics* (Eds. R. Lal and D.J. Greenland). Wiley, Chichester.

HALL D.G.M., REEVE M.J., THOMASSON A.J. & WRIGHT V.F. (1977) (Eds.) *Water retention, porosity and density of field soils.* Soil Survey of England and Wales, Technical Monograph No. 9, Harpenden.

KUBIENA W.L. (1938) *Micropedology.* Collegiate Press, Ames.

RUSSELL E.W. (1973) *Soil Conditions and Plant Growth*, 10th Ed. Longman, London.

SOIL SURVEY MANUAL (1951) United States Department of Agriculture Handbook No. 18.

STRUTT N. (1970) *Modern farming and the soil.* Report of the Agricultural Advisory Council on Soil Structure and Soil Fertility. HMSO, London.

TISDALL J.M. & OADES J.M. (1982) Organic matter and water-stable aggregates in soils. *Journal of Soil Science* **33**, 141–163.

FURTHER READING

BULLOCK P. & MURPHY C.P. (1976) The microscopic examination of the structure of subsurface horizons of soils. *Outlook on Agriculture* **8**, 348–354.

GREENLAND D.J. (1965) Interaction between clays and organic compounds in soils. I. Mechanisms of interaction between clays and defined organic compounds. *Soils and Fertilizers* **28**, 415–425.

GREENLAND D.J. (1965) Interaction between clays and inorganic compound in soils. II. Adsorption of soil organic compounds and its effect on soil properties. *Soils and Fertilizers* **28**, 521–532.

GREENLAND D.J. (1981) Soil management and soil degradation. *Journal of Soil Science* **32**, 301–322.

HARRIS R.F., CHESTERS G. & ALLEN O.N. (1966) Dynamics of soil aggregation. *Advances in Agronomy* **18**, 107–169.

HODGSON J.M. (1974) (Ed.) *Soil survey field handbook.* Soil Survey of England and Wales, Technical Monograph No. 5, Harpenden.

OADES J.M. (1984) Soil organic matter and structural stability: mechanisms and implications for management. *Plant and Soil* **76**, 319–337.

RUSSELL E.W. (1971) Soil structure: its maintenance and improvement. *Journal of Soil Science* **22**, 137–151.

QUIRK J.P. (1978) Some physico-chemical aspects of soil structural stability – a review, in *Modification of Soil Structure* (Eds. W.W. Emerson, R.D. Bond and A.R. Dexter). Wiley Interscience, Chichester.

Part 2
Processes in the
Soil Environment

'Soils are the surface mineral and organic
formations, always more or less coloured by
humus, which constantly manifest themselves as a
result of the combined activity of the following
agencies: living and dead organisms (plants and
animals), parent material, climate and relief.'
V.V. Dokuchaev (1879), quoted by
J.S. Joffe in *Pedology*

Chapter 5
Soil Formation

5.1 THE SOIL-FORMING FACTORS

In 1941 Jenny, in his book *Factors of Soil Formation*, presented an hypothesis that drew together many of the current ideas on soil formation, the inspiration for which owed much to the earlier studies of Dokuchaev and the Russian school. The hypothesis was that soil is formed as a result of the interaction of many factors, the most important of which are:

Parent material
Climate
Organisms
Relief
Time

Jenny's approach was to consider these *soil-forming factors* (section 1.2) as *control* variables, independent of the soil as it evolved, and also independent, but not necessarily so of each other. He then attempted to define the relationship between any soil property *s* and climate (*cl*), organisms (*o*), relief (*r*), parent material (*p*) and time (*t*), by a function of the form:

$$s = f (cl, o, r, p, t \ldots) \qquad (5.1)$$

The dots indicated that factors of lesser importance such as mineral accessions from the atmosphere, or fire, might need to be taken into account. An ensemble (*S*) of soil properties *s*, each being defined by a function such as equation 5.1, makes up an individual soil type, of which there is an almost infinite number. Subsequently, Jenny (1980) redefined the soil-forming factors as 'state' variables and extended the terms on the LHS of equation 5.1 to include total ecosystem properties, vegetation and animal properties, as well as soil properties.

To simplify the application of equation 5.1, it has been the practice to solve it for changes in a soil property *s* when only one of the control variables (e.g. climate) varies, the others being constant or nearly so. The relationship is then called a *climofunction*:

$$s \text{ or } S = f (cl)_{o, r, p, t \ldots} \qquad (5.2)$$

and the range of soils formed is called a *climosequence* (Figure 5.1). Biosequences, toposequences, lithosequences and chronosequences of soils have been recognized in various parts of the world. Indeed, the

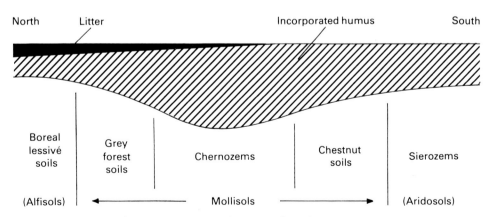

Figure 5.1. A climosequence of soils found along a north–south transect through the USSR showing soil depth and degree of humus incorporation (after Duchaufour, 1982).

main virtue of Jenny's attempt to quantify the relationship between soil properties and soil-forming factors lies not in the prediction of the exact value of s or S at a particular site, but rather in identifying *trends* in properties and soil types that are associated with readily observable changes in climate, parent material, etc. The concept of a causal relationship between S and the factors cl, o, r, p and t has been implicit, and at times explicit (see section 5.3), in attempts at imposing order on the apparently chaotic variation of soils in the field, i.e. *soil classification* (section 14.3). Furthermore, factors such as climate and organisms determine the rate at which chemical and biological reactions occur in the soil (the *pedogenic processes*, section 9.2); others such as parent material and relief define the *initial state* for soil development, and time measures the *extent* to which a reaction will have proceeded. In summary, therefore, one recognizes a logical progression: of *environment* (soil-forming factors) → *processes* → *soil properties* underlying soil formation. The following sections focus briefly on the role of the soil-forming factors individually.

5.2 PARENT MATERIAL

Soil may form directly by the weathering of consolidated rock *in situ* (a residual soil), or it may develop on superficial deposits, which may have been transported by ice, water, wind or gravity. These deposits originate ultimately from the denudation and geologic erosion of consolidated rock.

Weathering

Generally, weathering proceeds by physical disruption of the rock structure which facilitates chemical changes within the constituent minerals. Forces of expansion and contraction induced by diurnal temperature variations cause rock shattering and exfoliation, as illustrated by surface features in a desert soil (Figure 5.2). This effect is most severe when water trapped in the rock repeatedly freezes and thaws, resulting in irresistible forces of expansion and contraction.

Water is the dominant agent in weathering, not only because it initiates solution and hydrolysis (section 9.2), but also because it sustains plant life on the rock surface. Lichens play a special part in weathering, because they are rich in chelating agents which trap the elements of the decomposing rock in organo-metallic complexes. In general, the growth, death and decay of plants and other organisms markedly enhances the solvent action of rainwater by the addition of carbon dioxide from respiration. The plant roots also aid in the physical disintegration of rock.

Water carrying suspended rock fragments has a scouring action on surfaces. The suspended material can vary in size from the finest 'rock flour' formed by the grinding action of a glacier, to the gravel, pebbles and boulders moved along and constantly abraded by fast-flowing streams. Particles carried by wind also

Figure 5.2. Stones of a desert soil surface fractured due to extreme variations in temperature (J.J. Harris Teall in Holmes, 1978).

have a 'sand-blasting' effect, and the combination of wind and water to form waves produces powerful destructive forces along shorelines.

Mineral stability

The rate of weathering depends on:
(1) temperature;
(2) rate of water percolation;
(3) oxidation status of the weathering zone;
(4) surface area of rock exposed (largely determined by physical processes of fracturing and exfoliation);
(5) types of mineral present.

Much attention has been paid to the relationship between the crystalline structure of a rock mineral and its susceptibility to weathering. One scheme for the order of stability, or *weathering sequence*, of the more common silicate minerals in *coarse* rock fragments (> 2 mm diameter) is shown in Figure 5.3. This sequence demonstrates that resistance of a mineral to weathering increases with the degree of sharing of oxygens between adjacent silicon tetrahedra in the crystal lattice (section 2.3). The Si—O bond has the highest energy of formation, followed by the Al—O bond, and the even weaker bonds formed between O and the basic metal cations (Na^+, K^+, Ca^{2+}, Sr^{2+}, Ba^{2+}). Thus, olivine weathers rapidly because the silicon tetrahedra are only held together by O—metal cation bonds; quartz is very resistant because it consists entirely of linked silicon tetrahedra. In the chain (amphiboles and pyroxenes) and sheet (phyllosilicate) structures, the weakest points are the O—metal cation bonds. Isomorphous substitution (section 2.3) of Al^{3+} for Si^{4+} also contributes to instability because the proportion of Al—O to Si—O bonds increases and more O—metal cation bonds are necessary. This accounts for the decrease in stability of the calcium

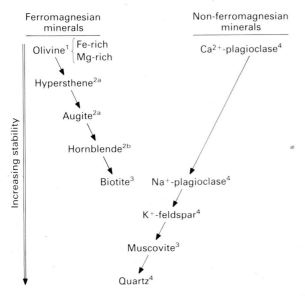

Figure 5.3. Ranking of the common silicate minerals in order of increasing stability. Note that the superscripts indicate the number of oxygens shared between adjacent SiO_4^{4-} tetrahedra: 1 = none; 2a = two (single chain); 2b = 2 and 3 (double chain); 3 = three (sheet structure); 4 = four (three-dimensional structure) (after Goldich, 1938 and Birkeland, 1984).

Table 5.1. Stages in the weathering of minerals in the < 2 mm fraction of soils.

Stage	Type mineral	Soil characteristics
Early weathering stages		
1	Gypsum	These minerals occur in the
2	Calcite	silt and clay fractions of young
3	Hornblende	soils all over the world, and in
4	Biotite	soils of arid regions where lack
5	Albite	of water inhibits chemical weathering and leaching.
Intermediate weathering stages		
6	Quartz	Soils found mainly in the
7	Muscovite (also illite)	temperate regions of the world, frequently on parent
8	Vermiculite and mixed layer minerals	materials of glacial or peri-glacial origin; generally fertile,
9	Montmorillonite	with grass or forest as the natural vegetation.
Advanced weathering stages		
10	Kaolinite	The clay fractions of many
11	Gibbsite	highly weathered soils on old
12	Hematite (also goethite)	land surfaces of humid and hot intertropical regions are
13	Anatase	dominated by these minerals; often of low fertility.

(After Jackson *et al.*, 1984.)

plagioclases when compared with the sodium plagioclases (Figure 5.3).

Weathering continues in the finely comminuted materials (< 2 mm diameter) of soils and unconsolidated sedimentary deposits, following a sequence from the *least* to the *most stable* of minerals, the stages of which are identified by certain type minerals. Table 5.1 presents one such sequence, together with the broad characteristics of the associated soils.

Rock types

Consolidated rocks are of igneous, sedimentary or metamorphic origin.

IGNEOUS ROCKS

These are formed by the solidification of molten magma in the Earth's crust, and are the ultimate source of all other rocks. They are broadly subdivided on mineral composition into *acidic* rocks, those relatively rich in quartz and the light-coloured Ca or K/Na silicates, and *basic* rocks, those low in quartz but high in the dark-coloured ferromagnesian minerals (hornblende, micas, pyroxene). The chemical composition of typical granite and basalt, rocks representative of these two groupings, is given in Table 5.2. The higher

Table 5.2. Chemical composition (per cent) of typical granitic and basaltic rock.

Element	Granite	Basalt
O	48.7	46.3
Si	33.1	24.7
Al	7.6	10.0
Fe	2.4	6.5
Mn	—	0.3
Mg	0.6	2.2
Ca	1.0	6.1
Na	2.1	3.0
K	4.4	0.8
H	0.1	0.1
Ti	0.2	0.6
P	—	0.1
Total	100.2	100.7

(After Mohr and van Baren, 1954.)

the ratio of Si to (Ca+Mg+K), the more probable it is that some of the Si is not bound in the silicates, and may remain as residual quartz as the rock weathers. Soils formed from granite are therefore usually higher in quartz grains (sand and silt size) than those formed from basalt.

SEDIMENTARY ROCKS

These are composed of the weathering products of igneous and metamorphic rocks and are formed after deposition by wind and water. Cycles of geologic uplift, weathering, erosion and subsequent deposition of eroded materials in rivers, lakes and seas have produced superimposed strata of sediments. Under the weight of overlying sediments, deposits gradually consolidate and harden—the process of *diagenesis*—to form sedimentary rocks. Faulting, folding and tilting movements, followed by erosion, can cause the sequence of rock types in the geological column to be revealed by their surface outcrops (Figure 5.4). *Clastic* sedimentary rocks are composed of fragments of the more resistant minerals. The size of fragments decreases in order from the conglomerates to sandstones to siltstones to mudstones. Other sedimentary rocks are formed by precipitation or flocculation from solution; most commonly limestone and chalks (Figure 5.5), which vary in composition from pure calcium carbonate to mixtures of calcium and magnesium carbonate (dolomite), or carbonates and detrital material, usually flinty silica and clay minerals.

Igneous and sedimentary rocks that are subjected to intense heat and great pressures are transformed into *metamorphic* rocks. Changes in mineralogy and rock structure generally render the metamorphosed rock more resistant to weathering.

Soils on transported material

SOIL LAYERS AND BURIED PROFILES

Most parts of the Earth's surface have undergone several cycles of submergence, uplift, erosion and denudation over the hundreds of millions of years of geological time. During the mobile and depositary

Column of sedimentary strata:

before faulting, tilting and erosion   after faulting, tilting and erosion

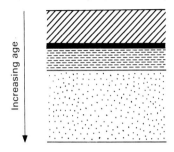

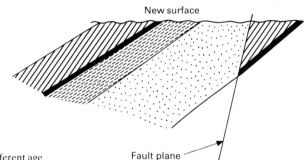

New surface

Increasing age

Fault plane

Figure 5.4. Diagram to show how sedimentary rocks of different age and type can provide the parent materials for present-day soil formation.

phases there is much opportunity for the mixing of materials from different rock formations. Thus, the interpretation of soil genesis on such heterogeneous parent material is often complex. When two or more layers of transported material are superimposed they form a *lithological discontinuity*, and the resultant profile is often difficult to distinguish from the pedogenically differentiated profile of a soil formed on

Figure 5.5. Electron micrograph of a sample of Middle Chalk (Berkshire, England) showing microstructure of calcified algae (courtesy of S.S. Foster).

Figure 5.6. 'Mottled zone' of a fossil laterite preserved under more recent transported material in southern New South Wales.

consolidated rock *in situ*. In other cases, a soil may be buried by transported material during a period of prolonged erosion of an ancient landscape: the old soil is then preserved as a *fossil soil*, as illustrated by the buried mottled-zone of a laterite (Oxisol) in Figure 5.6.

WATER

Transport by water produces alluvial, terrace and footslope deposits. During transport the rock material is sorted according to size and density and abraded, so that fluviatile deposits characteristically have smooth, round pebbles. The larger fragments are moved by rolling or are lifted by the turbulence of water flowing over the irregular stream bed, and carried forward some distance by the force of the water. This 'bouncing' process is called *saltation* (*cf.* saltation of wind-blown particles, section 11.5). The smallest particles are carried in suspension and salts move in solution. Colloidal material may not be deposited until the stream discharges into the sea, when flocculation occurs and estuarine deposits form. The modes of sediment transport in relation to the depth and velocity of the water are shown in Figure 5.7.

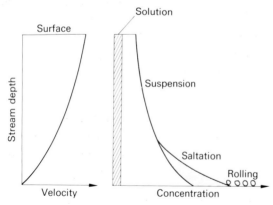

Figure 5.7. Sediment transport in relation to depth and velocity of water (after Mitchell, 1976).

ICE

Ice was an important agent in the transport of rock materials during the 2 million years of the Pleistocene.

During the colder *glacial* phases, the ice cap advanced from the North Pole and mountainous regions to cover a large part of the land surface in the Northern Hemisphere (the ice reached a line only a few kilometres north of the Thames Valley in southern Britain). The moving ice ground down rock surfaces and collected much debris on its surface by colluviation (see below). During the warmer *interglacial* phases, the ice retreated leaving extensive deposits of *glacial drift*, including boulder clay and till, which is very heterogeneous in nature. Streams flowing out of the glaciers produced glacifluvial deposits of sands and gravels (Figure 5.8). Repeated freezing and thawing in periglacial regions resulted in frost-heaving of the surface

Figure 5.8. Soil formation on glacifluvial sands and gravels of the Lleyn Peninsular, Wales.

and much swelling and compaction in subsurface layers. Such deposits may be distinguished from water-borne materials by the vertical orientation of many of the pebbles.

WIND

Wind moves rock fragments by processes of rolling, saltation and aerial dispersion (section 11.5). Material swept from dry periglacial regions has formed aeolian deposits called *loess* in central USA, central Europe, China and South America. These deposits, often cemented by the subsequent precipitation of calcium carbonate, are many metres thick in places and form the parent material of highly productive soils. Other characteristic wind deposits are the sand dunes of shorelines and desert areas.

GRAVITY

Gravity produces colluvial deposits, of which the scree formed by rock slides in mountainous regions is an obvious example. Less obvious, but widespread under periglacial conditions, are the *solifluction* deposits formed when wet thawed soil slips over frozen ground below, even on gentle hillslopes. *Creep* occurs when the surface mantle of soil moves by slow colluviation away from the original rock. The junction between creep layer and weathering rock is frequently delineated by a stone line.

5.3 CLIMATE

The key components of climate in soil formation are moisture and temperature.

Moisture

The effectiveness of moisture depends on several factors:
(1) the form and intensity of the precipitation;
(2) its seasonal variability;
(3) the evaporation rate (from vegetation and soil, see section 6.6);
(4) land slope;
(5) permeability of the parent material.

Various functions of meteorological variables have been proposed to measure *moisture effectiveness*. One such function is Thornthwaite's (1931) $P-E$ index, which is the summation over one year of the monthly ratios:

$$\frac{\text{Precipitation}, P}{\text{Evaporation}, E} \times 10 \qquad (5.3)$$

(The multiplier of 10 was introduced to eliminate fractions.) Based on the $P-E$ index, five major categories of climate and associated vegetation types are recognized, namely:

$P-E$ index	Category	Vegetation
128 and above	Wet	Rainforest
64–127	Humid	Forest
32–63	Subhumid	Grassland
16–31	Semi-arid	Steppe
< 16	Arid	Desert

When moisture effectiveness is high, as in wet or humid climates, there is a net downward movement of water in the soil for most of the year, which usually results in greater *leaching* of soluble materials, sometimes out of the soil entirely, and the *translocation* of clay particles from upper to lower horizons. More will be said about these processes, and profile differentiation, in Chapter 9.

Temperature

Temperature varies with lattitude and altitude, and the extent of absorption and reflection of solar radiation by the atmosphere. Temperature affects the rate of mineral weathering and synthesis, and the biological processes of growth and decomposition. Reaction rates are roughly doubled for each 10°C rise in temperature, although enzyme-catalysed reactions are sensitive to high temperatures and usually attain a maximum between 30 and 35°C.

Categories based on the thermal efficiency of the climate may be delineated by Thornthwaite's $T-E$ index, which is the sum for one year of the monthly values of:

$$9/20 \text{ (mean monthly temperature in °C)} \quad (5.4)$$

This empirical expression was chosen so that the cold margin of the Tundra had a $T-E$ index of zero and the cooler margin of the rainforest an index of 128. The $T-E$ indices and categories are:

$T-E$ index	Category
128 and above	Tropical
64–127	Mesothermal
32–63	Microthermal
16–31	Taiga
1–15	Tundra
0	Frost

Zonal soils

From Dokuchaev (*c.* 1870) on, many pedologists in Europe and North America regarded climate as preeminent in soil formation. The obvious relationship between climatic zones, with their associated vegetation, and the broad belts of similar soils that stretched roughly east–west across Russia inspired the zonal concept of soils (see Figure 5.1).

Zonal or normal soils are those in which the climatic factor, acting on the soil for a sufficient length of time, is so strong as to override the influence of any other factor.

Intrazonal soils, although showing well-developed soil features, are those in which some local anomaly of relief, parent material or vegetation is sufficiently strong to modify the influence of the regional climate.

Azonal or immature soils have poorly differentiated profiles, either because of their youth or because some factor of the parent material or environment has arrested their development. The zonal concept gave rise to systems of classifying soils reflecting the known or presumed genesis of like soils—the *Great Soil Groups*, listed in Table 5.3. Such a scheme was adopted, with minor modifications, over large continental areas, notably the USA and Australia, and

although the classification is now superseded, references to the old Great Soil Group names continue to appear in the soil science literature.

Predictably, the concept of soil zonality is not very helpful when applied to soils of the subtropics and tropics. There, where land surfaces are generally much older than in Europe and North America, and have consequently undergone many cycles of erosion and deposition associated with climatic change, the age of the soil (section 5.6) and its topographical relation to other soils in the landscape (section 5.5) are factors of major importance. The zonal concept is also of little use in regions such as Britain and Scandinavia where much of the parent material of present-day soils is young (Pleistocene deposits) and relief plays a powerful role in soil formation.

5.4   ORGANISMS

The soil, and the organisms living on and in it, comprise an *ecosystem*. The active components of the soil ecosystem are the plants, animals, microorganisms and man.

Vegetation

The primary succession of plants that colonize a weathering rock surface (section 1.1) culminates in the development of a *climax community*, the species composition of which depends on the climate and parent material, but which, in turn, has a profound influence on the soil that is formed. For example, in the mid-West and Great Plains of the United States, deciduous forest seems to accelerate soil formation compared to prairie (grassland) on the same parent material and under similar climatic conditions. In parts of England such as on the North York moors or the Bagshot Sands, the profiles preserved under the oldest barrows (burial mounds) of the Bronze Age people are typical of *brown forest soils* (Inceptisols) formed under the deciduous forests of some 4,000 years ago, in contrast to the *podzols* (Spodosols) on the surfaces of the barrows which have formed under the usurping

Table 5.3. Higher soil categories according to the Zonal Concept.

Order	Suborder	Great Soil Groups
Zonal soils	1. Soils of the cold zone	Tundra soils
	2. Light-coloured soils of arid regions	Desert soils Red desert soils Sierozems Brown soils Reddish-brown soils
	3. Dark-coloured soils of semiarid, subhumid, and humid grasslands	Chestnut soils Reddish chestnut soils Chernozems Prairie soils Reddish prairie soils
	4. Soils of the forest-grassland transition	Degraded chernozems Noncalcic brown or Shantung brown soils
	5. Light-coloured podzolized soils of the forest regions	Podzols Grey wooded or Grey podzolic soils Brown podzolic soils Grey–brown podzolic soils Red–yellow podzolic soils
	6. Lateritic soils of forested mesothermal and tropical regions	Reddish-brown lateritic soils Yellowish-brown lateritic soils Laterites
Intrazonal soils	1. Halomorphic (saline and alkali) soils of imperfectly drained arid regions and littoral deposits	Solonchak or saline soils Solonetz Soloth (solod)
	2. Hydromorphic soils of marshes, swamps, seep areas and flats	Humic gley soils Alpine meadow soils Bog soils Half-bog soils Low-humic gley soils Planosols Groundwater podzols Groundwater laterites
	3. Calcimorphic soils	Brown forest soils (Braunerde) Rendzinas
Azonal soils		Lithosols Regosols (includes Dry sands) Alluvial soils

(After Thorp and Smith, 1949.)

heathland vegetation. Differences in the chemical composition of leaf leachates can partly account for this divergent pattern of soil formation (section 9.2).

Fauna

The species of invertebrate animals in the soil and the type of vegetation are closely related. The acid litter of pines, spruce and larch creates an environment unfavourable to earthworms, so that litter accumulates on the soil surface where it is only slowly comminuted by mites and *Collembola* and decomposed by fungi. The litter of elm and ash, and to a lesser extent oak and beech, is more readily ingested by earthworms and becomes incorporated into the soil as faecal material. Earthworms are the most important of the soil-forming fauna in temperate regions, being supported to a variable extent by the small arthropods and the larger burrowing animals (rabbits and moles). Earthworms are also important in tropical soils, under good rainfall, but in general the activities of termites, ants and coprophagous (dung-eating) beetles are of greater significance, particularly in the subhumid to semiarid savanna of Africa and Asia.

Earthworms of the casting type build up a stone-free layer at the soil surface, as well as intimately mixing the litter with fine mineral particles they have ingested. The surface area of the organic matter that is accessible to microbial attack is then much greater. Termites are also important soil-formers in that they carry fine particles, water and dissolved salts from depths of 1 metre or more to their termitaria on the surface. On abandonment, the termitaria are reduced by weathering and the trampling of animals to form a thin mantle on the soil.

Man

Man influences soil formation through his manipulation of the natural vegetation, and its associated animals, his agricultural practices and his urban and industrial development. Indigenous populations of wild animals are regulated by natural controls; not so domestic stock which can, if too numerous, denude the vegetation so that the soil is exposed to severe erosion risk. Compaction by the traffic of domestic and wild animals decreases the rate of water infiltration into the soil, thereby increasing runoff and erosion. The subject of man and his management of soil resources is discussed in Part III.

5.5   RELIEF

Relief has an important influence on the local climate, vegetation and drainage of a landscape. Major topographical features are easily recognized: mountains and valleys; escarpments, ridges and gorges; hills, plateaux and floodplains. Changes in elevation affect the temperature (a decrease of approximately 0.5°C per 100 m increase in height), the amount and form of the precipitation and the intensity of storm events. These factors interact to influence the type of vegetation. More subtle changes in local climate and vegetation are associated with the *slope* and *aspect* of valley sides and escarpments.

Slope

Angle of slope and vegetative cover affect moisture effectiveness by governing the proportion of surface runoff to infiltration. Soils with impermeable subsoils, and those developing on slopes, may also show appreciable lateral sub-surface flow. Thus, at the top of a slope, the soils tend to be freely drained with the watertable at considerable depth, whereas at the valley bottom the soils are poorly drained, with the watertable near or at the soil surface (Figure 5.9a). The succession of soils forming under different drainage conditions on relatively uniform parent material comprises a *hydrological sequence*, an example of which is shown in Figure 5.9b. As the drainability deteriorates, the oxidized soil profile, with its warm orange-red colours due to the hydrated ferric oxides (section 2.4), is transformed into the mottled and gleyed profile of a reduced soil (iron in the ferrous state).

In addition to modified drainage, there is movement

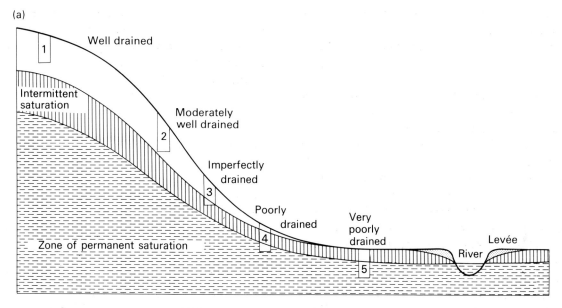

Hydrological sequence of soils from 1 to 5

Figure 5.9. (a) Section of slope and valley bottom showing a hydrological soil sequence.

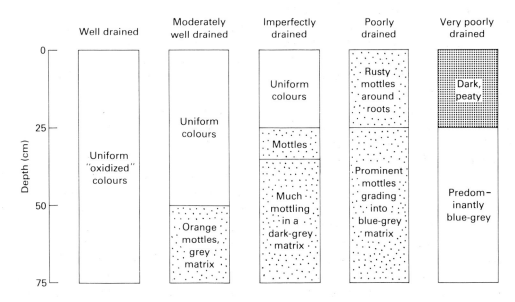

Figure 5.9. (b) Changes in profile morphology, principally colour, with the deterioration in soil drainage (after Batey, 1971).

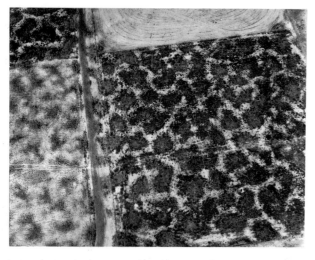

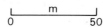

Figure 5.10. Aerial view of gilgai micro-relief (CSIRO photograph, courtesy of G.D. Hubble).

of soil on slopes; a mid-slope site is continually receiving solids and solutes from sites immediately up-slope by wash and creep, and continually losing material to sites below. Here the *form* of the slope is important— whether smooth or uneven, convex or concave, or broken by old river terraces. Minor variations in relief such as those caused by frost-heaving in very cold regions, or by faunal activity (rabbits, moles and termites) promote the slow creep of soils down even slight slopes.

Small variations in elevation can also be important in flat lands, especially if the groundwater is saline and rises close to the surface. A curious pattern of micro-relief of this kind, called *gilgai**, has been observed in the chernozem-like soils (Mollisols) of semiarid regions in Australia (Figure 5.10). The soils of the mound ('puff') and hollow ('shelf') are quite different, with calcium carbonate and gypsum occurring much closer to the surface in the yellow-brown crumbly clay of the puff than in the grey, compact cracking clay of

*From the Aboriginal word for a small waterhole.

the shelf. Comparable features are found in the black clay soils of Texas, the Orange Free State and in the Black Cotton soils of the Deccan. In the prairies of western Canada, glacial till covers much of the landscape and, depending on the extent to which it has been re-worked by water, forms a characteristic *knob* and *kettle* (depression) or *ridge* and *swale* topography. This markedly influences the local drainage and hence the type of soil formed, which ranges from being freely drained (even droughty) on the knobs to poorly drained (possibly saline) in the depressions.

The catena
The importance of relief was highlighted by Milne (1935) in central East Africa, who recognized a recurring sequence of soils forming on slopes in a generally undulating landscape. He introduced the term *catena* (Latin for a chain) to describe the suite of contiguous soils extending from hill top to valley bottom. The catena was restricted to soils of the one climatic zone, but uniformity of parent material was not a prerequisite, especially in the very old landscapes of the Tropics. However, in those temperate regions where parent materials have been rejuvenated during the Pleistocene glaciations, the term should be restricted to sequences of soils on uniform parent material. In this sense, the catena is synonymous with Jenny's *toposequence*, which defines the suite formed when the influence of relief is dominant over that of the other soil-forming factors.

Although Milne conceived of the catena as a mapping unit, Duchaufour (1982) and others have extended its meaning to include many denudational and hydrological sequences of soils on slopes. Figure 5.9 is one example, and others are given in Chapter 9.

5.6   TIME

Time acts on soil formation in two ways: (a) the *value* of a soil-forming factor may change with time (e.g. climatic change, new parent material, etc.); and (b) the *extent* of a pedogenic reaction depends on the time for which it has operated. *Monogenetic* soils are those that have formed under one set of factor values for a certain period of time; soils that have formed under more than one set of factor values are called *polygenetic*. The latter are much more common than the former.

It is well known that the world climate has changed over geological time: the most recent, large changes were associated with the alternating glacial and interglacial periods of the Pleistocene. Major climatic changes were accompanied by the raising and lowering of sea level, erosion and deposition, and induced isostatic processes in the Earth's crust, all of which produced radical changes in the distribution of parent material and vegetation, and in the shape of the landscape. On a much shorter timescale, covering a few thousand years, changes can occur in the biotic factor of soil formation, as witnessed by the primary succession of species on a weathering rock surface; and also in the relief factor through changes in slope form and the distribution of groundwater.

Rate of soil formation

Knowledge of a soil's age, together with information about any change in the factor values, enables rates of soil formation in different environments to be estimated. One criterion of a soil's age is its *profile morphology*, in particular, the multiplicity of horizons, their degree of differentiation and their depth. The use of profile criteria is feasible for a soil derived from uniform parent material. Frequently, however, the past history of climatic, geologic and geomorphological change is so complex that an accurate assessment of a soil's age using these criteria is impossible.

The organic matter in individual horizons can be dated using the natural radioactivity of the C and where the radio-C ages in a profile are not too dissimilar, an average age for the whole soil may be calculated. The relation between soil development and time can also be determined in special cases where the time that has elapsed since the exposure of parent material is known, and the subsequent influence of the other soil-forming factors has been reasonably steady. Such studies suggest that the rate of soil development is

extremely variable, ranging from very rapid (less than 100 years) on volcanic ash in the tropics to very slow (1 cm in 5,000 years) on chalk weathering in a cool temperate environment. Theoretically, the rate of soil formation sets the upper limit to the rate of soil loss by erosion that is acceptable (section 11.5). In some cases, however, this may be an unattainably low rate of loss for a soil under agriculture, so that some slow degradation of the soil profile with time must be expected.

The concept of the steady state

For soils forming on moderate slopes, the rate of natural erosion can, in certain circumstances, just counteract the rate of rock weathering so that the soil, although immature, appears to be in *steady-state equilibrium*, that is, processes of soil creation balance those of soil destruction. In many cases, however, because the various pedogenic processes do not operate at the same rate, all the properties that make a soil do not attain equilibrium at the same time. Soil formation is not linear with time, but tends to follow an exponential-type 'law of diminishing returns' (section 11.2). This has been demonstrated for the accumulation of C, N and organic P in soils of temperate regions (Figure 5.11). The time course of clay accumulation, on the other hand, is probably more sigmoidal in nature. When the rate of change of a soil property

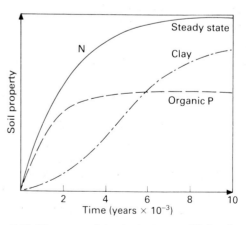

Figure 5.11. The approach to steady-state equilibrium for selected soil properties under temperate conditions (after Jenny, 1980).

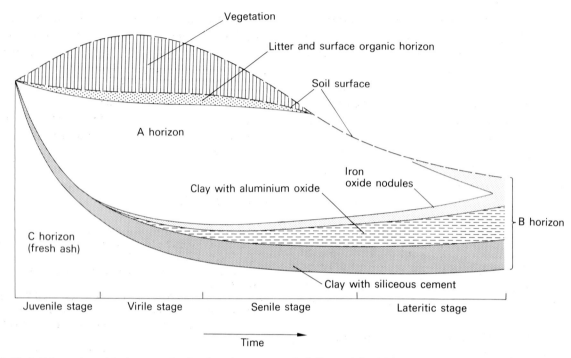

Figure 5.12. Soil formation with time on volcanic ash under a wet tropical climate (after Mohr and van Baren, 1954).

with time is negligibly small, the soil is said to be in steady state with respect to that property.

Despite the heterogeneity of pedogenic reaction rates, under a constant environment in temperate regions, it appears that a stable combination of soil and vegetation (the climax ecosystem) can form in 1,000 to 10,000 years. The system is stable in the sense that any changes occurring are immeasurably small within the time period since scientific observation began (*c.* 100 years). The soil component of such systems, the zonal soil, is sometimes referred to as *mature*. By contrast, in tropical regions, although the rate of many pedogenic processes (e.g. mineral weathering) is faster, the soil–vegetation system does not usually reach stability in a few thousand years but may take 10^5–10^6 years. Weathering occurs to much greater depths and the deeper profiles are less influenced by surface organic matter; crystalline clay minerals break down to mainly sesquioxides with the release of silica and the basic cations (section 9.2). This trend in soil formation with

time is illustrated in Figure 5.12 for a parent material of volcanic ash weathering in a wet tropical climate. Soil depth and vegetation biomass attain maxima in the 'virile' stage, only to decline in the 'senile' and final stages. Climate exerts less effect, when compared with parent material and relief, on the ultimate result of soil formation under such conditions.

5.7   SUMMARY

Jenny (1941) sought to quantify the effects of *climate*, *organisms*, *relief*, *parent material* and *time* on soil formation by writing the equation

$$\text{Soil property, } s = f(cl, o, r, p, t \ldots)$$

Lesser factors such as fire or atmospheric inputs are represented by the dots. Attempts have been made to solve this equation by examining the effect of one factor, such as climate, on s (or an ensemble of soil properties, *S*) when the other factors are constant or nearly so. This gives rise to suites of soils called *climosequences*, *biosequences*, *toposequences*, *lithosequences* and *chronosequences*.

Although quantification of *clorpt* has not been very successful, recognition of the role of these factors, and their various interactions, has helped scientists to understand why certain types of soil occur where they do. The concept has also been the basis of early attempts to impose order, through *soil classification*, on the apparently chaotic distribution of soils in the field. For example, the dominant effect of *climate* on soil formation in continental USSR gave rise to the *zonal system* of soil classification. Where only one set of factor values has operated during soil formation, the resultant soil is said to be *monogenetic*; *polygenetic* soils, which are more common, arise when one or more of the factors changes in value during the course of soil formation.

Parent material affects soil formation primarily through the type of rock minerals, their rate of weathering and the weathering products. These products are redistributed by the action of water, wind, ice and gravity, in many cases providing the parent material for another cycle of soil formation.

It is difficult to separate cause and effect in the case of *organisms*, because, although the range of colonizing species influences the pace and direction of soil formation, the soil that forms affects the competition for survival among the plants and organisms living in that soil. Man can exert the most profound effect, for good or bad, on the soil.

The importance of *relief* is obvious in the old weathered landscapes of the tropics, where the concept of the *catena* evolved to describe the recurring suites of soil that occur on slopes. On a smaller scale, the effect of relief on soil formation can be seen in the *gilgai* country of Australia and the *knob* and *kettle* country of western Canada.

Time is important because the value of a soil-forming factor may change with time (e.g. climatic change), and the extent of a pedogenic reaction depends on the length of time for which it has operated. A soil profile may attain *steady-state equilibrium* if the rate of natural erosion matches the rate of weathering of the parent material. The individual pedogenic processes, however, generally proceed towards equilibrium at different rates, following a type of 'law of diminishing returns' with respect to time. Steady state of the soil–vegetation system, and hence apparent stability, can be attained within 1,000–10,000 years in temperate regions, whereas it is more likely to take 10^5–10^6 years in the tropics.

REFERENCES

BATEY T. (1971) *Soil profile drainage*, in ADAS Advisory Papers, No. 9, Soil Field Handbook. Ministry of Agriculture, Fisheries and Food, London.

BIRKELAND P.W. (1984) *Soils and Geomorphology*. Oxford University Press, New York.

DUCHAUFOUR P. (1982) *Pedology, Pedogenesis and Classification*, (Translated by T.R. Paton). Allen & Unwin, London.

GOLDICH S.S. (1938) A study of rock weathering. *Journal of Geology* **46**, 17–58.

HOLMES A. (1978) *Principles of Physical Geology*, 3rd Ed. Nelson, London.

JACKSON M.L., TYLER S.A., WILLIS A.L., BOURBEAU G.A. & PENNINGTON R.P. (1948) Weathering sequence of clay-size minerals in soils and sediments. I. Fundamental generalizations. *Journal of Physical and Colloid Chemistry* **52**, 1237–1260.

JENNY H. (1941) *Factors of Soil Formation*. McGraw-Hill, New York.

JENNY H. (1980) *The Soil Resource, Origin and Behaviour*. Springer-Verlag, New York.

MILNE G. (1935) Some suggested units of classification and mapping, particularly for East African soils. *Soil Research* **4**, 183–198.

MITCHELL J.K. (1976) *Fundamentals of Soil Behaviour*. Wiley, New York.

MOHR E.C.J. & VAN BAREN F.A. (1954) *Tropical Soils*. N.V. Uitgeveriz, The Hague.

THORNTHWAITE C.W. (1931) The climates of North America according to a new classification. *Geographical Review* **21**, 633–656.

THORP J. & SMITH G.D. (1949) Higher categories of soil classification. *Soil Science* **67**, 117–126.

FURTHER READING

CLARKE G.R. (1971) *The Study of the Soil in the Field*. Clarendon Press, Oxford.

GREENLAND D.J. & HAYES M.H.B. (Eds.) (1981) *The Chemistry of Soil Processes*. Wiley-Interscience, Chichester.

THOMAS M.J. (1974) *Tropical Geomorphology*. Macmillan, London.

Chapter 6
Soil Water and the Hydrologic Cycle

6.1 THE FREE ENERGY OF SOIL WATER

Total water potential

In section 4.5 the concept of soil wetness measured by θ, the volumetric water content, was introduced. This is an extensive property of the water in a defined volume of soil because it relates to the *quantity* of water in that soil volume. To characterize its physicochemical state, it is necessary to know something of the forces acting on the water that affect its free energy (an intensive property of the water).

A small quantity of water added to a dry soil is distributed so that the force of attraction between soil and water is as large as possible (achieving a maximum reduction in free energy). Additional water is held by progressively weaker forces so that eventually, when the soil is saturated, there is little difference between the free energy of the soil water and water in the standard state, which is taken as pure, free water at the same temperature, pressure and height above sea level as the water in the soil. The difference in free energy may be calculated, per mole of water, from the equation:

$$\mu_w(\text{soil}) - \mu_w^{\circ}(\text{standard state}) = RT \ln e/e^{\circ} \quad (6.1)$$

where μ_w = chemical potential (free energy/mole) of water;
e/e° = ratio of vapour pressure of water in the soil to vapour pressure of water in the standard state;
R = gas constant (8.3143 J K^{-1}mol^{-1});
and T = absolute temperature (°K). (The reference state for pressure is taken as atmospheric pressure.)

This difference in chemical potential defines the *total water potential*, ψ. The upper value of ψ is obtained at $e/e^{\circ} = 1$, when $\psi = 0$. The units of ψ are energy per mole (J mol^{-1}), but to be compatible with earlier *pressure* measurements, ψ is often expressed as energy per unit volume (J m^{-3}) by dividing equation 6.1 by V_w, the volume per mole of water. The units J m^{-3} are equivalent to N m^{-2} or pascals (Pa).

The hydrostatic pressure due to a column of water h metres high (equivalent to ψ in units of energy per unit volume) is given by:

$$\psi = h \rho g \quad (6.2)$$

where ρ = density of water (kg m^{-3}) and
g = acceleration due to gravity (m s^{-2})

Thus, a pressure of 1 *bar* (10^5 Pa) is subtended at the base of a water column of height:

$$\frac{\psi}{\rho g} = \frac{10^5}{1000 \times 9.8} = 10.2 \text{ m}$$

Hydrologists make use of this relationship and usually express water potential in terms of the height of the equivalent water column, referred to as *hydraulic head* or simply as *head*. Note that head, h, is an expression of total water potential in units of energy per *unit weight*, i.e. J s^2/kg m. In summary, it can be seen that water potential, ψ, expressed as energy per unit volume, is related to head, h (energy per unit weight) by the equation:

$$h = \frac{\psi}{\rho g} \quad (6.3)$$

Soil water may be held at a positive hydrostatic pressure (Figure 6.1a) or a negative hydrostatic pressure (tension) due to an applied suction (Figure 6.1b).

In dry soils, the suction of water, when expressed in head units in the centimetre–gram–second system, was often a large and unwieldy number. It was therefore more convenient to use $\log_{10} h$ in cm, which defines the *pF value* of the soil water. However, plant

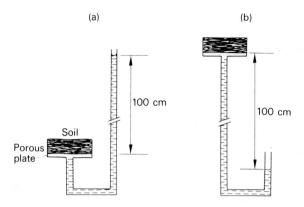

Figure 6.1. (a) Soil water under a hydrostatic pressure of 0.1 bar. **(b)** Soil water under a negative hydrostatic pressure of 0.1 bar.

physiologists traditionally measured the energy status of plant water in pressure units (ψ values) which are not readily compatible with soil water pF values. Thus, for studies of water movement through the soil–plant–atmosphere continuum (SPAC), the energy status of

Table 6.1. Water potential, hydraulic head and pF of a soil of varying wetness.

Soil moisture condition	Water potential ψ (bars)	Head, h (m)	pF	Equivalent radius of largest pores that would just hold water (μm)
Soil saturated or very nearly so	−0.001*	0.0102**	0	1500†
Moisture held after free drainage (field capacity)	−0.05	0.51	1.7	30
Approximately the moisture content at which plants wilt	−15	153	4.2	0.1
Soil at equilibrium with a relative vapour pressure of 0.85 (approaching air-dryness)	−220	2244	5.4	0.007

* Calculated from equation 6.1.
** Calculated from equation 6.2.
† Calculated from equation 6.5.

water is best expressed in terms of water potentials, ψ, and soil pF values are now rarely used.

Comparative values of water potential, head and pF for a soil at various states of dryness are given in Table 6.1.

Components of soil water potential

The free energy of water in the soil is reduced in the following ways.

(a) At low relative humidities (R.H. < 20 per cent), water is adsorbed on clay minerals of low permanent charge as a monolayer in which the molecules are hydrogen bonded to each other and to the surface. With an increase in R.H. more water is adsorbed and the influence of the surface on the structure of the liquid water extends further into the solution; yet the bulk of the effect occurs in the first three molecular layers, that is within approximately 1 nm of the surface.

(b) At the charged planar surfaces of the 2:1-type clay minerals, however, water is also more or less strongly attracted to the adsorbed cations, depending on whether the cations form inner- or outer-sphere complexes with the siloxane surface (section 2.4). The electric field of the cation orients the polar water molecules to form a solvation or hydration shell containing up to 6 water molecules for a monovalent cation and up to 12 for a divalent cation such as Mg^{2+} (Figure 6.2). The energy of hydration per mole of cation depends upon its ionic potential and polarizability, the energies for the common exchangeable cations falling in the order:

$$Al^{3+} > Mg^{2+} > Ca^{2+} > Na^+ > K^+ \simeq NH_4^+$$

Of course, the greater the hydration energy of the cation, the larger is the reduction in free energy of the

Figure 6.2. A hydrated exchangeable cation.

water molecules in its hydration shell. The relatively small amount of water present in dry soils (< 80 per cent R.H.) is therefore mainly associated with clay surfaces and their exchangeable cations.

(c) As water layers build up, the water begins to fill the finest pores and interstices between mineral particles, forming curved air–water menisci (Figure 6.3a).

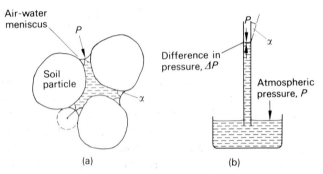

Figure 6.3. (a) Water held under tension between soil particles. **(b)** Rise of water in a capillary tube.

Equilibrium is attained when the surface free energy associated with the air–water–solid interface is at a minimum. The work done on the water is balanced by the increased potential energy of the water column, as for example in the rise of water in a capillary tube (Figure 6.3b). It follows from the latter figure that:

$$\pi r^2 \rho g h = 2\pi r \gamma \cos \alpha \qquad (6.4)$$

where π, r, ρ, g, h and α are as already defined, or indicated in Figure 6.3, and γ is the surface tension of water. Thus, the difference in pressure Δp across the meniscus is given by

$$\Delta p = \rho g h = \frac{2\gamma}{r} \qquad (6.5)$$

when the angle of wetting, $\alpha = 0$.

The combined effects of adsorption and capillarity on the free energy of soil water are expressed through the *matric potential*, ψ_m. Over the range of water contents for which capillary effects predominate, the matric potential, ψ_m, can be related directly to the pore radius through equation 6.5. Water held in pores of

diameter 1 μm, for example, has a matric potential of -3 bars.

(d) Solutes (ions and organic molecules) also reduce the free energy of soil water, the extent of the reduction being measured by the *osmotic potential*, ψ_s. Osmotic potential is numerically equal to the osmotic pressure of the soil solution, which is defined as the hydrostatic pressure necessary just to stop the inflow of water when the solution is separated from pure water by a semipermeable membrane. The osmotic pressure in bars (π) of soil solutions can be calculated from the equation:

$$\pi = \frac{RTC}{101.33} \qquad (6.6)$$

where R and T have been defined and C is the molar concentration of solute particles (molecules and dissociated ions) in solution.

ψ_m and ψ_s are *component potentials* of the soil water potential, ψ. In soil they are always less than 0 so that their contribution to ψ is always negative. Positive contributions to ψ can be made by the *gravitational potential*, ψ_g, if water is held at a height above that chosen for the standard state, or if soil water is subjected to a pressure greater than atmospheric, giving rise to a positive *pressure potential*, ψ_p.

The soil water potential, ψ, is therefore given by the sum:

$$\psi = \psi_m + \psi_s + \psi_p + \psi_g \qquad (6.7)$$

Although it is technically possible to increase the air pressure on a soil above normal atmospheric pressure, in practice variations in pressure potential due to changes in air pressure are negligible. The pressure potential, ψ_p, only becomes significant when the soil is saturated, in which case the pressure at some depth, z, below the surface of the saturated soil is directly proportional to z and defines the *piezometric head*. Obviously, ψ_m in the saturated soil must be zero so that ψ_m and ψ_p can be visualized as two subcomponents of a pressure-potential continuum. Above the water table, $\psi_p = 0$ (atmospheric pressure) and ψ_m is negative; at the water table $\psi_m = \psi_p = 0$, and below the water table, $\psi_m = 0$ and ψ_p is positive.

The retention and movement of water in the soil is discussed in terms of these potentials in the succeeding sections; for soil–plant relations in general, however, the important components of ψ are the matric, osmotic and gravitational potentials.

6.2 THE HYDROLOGIC CYCLE

A global picture

Water vapour enters the atmosphere by evaporation from soil, water and plants; it precipitates as rain, hail, snow or dew. Precipitation, P, equals evaporation, E, over the entire Earth's surface, but for land surfaces only, the average annual P of 0.71 m exceeds the average annual E by 0.24 m, the difference amounting to river discharge. If that part of the land surface which contributes little to evaporation—the deserts, ice caps and Tundra—is excluded from the calculation, the average rate of evaporation from the remainder is approximately 1 m per annum, which is not far short of the 1.2 m annual rate of evaporation from the oceans.

Evaporation depends on an input of energy, the energy available being much greater at the Equator

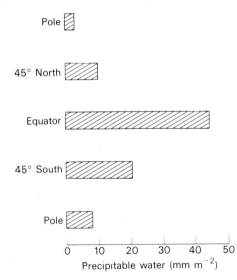

Figure 6.4. Amount of water vapour in the atmosphere by latitude (after Penman, 1970).

than at high latitudes. The capacity of the air to hold water also increases with temperature, so that the amount of precipitable water is greatest at the Equator, least at the Poles, and more in summer than in winter (Figure 6.4). It is less easy to estimate accurately the larger quantities of surface water, soil water and groundwater that sustain terrestrial plant and animal life: the average equivalent depth of soil water is *c.* 0.25 m, that of lake and river water *c.* 1 m and groundwater between 15 and 45 m.

Water balance on a local scale

As illustrated in Figure 6.5, of the gross precipitation on land, nearly all is intercepted by the vegetation with the remainder falling directly onto the soil. Water in excess of that required to wet the leaves and branches, the *canopy storage capacity* (usually 0.5–1 mm), drips from the canopy or runs down stems to the soil. Canopy drip, stem flow and direct rainfall constitute the *net precipitation* or *net rainfall*. Water retained by the canopy is lost by evaporation and is referred to as *interception loss*: it may be as much as one-quarter of the gross rainfall for broadleaf trees in summer.

Of the rain that reaches the soil, the difference between that soaking in and that running off comprises *surface runoff* or *overland flow*. The movement of water into the soil from above is called *infiltration*. When the soil is thoroughly wet down to the water-table, a steady rate of flow or *percolation* to the groundwater is maintained. Subsurface lateral flow or *interflow* also occurs through soils on slopes, or when vertical flow through the subsoil is impeded. Interflow eventually contributes to streamflow from the catchment, as does the lateral flow of ground water which is called *baseflow*. Surface runoff makes a much less predictable contribution to streamflow and may be accompanied by soil erosion.

Slow upward movement of water occurs in response to evaporation from the soil surface. Vegetation, however, can offer a much larger surface for the evaporation of water. The resultant decrease in water potential in the leaves provides a driving force for water to be sucked into the roots and through the plant,

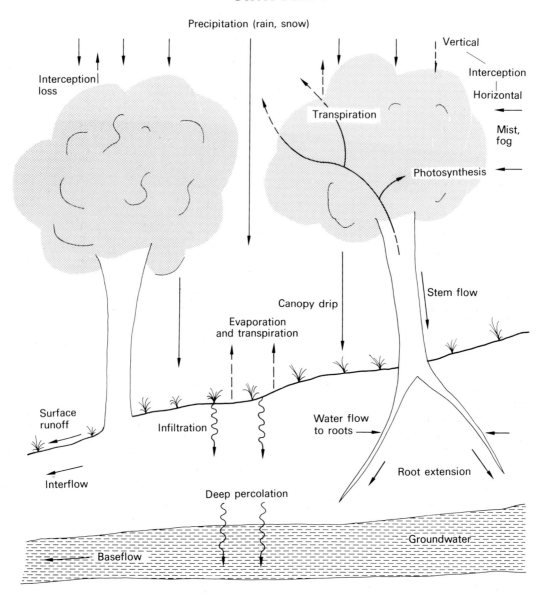

Figure 6.5. Hydrologic cycle in a small forested catchment.

to be lost by evaporation from the leaves. This is called *transpiration*. Some water flows into the water-depleted zones around absorbing roots while the remainder is intercepted by the roots as they grow through the soil.

The water balance equation

The interaction of all these processes within a regional hydrological unit or *catchment* is expressed by the equation:

$P = E + \text{Streamflow} \pm \text{Deep percolation} + \text{Change in soil storage}, \Delta S$ (6.8)

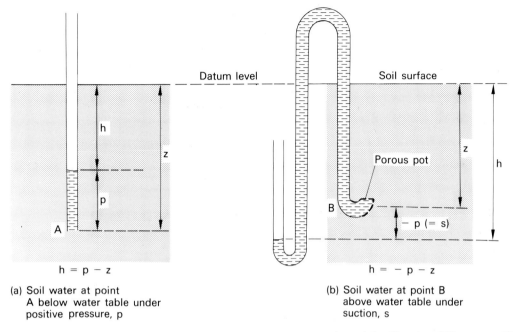

Datum level Soil surface

h

z

p

A

h = p - z

(a) Soil water at point
A below water table under
positive pressure, p

Porous pot

z

h

B

- p (= s)

h = - p - z

(b) Soil water at point B
above water table under
suction, s

Figure 6.6. Hydraulic head values for water in soil under a positive pressure, p, or suction, s (after Youngs and Thomasson, 1975).

For catchments in which deep percolation gains and losses are negligible, when periods of one year or more are considered, the changes in soil storage $\Delta S(+ \text{ or } -)$ tend to cancel out so that the difference between P and E is primarily due to streamflow. Over shorter periods, values of ΔS are nevertheless important for plant growth, and if ΔS is persistently negative irrigation may be necessary to supplement rainfall. Changes in ΔS can be measured by inserting sealed aluminium tubes in the soil into which a neutron probe is lowered. The soil water content θ is estimated by sensing the back-scatter from collisions with hydrogen nuclei of fast neutrons emitted from a radioactive source in the probe.

A knowledge of streamflow and deep percolation in a catchment is also important, first because streamflow provides water that can be conserved for human and animal use; second, because water in deep aquifers (permeable rocks that hold groundwater) is a valuable reserve from which to supplement surface supplies when E is consistently greater than P. Deep percolation is largely beyond man's control, as is precipi-

tation, despite attempts in the USA and Australia to induce rainfall by 'cloud seeding' with dry ice or silver iodide; there are, however, opportunities for manipulating the other variables in equation 6.8 which depend to a large extent on soil properties and land management in the catchment. It is therefore relevant to discuss infiltration, soil-water movement and evaporation in some detail.

6.3 INFILTRATION

Darcy's law
For water movement through porous materials, provided the velocity is low enough that flow is laminar and not turbulent, the rate of flow is directly proportional to the driving force and inversely proportional to the resistance, as expressed by Darcy's equation:

$$q = -K \operatorname{grad} \psi \qquad (6.9)$$

where q is the volume per unit time crossing an area, A, perpendicular to the flow; grad ψ denotes the change in water potential per unit distance in the direction of flow (which may be horizontal or vertical); and the proportionality coefficient K defines the *hydraulic conductivity* which is also the reciprocal of the flow resistance. (The negative sign accounts for the fact that flow occurs in the direction of decreasing potential and the gradient in ψ is negative).

The value of K depends on the amount of water in the pores and its viscosity, as well as on the pore geometry and surface roughness. The ideal conditions for applying Darcy's law are that both grad ψ and the resistance to flow are constant. Soils rarely satisfy these conditions because they are in a dynamic cycle of wetting and drying so that the parameters in equation 6.9 are continually changing in space and time (section 6.4).

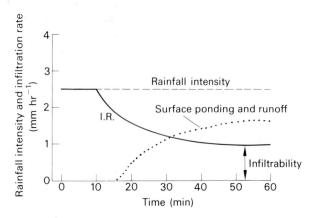

Figure 6.7. Infiltration, surface ponding and runoff during a storm.

For all except saline soils, ψ_m and ψ_g are the main component potentials governing water movement. Written in head units, the total water potential becomes the sum of the pressure head component, p, where $-p$ = suction, s, for soil water under tension, and the gravitational head component z, measured positively upwards from an arbitrary datum level, usually the soil surface, as illustrated in Figure 6.6. Thus, for flow downwards through a soil above the

water table, it can be seen that:

$$h = s - z \qquad (6.10)$$

and equation 6.9 becomes:

$$q = -K \operatorname{grad} h \qquad (6.11)$$

so that, combining equations 6.10 and 6.11 and writing in differential notation, the following is obtained:

$$q = -K\left(\frac{ds}{dz} - \frac{dz}{dz}\right)$$
$$= -K\left(\frac{ds}{dz} - 1\right) \qquad (6.12)$$

Note that to be consistent with the convention for signs introduced in equation 6.9, when s increases with depth, the product Kds/dz in equation 6.12 must be positive. Furthermore, because potential has been expressed in head units, K has the same dimensions as the flux density q ($L\,T^{-1}$).

Stages of infiltration

Rain falling on the soil surface is drawn into the pores under the influence of both a suction and gravitational head gradient, and if the rainfall intensity is less than the initial *infiltration rate* (*IR*) all the water is absorbed. In the early stages of wetting a dry soil, the suction gradient is predominant. However, as the depth of wet soil increases, the suction gradient decreases (the same difference in suction is spread over an ever-increasing depth interval Δz) and the gravitational head gradient becomes the main driving force. The flux density, q, then approaches the hydraulic conductivity, K, under which circumstances the soil may not be able to accept water rapidly enough to prevent ponding on the surface. For bare soil, structural changes in the surface following raindrop impact may also reduce the rate of acceptance so that the *IR* falls below the rainfall intensity within a few minutes to a few hours. Initially, ponding may be confined to small depressions, but once the surface detention reaches a critical value, runoff occurs. This sequence of events for rainfall of moderate intensity is illustrated in Figure 6.7.

6.4 REDISTRIBUTION OF SOIL WATER

The wetting front

If rain continues to fall on the soil surface, as in Figure 6.7, and the soil is structurally homogeneous, water travels downwards at a constant rate, q, which determines the *infiltration capacity* or *infiltrability* of the soil. As water penetrates more deeply a zone of uniform water content, the *transmission zone*, develops behind a narrow *wetting zone* and well defined *wetting front*. The plot of water content against depth for such a soil, illustrated in Figure 6.8, shows a sharp change in water content at the wetting front. This occurs because water at the boundary takes up a preferred position of minimum potential in the narrowest pores, for which the hydraulic conductivity is very low, and does not move at an appreciable rate until the macropores begin to fill.

Figure 6.9. Vertical infiltration into a soil in the field.

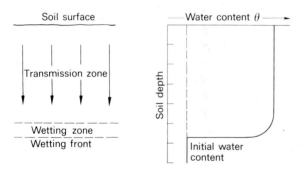

Figure 6.8. Vertical infiltration into a homogeneous column of dry soil.

For a constant rainfall intensity, q, and soil water content, θ, the mean *pore* water velocity, $\bar{v}$, in the wetted soil is given by:

$$\bar{v} = \frac{q}{\theta} \qquad (6.13)$$

and the distance, z, travelled by the wetting front in time, t, is given by:

$$z(t) = \frac{qt}{\theta} \qquad (6.14)$$

In practice, field soils are rarely structurally homogeneous due to the presence of old root channels, worm holes and planes of weakness between aggregates, so the downward movement of water is more erratic than Figure 6.8 would imply. A wetting front in a natural soil profile is shown in Figure 6.9. In this case, there may be a wide range of pore water velocities because some of the infiltrating water flows rapidly down channels and cracks (sometimes referred to as 'bypass' flow), whereas the bulk of the water penetrates slowly into the micropores within aggregates ('matrix' flow).

Within a well-developed transmission zone, the soil is saturated or near saturated (allowing for some trapped air pockets) and the suction gradient ds/dz is negligible. Therefore, equation 6.12 simplifies to:

$$q = K_s \qquad (6.15)$$

where K_s is the *saturated hydraulic conductivity* of the soil (which is also the infiltrability, see Figure 6.7). Values for K_s range from 1 to 500 mm day^{-1}, depending on the texture and structure of the soil.

Field capacity

When infiltration ceases, redistribution of water occurs at the expense of the initially saturated zone of soil. If the soil is completely wet to the watertable, or an artificial outlet such as a drain, drainage ceases when the suction gradient acting upwards balances the gravitational head gradient acting downwards. Even when the soil is not completely wet, depthwise, the drainage rate often becomes very small one to two days after rain or irrigation. This is possible because of the sharp fall in the hydraulic conductivity of the soil when most of the macropores have drained of water. The water content then defines the *field capacity* (*FC*), which for well-structured British soils is in equilibrium with a suction of about 50 mbars (section 4.5). As will be seen in section 6.5, the concept of field capacity is useful for setting an upper limit to the amount of 'available' water in the soil; but it is imprecise for two reasons: the soil may remain above the *FC* due to frequent showers of rain for several days; and soils with a high proportion of micropores continue to drain slowly for several days after rain.

Unsaturated flow

Flow under saturated or near-saturated conditions is usually in *steady state*, i.e. the flux density into a given volume is equal to the flux density out. However, as the larger, more continuous pores drain and the soil becomes *unsaturated*, the resistance to flow increases and the value of *K* decreases accordingly. Darcy's equation can be applied to unsaturated flow, and provided the suction gradient can be measured, it is possible to evaluate *K* using equation 6.12. The unsaturated conductivity is a function of the average suction or matric potential, i.e. $K(s)$; alternatively it may be written as a function of θ, if the relation between suction and θ is known (section 6.5).

For drainage to a very deep water table ds/dz is much smaller than 1, and the water content throughout the profile adjusts so that $q \simeq K$ (compare with saturated flow through a transmission zone during infiltration). Unsaturated flow of this kind occurs in deep permeable soils below the roots of the vegetation. Alternatively, flow is sometimes impeded by a layer of less permeable soil or parent material deep in the profile, resulting in localized saturation and the development of a *perched water table* (Figure 6.10). The increase in hydraulic head immediately above the impeding layer induces

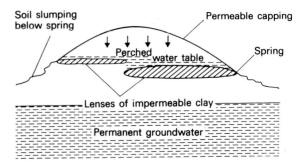

Figure 6.10. A landscape profile showing perched water tables and spring lines.

lateral flow which may eventually appear as a line of springs at another position in the landscape.

NON-STEADY STATE CONDITIONS

Soil water is rarely in steady-state equilibrium in the root zone. Intermittent rainfall and surface evaporation combined with the variable plant uptake of water cause the water content, θ, to fluctuate with time at any particular point in the soil. Flow can then only be described by combining Darcy's equation with an equation for the conservation of mass—the *continuity equation*. The latter relates the rate of change of θ in a small soil volume to the change in flow rate into and out of that volume, allowing for any water removed by evaporation and transpiration (*ET*) (Figure 6.11). Accordingly,

$$\frac{\partial \theta}{\partial t} = -\frac{\partial q}{\partial z} - ET \qquad (6.16)$$

Substituting for q from equation 6.12 we have

$$\frac{\partial \theta}{\partial t} = \frac{\partial}{\partial z}\left\{K(s)\left(\frac{\partial s}{\partial z} - 1\right)\right\} - ET \qquad (6.17)$$

Values of $K(s)$ can be measured in the field, for example using the 'instantaneous profile' method. Porous-cup tensiometers, based on the principle illustrated in Figure 6.6b, are installed at different depths to measure suctions up to 0.85 bars. A neutron probe can be used to measure change in θ. After the soil has

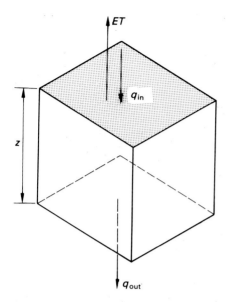

Figure 6.11. Water flow through a soil volume under non-steady state conditions.

been thoroughly wetted and covered to prevent evapotranspiration (ET), the changes in θ with time for measured values of ds/dz can be substituted in equation 6.17 and the equation solved for $K(s)$. Other field and laboratory methods for measuring $K(s)$ are discussed by Hillel (1982).

SOLUTE TRANSPORT

Infiltration of rain and the subsequent internal drainage of water in the soil can lead to the redistribution of solutes. The net downward movement of solutes is called *leaching* (section 5.3). On the assumption that the mechanisms involved are mass flow of solution (convection) and diffusion between regions of different concentration in the solution, the solute flux density, J, vertically downwards through the soil can be written as:

$$J = J_v + J_d \qquad (6.18)$$

where the convective flux, J_v, is given by:

$$J_v = qC \qquad (6.19)$$

and the diffusive flux, J_d, is given by Fick's first law:

$$J_d = -D\theta\frac{\partial C}{\partial z} \qquad (6.20)$$

C is the concentration of the solute in the soil solution and D is its diffusion coefficient in a porous medium.

Recalling the principle of mass conservation embodied in the continuity equation 6.16, for solute flux through a defined volume of soil, we may write:

$$\left\{\frac{\partial(\theta C)}{\partial t}\right\} = -\frac{\partial J}{\partial z} \qquad (6.21)$$

Substituting for J from equations 6.18, 6.19 and 6.20 we have:

$$\theta\frac{\partial C}{\partial t} = -\frac{\partial}{\partial z}\left\{qC - D\theta\frac{\partial C}{\partial z}\right\}$$

and:

$$\frac{\partial C}{\partial t} = D\frac{\partial^2 C}{\partial z^2} - \frac{q}{\theta}\frac{\partial C}{\partial z} \qquad (6.22)$$

In practice, if the substitution $q/\theta = \bar{v}$ is made in equation 6.22 on the assumption that equation 6.13 is valid, it is found that the value of D obtained by solving equation 6.22 is invariably greater than the porous medium diffusion coefficient of the solute. This is due to the coupled effects of molecular diffusion and hydrodynamic dispersion on the solute as water flows through the soil, a subject outside the scope of this book. For this reason, equation 6.22 is written in the form:

$$\frac{\partial C}{\partial t} = D_h\frac{\partial^2 C}{\partial z^2} - \bar{v}\frac{\partial C}{\partial z} \qquad (6.23)$$

where D_h is called the *dispersion coefficient* and (6.23) is known as the convection–dispersion equation for solute transport.

6.5 THE MOISTURE CHARACTERISTIC CURVE

Pore-size distribution

If a saturated soil core is placed on a porous plate

attached to a hanging water column and allowed to attain equilibrium (Figure 6.1b), the soil water content at the applied suction can be determined. For suctions > 0.85 bars, the soil is usually brought to equilibrium with an air pressure which is then equal but opposite to the water suction in the soil pores. A graph of θ against soil suction is constructed, which is called the *moisture characteristic* or *moisture retentivity curve*, illustrated by curve A in Figure 6.12. The rate of change of θ with s ($d\theta/ds$) is the *specific* or *differential water capacity*.

Since s is inversely proportional to the radius of the largest pores holding water at that suction (equation 6.5), the volume of pores having radii between r and $r-\delta r$, where δr is a very small decrement in r corresponding to an increase in suction δs, can be determined from the slope of the moisture characteristic curve, provided that shrinkage of the soil on drying is negligible. Furthermore, the larger the value of $d\theta/ds$, the greater is the volume of pores holding water within the size class defined by δr. Therefore, the value of s when $d\theta/ds$ is at a maximum defines (through the

relationship $s \propto 1/r$) the most frequent pore-size class, as illustrated by curve B in Figure 6.12. The example chosen is typical of a well-structured soil showing an adequate volume of large pores between peds, which promote drainage and aeration, relative to fine pores within peds that hold water in the so-called available range (see below).

HYSTERESIS

The shape of the moisture characteristic curve depends on whether the soil is drying or wetting—a phenomenon known as *hysteresis*, which occurs for two main reasons:

(a) Large pores do not empty at the same suction as they fill, if access to the pore is restricted by narrow necks. Take, for example, a large pore of diameter, D, which is connected to two smaller pores of diameters, d_1 and d_2. On drying, the large pore will not empty until the suction is great enough to break the meniscus across the larger of the two connecting pores, diameter, d_2 (Figure 6.13a). On wetting, water will advance into the large pore through pore d_1 or d_2 at roughly the same suction because they are of similar size. However, the large pore will not fill completely until the suction is small enough to sustain a meniscus across the distance, D (Figure 6.13b). Because $D > d_1$ or d_2, the suction at which the large pore fills on wetting is much less than the suction at which it empties on drying.

(b) For a given pore, the contact angle between the water meniscus and the pore walls is greater when the meniscus is advancing (wetting) than when it is retreating (drying). This means that the radius of curvature is greater for the advancing meniscus and that the suction developed is correspondingly less for the same water content.

The combined effect of (a) and (b) is illustrated in Figure 6.13(c) which shows that the soil comes to equilibrium at a higher water content for a given suction on the drying than the wetting phase of the cycle. A minor cause of hysteresis applies to clays that hold appreciable water at high suctions. It is thought to be due to changes in the size and arrangement of the

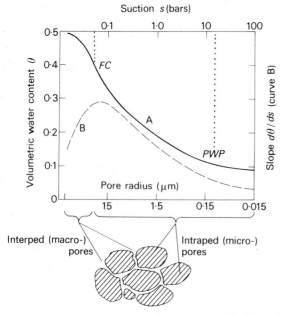

Figure 6.12. Graphs of the moisture characteristic (A) and specific water capacity (B) of a clay soil.

smallest pores as water films contract and clay domains become re-oriented. Such changes may not be reversible. Similarly, the water content attained at zero-applied suction at the end of the wetting phase may differ slightly from that at the start of the original drying phase because of air entrapment in the soil (Figure 6.13c).

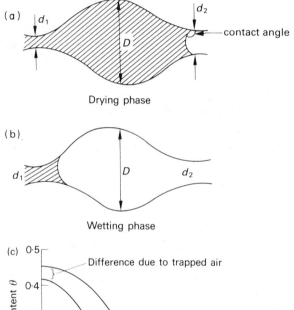

Figure 6.13. (a) and **(b)** Differences in pore water content at similar suctions during wetting and drying. **(c)** Hysteresis in the soil moisture characteristic curve.

Available water capacity

The term available water capacity (*AWC*) was introduced to define the amount of water in a soil that is available for plant growth. The upper limit is set by the field capacity (*FC*) and the lower limit as the value of θ at which plants lose turgor and wilt, that is, the *permanent wilting point* (*PWP*). When related to the moisture characteristic curve, these water contents are found to correspond to suctions of around 0.05 bars for *FC* and 15 bars for *PWP* for a number of soil–plant associations. It must be recalled from section 6.5, however, that *FC* is an inexact concept for many soils; further, the onset of permanent wilting depends as much on the plant's ability to lower the water potential within its tissues (to maintain a favourable gradient for water inflow from the soil) as it does on the soil water potential. For this reason, the choice of 15 bars as the suction at the *PWP* is essentially arbitrary. Nevertheless, the use of *AWC* as a measure of the soil's reserve of plant-available water has been widespread, particularly for assessing the growth potential of crops under dryland farming and for calculating the frequency of application of irrigation water (section 13.2). As the soil water content falls below the *FC*, the difference between the two is referred to as the *soil moisture deficit* (*SMD*).

It is sometimes assumed that water held between *FC* and *PWP* is equally available for growth, in the sense that a plant will *transpire at a constant rate* over the whole range. This is approximately true provided that:
(a) The soil moisture characteristic curve is of a type that the bulk of the available water is held at low suctions at which the hydraulic conductivity of the soil is relatively high—this is true of many sandy and sandy loam soils but not of clays.
(b) The evaporative power of the atmosphere is low enough that the rate of water movement from soil to root can satisfy the plant's transpiration. This is clearly demonstrated by Figure 6.14, which shows that when the evaporative power of the atmosphere is low (curve C), almost all the available water is transpired before the plant begins to suffer moisture stress and the transpiration rate falls. By contrast, when the evaporative power of the atmosphere is high (curve A), the transpiration rate drops markedly after only a small fraction of the *AWC* has been consumed. The plant

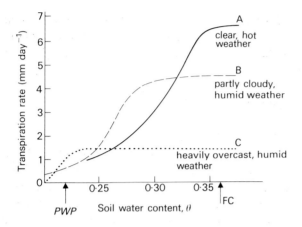

Figure 6.14. The effect of atmospheric conditions and soil water content on plant transpiration rate (after Denmead and Shaw, 1962).

then shows signs of wilting at θ values well above the *PWP*, but turgor is restored if cool, humid weather conditions supervene.

These observations illustrate the complexity of the interaction between soil properties (θ–suction and K–suction relationships), plant properties (rooting depth, root morphology, salt accumulation, osmotic adjustment) and atmospheric factors affecting the plant's transpiration rate. The strong link between atmospheric factors and transpiration rate, which determines the rate at which the plant must withdraw water from the soil to remain turgid, focuses attention on the physically based measurements of evapotranspiration that are discussed in section 6.6.

6.6 EVAPORATION AND EVAPOTRANSPIRATION

The energy balance

Evaporation from open water surfaces E_o is measured using evaporation pans, which are shallow tanks of water exposed to wind and sun (Figure 6.15a); evapotranspiration, *ET*, is measured using weighing lysimeters, which are large tanks filled with soil and vegetation (Figure 6.15b). Where such direct measurements are not possible, and especially for the estimation of *ET* losses on a regional scale, the *potential* rate of evaporation from any surface—open water, plant or wet soil—can be calculated from:

(a) the input of energy into the system;

(b) the capacity of the air to take up and remove water vapour.

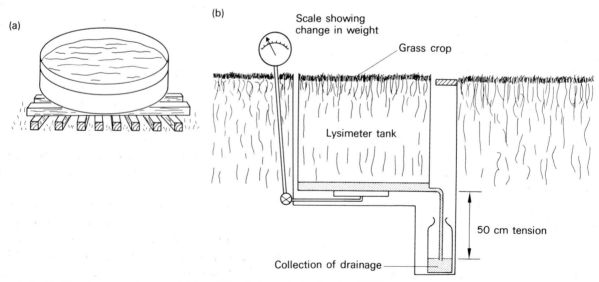

Figure 6.15. (a) Class A evaporation pan. **(b)** A weighing lysimeter (after Pereira, 1973).

The main source of energy is solar radiation of which a fraction is directly reflected, depending on the *albedo* of the surface (ranging from 0.15 for forests to 0.30 for agricultural crops), and some is lost as a net efflux of long-wave radiation from the surface to the atmosphere. Of the energy absorbed H, part is dissipated by evaporation (component LE, where L is the latent heat of vaporization and E the rate of evaporation), part by the transfer of sensible heat to the air (Q) and part stored as heat in the receiving system. Over a 24-hour period, the last component is negligible except for aquatic ecosystems, so we may write:

$$H = LE + Q \qquad (6.24)$$

The transport capacity of the air is estimated from the vapour pressure gradient over the surface and the average wind speed, with an added empirical factor to account for surface roughness when evaporation from vegetation, rather than a free water surface, is considered.

Q/LE is the ratio of sensible heat transported to latent heat transported and defines the *Bowen ratio*. Under humid temperate conditions, Q/LE is small so to a reasonable approximation, Q may be neglected in equation 6.24 and we may write:

$$E \simeq \frac{H}{L} \qquad (6.25)$$

Penman (1970) suggested that approximately 40 per cent of the incoming radiation, R_1, is dissipated in the evaporation of water from a short grass crop completely covering the soil surface and for which moisture is not limiting. Thus, substituting $H = 0.4\,R_1 = 219$ J m^{-2} s^{-1} and $L = 2.4 \times 10^9$ J m^{-3} in equation 6.25 gives an evaporation rate, E, of 3 mm day^{-1}: this is an estimate of *potential ET*, which for the conditions specified, is comparable with the actual rates of *ET* under mild cloudy conditions shown in Figure 6.14. However, when the soil moisture supply becomes limiting, actual rates of *ET* fall below potential rates by as much as 20–40 per cent.

It is clear that under arid conditions when plants may suffer water stress, the leaf surfaces will warm up and a much greater share of R_1 is lost to the air as sensible heat. In this case the Bowen ratio becomes very large and the approximation used in equation 6.25 is no longer valid.

Evaporation from soil

STAGES OF DRYING

While the soil surface remains wet, the rate of evaporation is determined by the energy balance between the soil and the atmosphere and the transport capacity of the air. Depending on the albedo of the soil, the rate can approach that of evaporation from an open water surface. This stage is called 'constant rate' drying.

With high evaporation rates from a soil, unaffected by a water table, the rate of water loss soon exceeds the rate of supply from below and the surface 1–2 cm become air-dry. As the macropores empty of water first, the unsaturated conductivity for water flow across the dry zone from below decreases sharply and more than offsets the steep rise in suction gradient. The rate of evaporation is now controlled by soil properties and decreases with time, a stage known as 'falling rate' drying. The dry surface layer has a *self-mulching* effect because it helps to conserve moisture at depth, especially under conditions of high insolation in soils with a fine granular or crumb structure.

Water movement in dry soil

Despite the self-mulching effect of an air-dry surface, if evaporation is prolonged and uninterrupted by rainfall, the zone of dry soil gradually extends downwards. This is because there is a slow flow of liquid water through soil wetter than the wilting point, to the drying surface, and there is also some diffusion of water vapour.

The flow of liquid water through relatively dry soil can be demonstrated by the simultaneous transfer of water and its dissolved salts in soil subjected to controlled, isothermal, evaporative conditions. As Figure 6.16 illustrates, the movement of water and chloride from the lower half of a series of soil blocks at initially

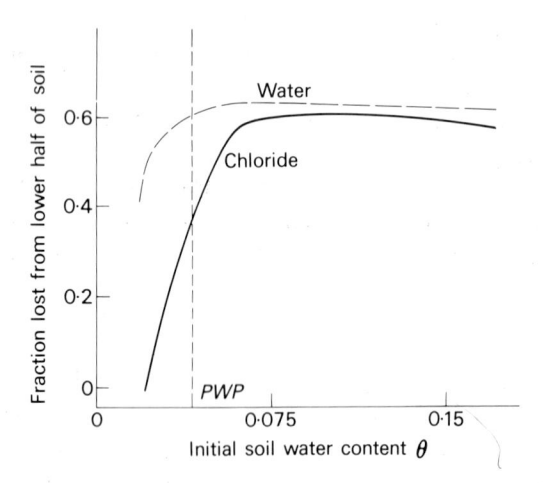

Figure 6.16. The movement of water and chloride through soil blocks exposed to constant evaporative conditions (after Marshall, 1959).

uniform water contents is identical down to a water content only slightly above the wilting point. Below that it is mainly water that moved, by vapour diffusion, which occurs in response to vapour pressure gradients created by differences in matric suction, salt concentration or temperature. The effect of matric or osmotic suction on vapour pressure is normally small because over the suction range 0–15 bars, the relative vapour pressure merely changes from 1.0 to 0.989. Only when the soil becomes air-dry and the relative vapour pressure of the soil water drops below 0.85 does the vapour pressure gradient under *isothermal* conditions become appreciable. Temperature, however,

has a large effect on vapour pressure, producing a three-fold increase between 10 and 30°C.

An exaggerated diurnal variation in temperature can cause redistribution of salt within the top 15 cm or so of a dry soil through the alternation of vapour and liquid phase transfer of water (Figure 6.17). As the surface heats up during the day, water vapour diffuses into the cooler layers and condenses. Transfer and condensation of water means heat transfer so that the temperature gradient, and hence the vapour pressure gradient, gradually diminishes until the evening when, with concurrent surface cooling, the temperature differential disappears. At this point, the increased water content and matric potential in the condensation zone induces a return flow of water to the surface, carrying dissolved salts that are deposited when the water subsequently evaporates.

CAPILLARY RISE

Soil water above the water table is drawn upwards in continuous pores until the suction gradient acting upwards is balanced by the gravitational potential gradient. This rise of groundwater, known as *capillary rise*, depends very much on the pore size distribution of the soil, being usually not greater than 1 m in sandy soils but as much as 2 m in some silt loams. Thus, a water table within 1–2 m of the soil surface can lead to excessive evaporative losses, the actual rate of loss being governed by the external atmospheric conditions up to the point where the conductivity of the soil becomes limiting. Accumulation of salts at the surface, as occurs in mismanaged irrigated soils, encourages capillary rise because the high osmotic suction maintains an upward head gradient even in moist soils, when the unsaturated hydraulic conductivity is high and appreciable water movement can therefore occur. The rise of groundwater often exacerbates the problem of soil salinity, which is discussed in Chapter 13.

6.7 SUMMARY

The interaction between soil water and organo-mineral surfaces, exchangeable cations and dissolved salts

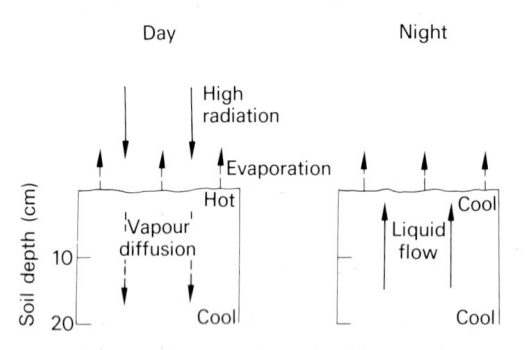

Figure 6.17. Diurnal alternation of vapour and liquid water movement at the surface of a drying soil.

reduces its partial molar free energy (or chemical potential μ_w) relative to pure water outside the soil, at the same height, temperature and pressure. The *total water potential*, ψ, is therefore defined as:

$$\psi = \mu_w(\text{soil}) - \mu_w^o(\text{standard state}) = RT \ln e/e^o$$

where e/e^o is the relative vapour pressure of the soil water. The value of ψ may be apportioned between several component potentials attributable to matric, osmotic, pressure and gravitational forces, respectively:

$$\psi = \psi_m + \psi_s + \psi_p + \psi_g$$

Usually, ψ_m and ψ_p are regarded as parts of a continuum such that at normal atmospheric pressure in an unsaturated soil, $\psi_p = 0$ and ψ_m is negative, whereas in saturated soil ψ_p is positive and $\psi_m = 0$. ψ can be expressed as energy per unit volume, equivalent to pressure in bars or pascals; or as energy per unit weight, which is equivalent to *hydraulic head* in units of length. Head, h, and potential, ψ, are related through:

$$h = \psi/\rho g$$

Broadly, the interchange of water between the atmosphere and Earth's surface by precipitation, P, and evaporation, E, from soil, water and plants comprises the *hydrologic cycle*. Within a defined catchment, $P = E + \text{Streamflow} \pm \text{Deep percolation} + \text{Change in soil storage}$, ΔS. Over a year or longer, ΔS is negligible so that streamflow and deep percolation determine the potential water storage, either in surface dams or as groundwater. Nevertheless, changes in ΔS are crucial in the short term and may decide crop success or failure.

Changes in soil volumetric water content, θ, can be measured by a *neutron probe*; changes in soil water suction, s (equivalent to $-\psi_m$), up to $c.$ 0.85 bars are measured by a *tensiometer*. Under static equilibrium conditions, the relationship between θ and s defines the *moisture characteristic curve*, the slope of which ($d\theta/ds$) depends on the pore-size distribution in non-swelling soils. Two points identified on this curve, the FC corresponding to suctions $c.$ 0.05 bars and the PWP

corresponding to a suction of 15 bars, set the approximate upper and lower limits for the *available water capacity* (AWC) of the soil. However, the rate of water uptake by plants also depends on plant properties and atmospheric factors, the latter exerting an overriding effect on evaporation and transpiration (*evapotranspiration*) rates when soil moisture is non-limiting.

The flux density q of water through the soil is given by Darcy's law:

$$q = -K \operatorname{grad} h$$

where K is either the saturated or unsaturated *hydraulic conductivity*. For predominantly vertical flow in non-saline soils, the main driving forces are the gradients in matric and gravitational potentials, represented in terms of head by s and depth, z, respectively. During saturated flow, $ds/dz = 0$ and $q = K_s$, the saturated hydraulic conductivity. For unsaturated flow, ds/dz must be measured, and if flow is unsteady, K can be determined only by combining Darcy's equation with an equation for the conservation of mass (*the continuity equation*). Solute transport through structurally homogeneous soils can be described by combining equations for mass flow and diffusion, which give rise to the *convection-dispersion equation*.

The hydraulic conductivity falls markedly with a decrease in θ (or increase in s). This accounts for the fall in evaporation rate from a bare soil surface once the top 1–2 cm has become air-dry. Vapour phase transport is insignificant in non-saline soils wetter than the PWP unless a temperature gradient exists between different layers in the soil.

REFERENCES

DENMEAD O.T. & SHAW R.H. (1962) Availability of soil water to plants as affected by soil moisture content and meteorological conditions. *Agronomy Journal* **54**, 385–390.

HILLEL D. (1982) *Introduction to Soil Physics*. Academic Press, New York.

MARSHALL T.J. (1959 *Relations between water and soil*. Commonwealth Bureau of Soils, Technical Communication No. 50.

PENMAN H.L. (1970) The water cycle. *Scientific American* **223**, 98–108.

PEREIRA H.C. (1973) *Land Use and Water Resources in Temperate and Tropical Climates.* Cambridge University Press.

YOUNGS E.G. & THOMASSON A. (1975) Water movement in soil, in *Soil Physical Conditions and Crop Production.* MAFF Bulletin No. 29. pp 228–239.

FURTHER READING

CHILDS E.C. (1969) *An Introduction to the Physical Basis of Soil Water Phenomena.* Wiley, New York.

HANKS R.J. & ASHCROFT G.L. (1980) *Applied Soil Physics.* Advanced Series in Agricultural Sciences 8, Springer-Verlag, Berlin.

HILLEL D. (1980) *Applications of Soil Physics.* Academic Press, New York.

HILLEL D. (1980) *Fundamentals of Soil Physics.* Academic Press, New York.

MARSHALL T.J. & HOLMES J. (1979) *Soil Physics.* Cambridge University Press, Cambridge.

PENMAN H.L. (1963) *Vegetation and hydrology.* Commonwealth Bureau of Soils, Technical Communication No. 53.

Chapter 7
Reactions at Surfaces

7.1 CHARGES ON SOIL PARTICLES

Some definitions

Chapter 2 introduced the concept of *permanent charge* on soil minerals arising out of the substitution of elements of similar size but different valency within the crystal lattice, and *pH-dependent charge* arising through the dissociation of protons from surface groups according to their Bronsted acid strength and the activity of H^+ ions in the bathing solution.

The net permanent charge created by isomorphous substitution in clay minerals (mainly the 2:1 clays) is invariably negative and is unaffected by the concentration and type of ions in the soil solution, unless the pH is so low that it induces the decomposition of the crystal lattice. With some of the hydrated oxides, there can be a small positive permanent charge due to the substitution of cations of higher valency within the crystal lattice. The result is a mineral of *constant* surface charge density (σ_0) and variable electrical potential, ϕ. On the other hand, the reversible dissociation of protons from carboxyl and phenolic groups in organic polymers, or at O and OH groups in the surfaces of oxides and edge faces of kaolinite crystals, results in a *variable* surface charge density (σ_H). H^+ (and OH^-) are called *potential-determining* ions for such surfaces, and the electrical potential of the surface remains constant provided the pH of the solution does not change, hence the expression 'constant potential' surface which is sometimes used. When surfaces of constant and variable charge occur on the one mineral, such as kaolinite and some allophanes (section 2.4), the resultant surface charge density is called the *intrinsic* surface charge density (σ_{in}), defined by:

$$\sigma_{in} = \sigma_0 + \sigma_H \qquad (7.1)$$

Methods of measuring or calculating σ_0, σ_H and σ_{in} are

discussed by Sposito (1984). σ_{in} is an operationally defined parameter because its value will depend on the experimental conditions under which it is measured.

The charged particle–solution interface

The existence of *immobile* charges on a mineral surface produces a change in the distribution of *mobile* charges (ions) in the soil solution in contact with the surface. A simple example is provided by the group of alumino-silicates called the *zeolites*, in which the permanent negative charge is distributed throughout the crystal lattice to give a uniform volume charge. Ions dissolved in the solution permeating the mineral can diffuse through a network of interconnecting pores, under-going frequent changes in electrical potential as they move from the influence of one lattice charge to another. However, when equilibrium is reached between the exchanger and the bathing salt solution, each ion suffers an *average* change in electrical potential on passing into or out of the exchanger, the value of which measures the *Donnan membrane potential* or simply the *Donnan potential*. Cations tend to be retained, or positively adsorbed, within the exchanger and anions are excluded or negatively absorbed. The volume from which anions are effectively excluded is called the *Donnan free space* (Figure 7.1). This model applies equally well to soil humic polymers and plant constituents, cell walls and cyto-plasmic proteins, except that the charges on these are not permanent but pH-dependent.

NON-UNIFORM VOLUME CHARGE

Unlike the zeolites, clay minerals have thin laminar structures in which the permanent negative charge acts as if it were spread over the planar surfaces of the crystals. The negative charge creates a surplus of cations (the counterions) and deficit of anions (the

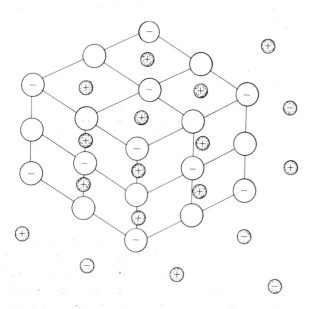

Figure 7.1. Concentration of mobile cations (+) within the Donnan Free Space and exclusion of mobile anions (−).

co-ions) in the solution immediately adjacent to the surface. The simplest case is one in which each surface charge is neutralized by the close proximity of a mobile charge of opposite sign—the parallel alignment of charges in two planes is called a *Helmholtz double layer* (Figure 7.2).

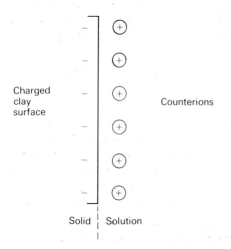

Figure 7.2. The Helmholtz double layer.

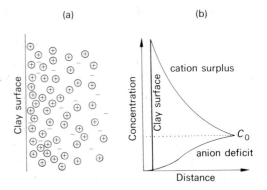

Figure 7.3. (a) and **(b)** The ionic distribution at a negatively charged clay surface.

However, the Helmholtz model is unrealistic: in aqueous solution the high dielectric constant of water reduces the Coulombic force of attraction between the fixed and mobile charges, which is also opposed by a diffusive force tending to drive the cations out of the region of high concentration near the charged surface. At equilibrium, the Coulombic force across a plane of unit area near to and parallel to the surface is just balanced by the difference in osmotic pressure between the plane and the external solution far removed from the surface. The result is a diffuse distribution of cations and anions in solution (Figure 7.3a), which together with the surface charge comprises the *Gouy–Chapman double layer*, so named after the two scientists who independently developed the mathematical description of the diffuse space charge. Such a diffuse double layer (DDL) can develop at surfaces of constant charge or variable charge.

Characteristics of the Gouy–Chapman double layer
The distribution of ions in solution in the direction normal to the charged planar surface (Figure 7.3a) is given by the Boltzmann equation:

$$C(x) = C_0 \exp[-zF\phi(x)/RT] \qquad (7.2)$$

$C(x)$ is the concentration of an ion of valency, z (cation or anion) at a distance, x, from the surface where the potential is $\phi(x)$; C_0 is the ion concentration at a plane in the bulk solution, F is Faraday's constant and R and T have their usual meaning (see equation 6.1). It is implicit in equation 7.2 that the potential in the bulk solution (concentration, C_0) is zero. Thus, for a negatively charged surface, as x becomes smaller and the absolute value of $\phi(x)$ increases, the concentration $C(x)$ of cations increases exponentially relative to C_0. At the same time, the concentration of anions decreases exponentially, becoming vanishingly small as the surface is approached. The concentration profiles for cations and anions in solution near a negatively charged surface are illustrated in Figure 7.3b.

At a clay mineral planar surface, the volume density of charge in the diffuse layer $\sigma(x)$ must remain constant irrespective of changes in the ionic composition of the solution. For each square cm of surface, part of the charge is neutralized by the attraction of cations and part by the repulsion of anions so that

$$\sigma(x) = \sigma(+)+\sigma(-) \tag{7.3}$$

where $\sigma(+)$ is the cationic charge in the diffuse layer (cmol$(+)$ cm^{-2}) and $\sigma(-)$ is the anionic charge repelled from the diffuse layer (cmol$(-)$ cm^{-2}). (Note that when $x = 0$, $\sigma(x) = \sigma_0$, the density of permanent negative charge on the planar clay surfaces.) As in a parallel plate condenser, the capacity of the DDL increases as the thickness of the diffuse layer decreases. The *effective thickness* of the diffuse layer is measured by the parameter d_{ex}, which is the mean distance over which an ion is depleted at the charged surface. It can be shown that d_{ex} depends on the valency of the counterion and the concentration, C_0 of the bulk solution, according to the equation:

$$d_{ex} \simeq \frac{2}{z\sqrt{\beta C_0}} \tag{7.4}$$

where β, the DDL constant, has the value 1.084×10^{16} cm cmol^{-1}. Calculated values of d_{ex} for different concentrations of NaCl and CaCl$_2$ solutions in contact with a clay surface are given in Table 7.1. The figures illustrate how, as the valency of the cation increases or the

solution concentration increases, the diffuse layer is compressed.

Table 7.1. Calculated diffuse layer thicknesses for different concentrations of 1:1 and 2:1 electrolytes.

Electrolyte concentration C_0 (mmol cm^{-3})	Effective diffuse layer thickness, d_{ex} (nm)	
	NaCl	CaCl$_2$
0.1	1.94	1.0
0.01	6.2	3.2
0.001	19.4	10.1

For a surface of *constant* charge (negative), as the DDL is compressed and its capacity increases, the potential $\phi(x)$ in the inner region of the DDL must decrease. In these circumstances, the fraction of the surface charge that is balanced by anion repulsion within the double layer increases, but at low salt concentrations this component is negligible when compared with the volume charge contributed by cation accumulation in the double layer. For a surface of *variable* charge (e.g. positive), as the DDL is compressed and its capacity increases, the charge on the surface must also increase. This increase in surface charge requires that protons be adsorbed from the solution, so that the pH rises and the surface potential, ϕ_0, which is controlled by the solution pH (see equation 7.7), decreases accordingly.

MODIFICATIONS TO THE GOUY–CHAPMAN DOUBLE LAYER

The Gouy–Chapman theory works well for cation solutions of intermediate concentrations (0.1–0.0001 M) and surface charge densities of 1–4×10^{-4} cmols m^{-2}. It breaks down once certain limits of cation valency and solution concentration are reached: first, because the ions are assumed to be point charges, which leads to absurdly high concentrations of cations in the inner region of the diffuse layer; second, because the theory ignores the existence of forces between the surface and counterions other than simple Coulombic forces. It predicts that cations of the same valency will be absorbed with the same energy, whereas it is known from cation exchange measurements that Li$^+$ and K$^+$, or Ca^{2+} and Sr^{2+}, for example, are not held with the same tenacity by clays in contact with solutions of identical normality.* These factors are accommodated in the *Stern model*—essentially a combination of the Helmholtz and Gouy-Chapman concepts—which splits the solution component of the double layer into two parts.

(a) A plane of cations of finite size located within a few Angstroms† of the surface, the Stern layer, across which the electrical potential decays linearly.

(b) A diffuse layer of cations, across which the potential falls almost exponentially, the inner surface of which abuts onto the Stern layer and is called the outer Helmholtz plane (*OHP*) (Figure 7.4). Allowance is made in the Stern model for the reduction in the dielectric constant of water due to its ordered structure very near the surface, and for specific adsorption forces between the surface and Stern layer cations.

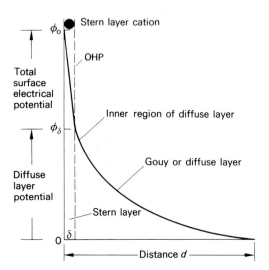

Figure 7.4. Electrical potential gradients at a planar clay surface (after van Olphen, 1977).

*A normal (N) solution is one containing 1 mmol of cationic or anionic charge per cm³.
†1 Angstrom (Å) = 10^{-10} m

The Stern model predicts that an increase in solution concentration not only compresses the double layer but also increases the proportion of ions in the Stern layer. The diffuse layer potential ϕ_δ decreases relative to the total surface potential ϕ_0 and the repulsion of anions from the inner region of the diffuse layer is diminished.

SPECIFIC ADSORPTION OF CATIONS
At a distance from the surface less than 4 Å the assumption of a plane of uniform surface charge does not hold and cations in the Stern layer experience a polarization force in the vicinity of each charged site. Depending on the balance of forces—the polarizing effect of the cation on the water molecules in its hydration shell compared with the polarization of the cation by the surface charge—the cation may shed its water of hydration to enter the Stern layer where it forms an *inner-sphere* complex (section 2.3) with surface groups of a siloxane trigonal cavity. The cations most likely to do this belong to elements of Group I in the Periodic Table. Li, Na, K, Rb and Cs form a lyotropic series in which the ionic potential and hence hydration energy of the monovalent cation decreases with increasing atomic size; the small cation Li$^+$ tends to remain hydrated at the surface whereas the large Cs$^+$ ion tends to dehydrate and become tightly adsorbed.

The attractive force between the charged surface and the solution cations, over and above the simple electrostatic force, is called a *specific adsorption* force and the cations are said to be specifically adsorbed. Cations held in the Stern layer are continually exchanging with cations in the diffuse layer, but at any instant the proportion held in the Stern layer increases from about 16 per cent for Li$^+$, to 36 per cent for Na$^+$ and 49 per cent for K$^+$, and the overall strength of adsorption increases in the sequence Li < Na < K < Rb < Cs.

Divalent cations, such as those formed by the Group II elements Mg, Ca, Sr, Ba and Ra, also form inner-sphere complexes at clay surfaces, but because of their generally higher ionic potentials they may retain some of their hydration water even when adsorbed in the Stern layer. For example, when Ba^{2+} is adsorbed on vermiculite, a monolayer of water is retained between the clay layers (but not between the Ba^{2+} ions and the nearer siloxane surface). However, K^+, which has almost the same ionic radius as Ba^{2+}, is adsorbed without any hydration water. The divalent cations also have a marked tendency to form *outer-sphere* complexes (section 2.4) which leads to quasi-crystal formation (section 4.4). The degree of dissociation of the cations from the interlayer surfaces in quasi-crystals is small and consequently differences between the divalent cations in their affinity for the surface are small. Nevertheless, the strength of adsorption of the divalent cations of Group II increases in the order Mg < Ca < Sr < Ba < Ra.

Following from equation 7.1, the density of *net* charge on a clay particle (σ_p) may be written as:

$$\sigma_p = \sigma_0 + \sigma_H + \sigma_{IS} + \sigma_{OS} \qquad (7.5)$$

where σ_{IS} is the charge density due to ions forming inner-sphere complexes (other than H^+ and OH^-) and σ_{OS} is the charge density due to outer-sphere complexes. For clays, σ_p is almost invariably negative and the balance of charge for the particle and its ionic atmosphere (the whole double layer) is made up by σ_d, the charge density of cations not complexed with the surface in any way, that is:

$$\sigma_p + \sigma_d = 0 \qquad (7.6)$$

It can be seen that even the Stern model of the DDL is incomplete, because although inner-sphere complexes (Stern layer cations) are accounted for, outer-sphere complexes are not. This has led to the development of models based on the molecular properties of the interface (surface complexation models), but such models are outside the scope of this book.

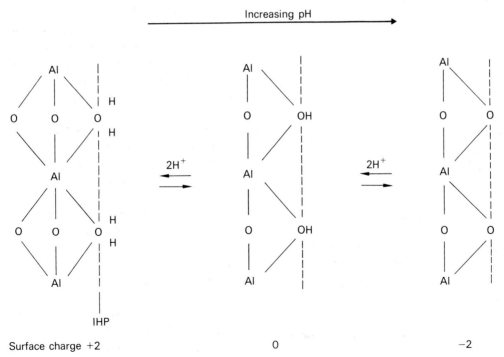

Figure 7.5. Changes in surface charge on an aluminium oxide.

Surfaces of variable charge

OXIDE SURFACES

Free oxides in soil, especially the hydrated Fe and Al oxides, exhibit pH-dependent charges due to the reversible adsorption of potential-determining H^+ ions. The change in surface charge with pH change is illustrated for hydrated Al_2O_3 in Figure 7.5. For pure oxides, $\sigma_{in} = \sigma_H$ and the pH at which the net charge on the surface is zero defines the *point of zero charge* (*PZC*). The greater the polarizing effect of the metal atom on the O—H bond, the more acidic is the surface; that is, the lower is the *PZC*. Silica, for example, has a *PZC* < 2 whereas the *PZC* of ferric oxide is ~ 9.

The oxygen, OH and OH_2^+ groups of the oxide surface lie in the inner Helmholtz plane (*IHP*) (Figure 7.5); naturally, when the surface has a net positive charge, anions are attracted from the solution to form a double layer. If the charge on the surface is all balanced by counterions forming inner- and outer-sphere complexes, so that the diffuse layer potential is zero, then the solution pH defines the *isoelectric point* (*IEP*) of the solid.

The potential ϕ_0 of a surface where protons are potential-determining is related to the *PZC* by the Nernst equation:

$$\phi_0 = \frac{2.303\,RT}{zF}(\mathrm{pH}_s - PZC) \qquad (7.7)$$

where pH_s is the pH of the solution in equilibrium with the surface. Clearly, ϕ_0 changes from positive through zero to negative as pH_s changes from being less than, equal to or greater than the *PZC*. The *PZC* is therefore a useful reference point from which to measure changes in the potential of surfaces of variable charge.

CLAY MINERAL EDGE FACES

Aluminium atoms coordinated to O and OH groups are also exposed at the edges of clay crystals. This structure differs from that of the free oxides only in that the oxygens are also coordinated to Si atoms in at least one contiguous silica layer. The full negative charge of 2 electrons per 0.4 nm² of edge area is only developed above pH 9 when the $\equiv$Si—OH groups dissociate (see Figure 2.17). The association of H^+ ions with the O and OH coordinated to the aluminium increases as the pH decreases from 9 to 5 in the manner illustrated in Figure 7.6, the *PZC* of the edge face occurring between pH 6 and 7 for kaolinite. Thus, the edge faces of the clay minerals show a pH-dependent charge.

ORGANIC MATTER

Although the charge of humified organic matter is completely dependent on pH due to the dissociation of carboxyl and phenolic groups (section 3.4), the charge is negative from pH 3 upwards and so augments the permanent negative charge of the clay minerals. In

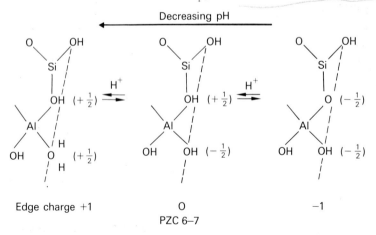

Figure 7.6. Charge development at the edge face of kaolinite with change in pH.

soils of high organic content, however, it has the effect of considerably increasing the pH-dependence of the soil's *CEC*.

The significance of pH-dependent charges in soil

Edge charges are of greatest significance in the kaolinites which have low permanent charges (2–5 cmol (+) per kg) and high edge:planar area ratios (1:10 to 1:5). The development of one positive charge per $0.4 \, nm^2$ of edge area amounts to $4 \times 10^{-4} \, cmol \, m^{-2}$, or 0.4–$4 \, cmol \, kg^{-1}$ of kaolinite ranging in edge face

area from 1 to $10 \, m^2 \, g^{-1}$. Conversely, edge charges are unimportant in the 2:1 clay minerals which have high permanent negative charges (40–150 cmol kg^{-1}) and exist as smaller crystals, especially in the direction of the *c* axis, thereby offering a smaller edge:planar ratio than the kaolinites. The charge characteristics of soil clays may, however, be considerably modified from that of the pure clay minerals by the presence of sesquioxides in variable amounts and spatial distribution.

The formation and occurrence of iron and aluminium oxides was discussed briefly in section 2.4. With *PZC* values in the range 6.5–9.5, they are normally positively charged in soil and occur as thin films adsorbed on the planar surfaces of clays, as illustrated by the $Fe(OH)_3$ film precipitated on kaolinite at pH 3 (Figure 7.7). $Al(OH)_3$ films occur preferentially within the interlamellar spaces of mica-type clay minerals to form mixed layer minerals. Positive charge densities of 30–50 cmol kg^{-1} have been recorded for soil oxides, so that soils that are low in organic matter and contain mainly kaolinite and sesquioxides can develop a *net positive* charge at low pH. Such behaviour is prevalent in highly weathered

Figure 7.7. Fe(OH)$_3$ deposit on a kaolinite cleavage face (courtesy of D.J. Greenland).

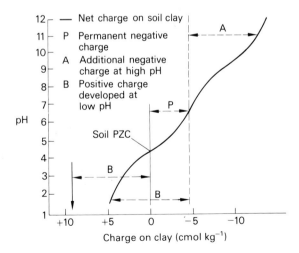

Figure 7.8. Charge *vs* pH curve for an acid red loam (Oxisol) (after Schofield, 1939).

soils of the tropics, exemplified by an acid subsoil from South Africa, the charge *vs* pH curve of which is illustrated in Figure 7.8. The presence of significant positive-charge densities, wholly dependent on pH, has profound implications not only for the ability of the soil to retain cations, but also for the adsorption of anions, two topics that are discussed in the following sections.

7.2 CATION EXCHANGE

Exchangeable cations

A high proportion of the cations released by rock weathering and organic decomposition are adsorbed by clay and organic colloids, leaving a low concentration in the soil solution to be balanced by mineral anions, and bicarbonate generated from the respiration of soil organisms. Apart from those forming inner-sphere complexes, cations held in the double layer are called *exchangeable* because they can exchange rapidly with cations in solution. In all but the most acid or alkaline of soils, the major cations are Ca^{2+}, Mg^{2+}, K^+ and Na^+, in roughly the proportions of 80 per cent Ca: 15 per cent Mg: 5 per cent (Na+K) with variable amounts of NH_4^+, depending on the efficiency of microbial nitrification (section 8.3). Trace amounts of other cations such as Cu^{2+}, Mn^{2+} and Zn^{2+} are also adsorbed, although chelation with organic compounds and precipitation play a more prominent part in the retention of these elements (section 10.5).

Cation exchange capacity and exchange acidity

One method of determining soil *CEC*, as defined in section 2.5, is to displace all the exchangeable cations with NH_4^+ ions provided by a 1 M solution of ammonium acetate at pH 7, and to measure the amount of NH_3 which can be distilled off by steam at pH > 10. The pH of the displacing solution is specified because of the pH-dependence of the *CEC*. In calcareous soils the sum of the cations (ΣCa, Mg, K, Na) is invariably equal to the *CEC* since any deficit of the cations on the exchanger can be made up by Ca^{2+} ions

from the dissolution of $CaCO_3$; in non-calcareous soils, however, ΣCa, Mg, K, Na is frequently less than the *CEC*, the difference being referred to as the *exchange acidity* (in cmol H^+ per kg soil). The ratio:

$$\frac{\Sigma(Ca, Mg, K, Na)}{CEC} \times 100$$

defines the *per cent base saturation*, which is crudely correlated with the pH in many mildly acid and neutral soils.

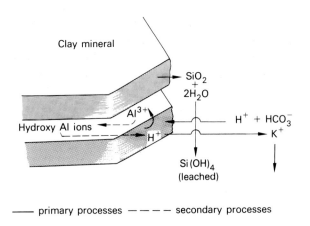

Figure 7.9. Clay mineral weathering in an acidic environment.

The exchangeable acidity is not primarily due to H^+ ions. As the soil pH drops below 5, increasing amounts of Al^{3+} ions can be displaced by leaching with strong salt solutions. Under acid conditions the weathering of clay minerals is accelerated, with the release of Al, SiO_2 and smaller amounts of Mg, K, Fe and Mn (Figure 7.9). While the silicic acid formed is removed by leaching, the Al, Mg, K and Mn are retained initially as exchangeable cations. The hydrated ferric ion $(Fe6H_2O)^{3+}$ is hydrolysed at pH < 3 and therefore readily precipitates, in the pH range 4–5, as ferric hydroxide or as the insoluble carbonate or phosphate if the appropriate anions are present. With the exception of illitic clay soils, K^+ is also lost by leaching so that an acid clay remains which is dominated by Al^{3+} with some Mg^{2+} (and H^+ ions produced by the hydrolysis of the Al^{3+}).

pH buffering capacity

High levels of exchangeable aluminium greatly increase the soil's capacity to neutralize OH^- ions, which is a measure of the soil's *pH buffering capacity*. The hydrated Al^{3+} ion hydrolyses on the addition of alkali according to the reaction:

$$(Al6H_2O)^{3+} + H_2O \rightleftharpoons [Al(OH) 5H_2O]^{2+} + H_3O^+ \quad (7.8)$$

there being equal concentrations of the di- and trivalent species at pH 5. The hydroxyaluminium ions $(AlOH)^{2+}$ show a pronounced tendency to combine through bridging OH groups and build up into polymeric units of 6 and more Al atoms. The consequent release of H^+ contributes to the buffering capacity

until eventually neutralization is complete and amorphous $Al(OH)_3$ precipitates on the surfaces and interlamellar spaces of the clay.

Three stages are recognized in the neutralization of acid clays and these are illustrated in Figure 7.10 where representative curves for kaolinite, vermiculite and montmorillonite titrated in a neutral salt solution are plotted. Provided that sufficient time is allowed for near-equilibrium pH values to be attained, the three clays show markedly different buffering within the ranges:

I   < pH 4   exchangeable H^+
II   pH 4–5.5   exchangeable Al^{3+}
III   pH 5.5–7.5   hydroxyaluminium ions

Some 'weak' acidity due to proton release from hydroxy–Al polymers continues to be neutralized at pH > 7.5, and the quantity of alkali required to raise the soil pH to 8.2 is usually taken as a measure of the

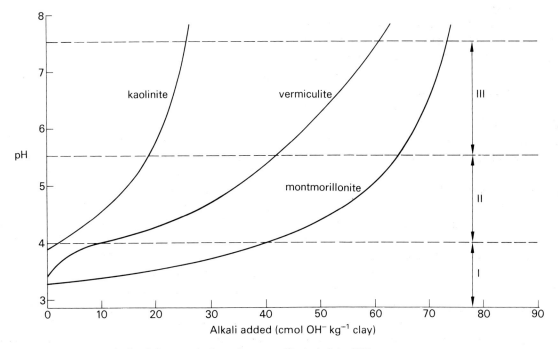

Figure 7.10. pH-titration curves for kaolinite, vermiculite and montmorillonite in 0.1 M KCl.

soil's *titratable acidity* (cmol H^+ kg^{-1}). The buffer solution used is $BaCl_2$ (0.1–0.3 M) containing half-neutralized 0.2 M triethanolamine. In practice, titration of a soil to a specific pH is used to estimate the soil's *lime requirement* (section 11.3). In soils with much interlayer and surface hydroxy-Al films, the titratable acidity is always greater than the exchange acidity because part of the *CEC* is blocked by the hydroxy–Al polymers which are not displaced in neutral salt solutions.

Al is extensively hydrolysed when adsorbed by organic matter, the average charge per mole of adsorbed Al being +1. The Al^{3+} ions react preferentially with the stronger acid groups and the H^+ ions released by hydrolysis suppress the dissociation of weaker acid groups on the organic matter. The apparent pK of Al-free organic matter is between 4 and 5 but rises to *c.* 6 when Al is added. Thus, the adsorption of Al reduces the acidity of soil organic matter and increases its capacity to adsorb anions such as phosphate (section 7.3).

Cation exchange reactions

EXCHANGE EQUATIONS

The technique of using a strong solution of NH_4^+ ions to displace the exchangeable cations and determine the *CEC* is an example of *cation exchange*. Cation exchange is a reversible process in which one mole of cation charge (or equivalent) in solution replaces one mole of cation charge on the exchanger, as for example:

$$(NH_4^+) + (\text{Ca-exchanger}) \rightleftharpoons (NH_4\text{-exchanger})$$
$$+ \tfrac{1}{2}(Ca^{2+}) \qquad (7.9)$$

where (NH_4) and (Ca^{2+}) are the molar activities of the ions in solution, and (Ca-exchanger) and $(NH_4$-exchanger) represent the activities of the ions on the exchanger. Much has been written about the validity of cation exchange equations, which all founder on the problem of measuring the *activity* of ions in the adsorbed state. The best approach is probably an

empirical one in which it is assumed that NH_4^+ and Ca^{2+} are adsorbed in proportion to: (a) the 'reduced' ratio of their equilibrium activities in solution; and (b) the relative affinity of the exchanger for NH_4^+ compared with Ca^{2+}, as measured by a selectivity coefficient, k_{NH_4-Ca}. The resultant equation:

$$\frac{NH_4\text{-exchanger}}{Ca\text{-exchanger}} = k_{NH_4-Ca} \frac{(NH_4^+)}{(Ca^{2+})^{1/2}} \qquad (7.10)$$

is called the *Gapon equation* and has proved of wide applicability to cation exchange in soils. Provided that the monovalent cation (Na^+ or K^+) amounts to < 50 per cent of the exchangeable Ca^{2+} and Mg^{2+} combined, reasonably constant values of the Gapon coefficients $k_{Na-Ca, Mg}$ and $k_{K-Ca, Mg}$ are obtained for soils of similar clay mineral type from the equations:

$$\frac{Na\text{-exchanger}}{(Ca+Mg)\text{-exchanger}} = k_{Na-Ca, Mg} \frac{(Na^+)}{(Ca^{2+}+Mg^{2+})^{1/2}}$$
$$(7.11)$$

and

$$\frac{K\text{-exchanger}}{(Ca+Mg)\text{-exchanger}} = k_{K-Ca, Mg} \frac{(K^+)}{(Ca^{2+}+Mg^{2+})^{1/2}}$$
$$(7.12)$$

(Ca^{2+} and Mg^{2+} have sufficiently similar adsorption affinities, relative to the monovalent cations, to be regarded as interchangeable). An equation similar to equation 7.11 has been of great value in predicting the likely 'sodium hazard' of irrigation waters (section 13.2), and equation 7.12 has been applied to the study of potassium availability in soil.

For example, equation 7.12 may be rewritten as:

$$\frac{\text{exchangeable K}}{CEC\text{-exchangeable K}} = k' \frac{K}{\sqrt{Ca+Mg}} \qquad (7.13)$$

where the quantities on the left-hand side of the equation are in cmol per kg soil. When exchangeable K is small (< 10 per cent of the *CEC*), equation 7.13 reduces to the simple form:

$$\text{exchangeable K} = k'' \frac{K}{\sqrt{Ca+Mg}} \qquad (7.14)$$

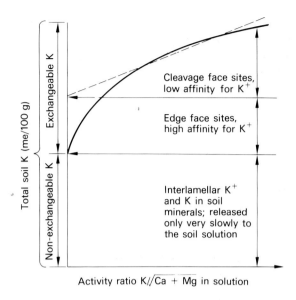

Figure 7.11. Schematic Q/I relation for K in a micaceous clay soil (after Beckett, 1971).

where k'' includes the Gapon constant and the CEC of soil. Usually the *change* in exchangeable K (ΔK) is plotted against the activity ratio (AR) to give a *quantity/intensity* relation (Q/I) for the soil's exchangeable potassium (Figure 7.11). However, the reactions of K in soil are complex, especially in soils containing much partially-expanded clay mineral. As

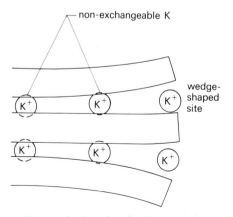

Figure 7.12. The weathering edge of a micaceous clay crystal.

described in section 2.4, K^+ ions form inner-sphere complexes in the unexpanded interlayer spaces of illite and micaceous clays and are therefore non-exchangeable. (NH_4^+, having a similar ionic potential to K^+, may also be held as a non-exchangeable cation in this way.) However, with weathering and the depletion of K in solution, the clay crystals slowly exfoliate from the edges exposing the interlayer K to exchange by other cations in solution (Figure 7.12). The exposed inter-layer sites at the crystal edges (which have been likened to 'wedge-shaped' sites) have a higher affinity for K^+ than the cleavage faces so that the whole crystal exhibits a changing affinity for K^+ relative to Ca^{2+} and Mg^{2+}, as first the edge sites and then the cleavage face sites are occupied. This is reflected in a gradual change in the slope of the Q/I plot, and hence the soil's buffering capacity for K, in going from low to high AR values.

The ratio law

The Gapon equation indicates that if the proportions of Na^+ and K^+ on the exchanger are unchanged, the ratio Na^+/K^+ in the equilibrium solution remains constant, and similarly, if the proportions of K^+ and Ca^{2+} on the exchanger are constant, the ratio $K^+/\sqrt{Ca^{2+}}$ is also constant. Schofield (1947) was the first to formulate such observations into a general hypothesis called the *ratio law*. It states that 'when cations in solution are in equilibrium with a larger number of exchangeable cations, a change in the concentration of the solution will not disturb the equilibrium if the concentrations of all the monovalent ions are changed in the one ratio, those of all the divalent ions in the square of that ratio and those of all the trivalent ions in the cube of that ratio'. For 'concentration' one may read 'activity'.

The constancy of the cation activity ratios depends on the effective exclusion of the accompanying anions such as Cl^- from the adsorption sites, since only then will the activities of the exchangeable cations be unaffected by changes in the Cl^- concentration of the external solution. Chloride exclusion is determined by the thickness of the double layer (equation 7.4) so that

the ratio law is only valid provided that:
(1) there is a preponderance (> 80 per cent) of negative over positive charge on the surface, and
(2) the total solution concentration is not too high. As theory predicts, the upper limit of concentration for which the ratio law applies varies with the valency of the cation, being approximately 0.1 M for Na and K, 0.01 M for Ca and 0.002 M for Al. Despite these limitations, the ratio law is very useful in interpreting ion exchange in soils as the following examples illustrate.

LEACHING OF CATIONS

Percolation of water through the soil leads to losses of solutes by leaching. If the concentration of cations in the leaching solution is adjusted according to the ratio law, the equilibrium between exchangeable and solution cations will be undisturbed despite the overall reduction in concentration. This is confirmed by the results of percolation tests on an unlimed Rothamsted Park Grass soil (Table 7.2), in which the concentrations of cations in the second leaching solution (B) were reduced in accordance with the ratio law. With percolating rain water, which is naturally very dilute, all the cations are in low concentration, and exchange occurs between cations of higher valency in solution and adsorbed cations of lower valency so that the activity ratios Na:$\sqrt{}$Ca, K:$\sqrt{}$Ca and so on tend to remain constant. Gradually Na followed by K is lost,

the exchange surfaces becoming dominated by Ca and Mg and ultimately by Al. The result is the genesis of an acid soil of low-base saturation.

pH MEASUREMENT

Soil pH is usually measured after shaking one part of the soil by weight with 2.5 or 5 parts of distilled water by volume and immersing the tips of a glass and a calomel reference electrode in the supernatant solution. For soils containing predominantly 2:1 and 1:1 clays (negatively charged), dilution of the soil solution by distilled water increases the absolute value of the surface potential and therefore changes the distribution of H^+ ions between the DDL and the bulk solution. The measured pH, which is the pH of the bulk solution, is higher than that of the undisturbed soil, an effect especially noticeable in saline soils. However, if the soil is shaken with a solution containing the most abundant cation at a concentration approximating to the total concentration of the soil solution, the measured pH changes little with dilution. For many soils a solution of 0.01 M $CaCl_2$ is found to satisfy these requirements and the pH so measured is very close to the natural pH of the soil, as determined in a solution which has been equilibrated with successive samples of the soil (Table 7.3).

Table 7.3. pH measurements on a basaltic red loam.

Soil/liquid ratio (g ml⁻¹)	pH in H₂O	pH in 0.01 M CaCl₂	pH in an equilibrium solution*
1:2	5.08	4.45	4.45
1:5	5.29	4.45	4.45
1:10	5.43	4.46	4.45
1:50	5.72	4.52	4.45

*0.01 M $CaCl_2$ after successive equilibrations with four separate samples of soil (after White, 1969).

7.3   ANION ADSORPTION

Adsorption sites and mechanisms
Protonated hydroxyl groups on the edge faces of clay

Table 7.2. Rothamsted Park Grass Soil, at equilibrium with solution A and percolated with solution B, the concentration of which was adjusted according to the ratio law.

Solution	Percolate no.	Na⁺	K⁺	Mg²⁺	Ca²⁺	Total cations
				(me l⁻¹)		
A	3	0.0	3.9	1.3	5.1	10.3
			(at equilibrium)			
B	1	0.0	2.1	0.3	1.5	3.9
B	2	0.0	2.0	0.3	1.3	3.6
B	3	0.0	2.0	0.4	1.2	3.6
			(at equilibrium)			

(After Schofield, 1947.)

minerals or on oxide surfaces will adsorb anions from solution according to the reaction:

$$M—OH_2^+ + A^- \rightleftharpoons M—OH_2^+A^- \qquad (7.15)$$

The anion A^- may exchange with other anions in the soil solution, in accordance with the ratio law, that is the ratio of adsorbed $H_2PO_4^-$ to SO_4^{2-} varies with the activity ratio $H_2PO_4^-/\sqrt{SO_4^{2-}}$ in the solution provided that the concentration of cations at the surface is vanishingly small. However, when the positive sites are only isolated spots on a clay-mineral surface, or the soil solution concentration is low enough for the diffuse layer at planar surfaces to envelope the positive sites at the edges, anion adsorption is determined by the characteristics of the double layer at the negatively charged surfaces (Figure 7.13a). In consequence, the concentration of SO_4^{2-} at the inner surface of the double layer is less than the concentration of $H_2PO_4^-$, and as the soil solution becomes more dilute, the increase in the concentration of SO_4^{2-} in the bulk solution is relatively greater than the increase in $H_2PO_4^-$ concentration. Under these conditions the cation-anion *activity products* $(Ca^{2+}) \times (SO_4^{2-})$ and $(Ca^{2+}) \times (H_2PO_4^-)^2$ measured in solution should have constant values.

Non-specific adsorption, as represented by equation 7.15, is typical of anions like ClO_4^-, Cl^- and NO_3^-. But oxyanions such as $H_2PO_4^-$, HCO_3^-, SO_4^{2-}, MoO_4^{2-}, $H_3SiO_4^-$ and $B(OH)_4^-$ can be adsorbed by a *ligand exchange* mechanism in which an OH or OH_2^+ group in the mineral surface is displaced by an O of the oxyanion, e.g.

(a)   (b)

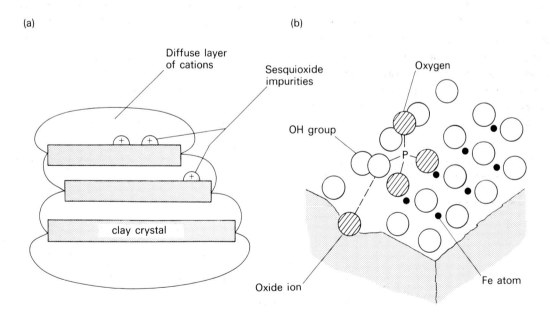

Sites for anion adsorption on clay crystals

Model of $HPO_4^=$ adsorption on a goethite crystal surface after Parfitt and others (1976)

Figure 7.13. (a) and (b) Physical models of the interrelations between clays, sesquioxides and anions in solution.

$$M—OH+H_2PO_4^- \rightleftarrows M—O—\overset{\displaystyle OH}{\underset{\displaystyle OH}{\overset{|}{\underset{|}{P}}}}{=}O+OH^- \qquad (7.16)$$

or:

$$M—OH_2^++H_2PO_4^- \rightleftarrows M—O—\overset{\displaystyle OH}{\underset{\displaystyle OH}{\overset{|}{\underset{|}{P}}}}{=}O+H_2O \qquad (7.17)$$

In these reactions, the anion forms an ionic–covalent bond with the cation (M) by entering into its co-ordination shell. The anion is therefore very strongly adsorbed, a condition described as *chemisorbed*. If both types of site M—OH and M—OH$_2^+$ occur on the mineral surface, then as phosphate is adsorbed, the solution pH can rise *and* the positive charge on the surface can decrease. The latter effect decreases the *PZC* of the surface.

Experimentally, decreases in surface charge from 1 to 0 have been observed per mole of anionic charge adsorbed onto pure Fe and Al oxides. Studies of phosphate adsorption on goethite using infra-red (IR) spectroscopy suggest that phosphate displaces two singly coordinated OH groups of adjacent Fe atoms to form a binuclear HPO$_4$ bridging group (Figure 7.13b). Although ligand exchange is probably the principal mechanism by which phosphate, molybdate, silicate and borate are adsorbed, HCO$_3^-$ and SO$_4^{2-}$ are intermediate in that they can be adsorbed by this mechanism, or by simple electrostatic attraction as in equation 7.15. Phosphate and other anions non-specifically adsorbed at positive sites on oxide surfaces can be desorbed by reducing the ion's concentration in the ambient solution, but anions coordinated to the surface can only be desorbed by raising the pH or introducing another anion with a higher affinity for the metal of the oxide.

Precipitation of insoluble salts

Phosphate also forms insoluble salts with Fe, Al and Ca. The crystalline minerals, *strengite* (FePO$_4$.2H$_2$O) and *variscite* (AlPO$_4$.2H$_2$O), are only stable below *c.* pH 1.4 and 3.1, respectively, and are most unlikely to exist in soil. However, in less acid soils, slightly more soluble compounds with an Al:P or Fe:P mole ratio $\simeq$ 1 (amorphous AlPO$_4$ and FePO$_4$) can form, especially in the vicinity of a dissolving granule of superphosphate (section 12.3). Further, the hydroxy–Al polymers that form on clay surfaces in the pH range 4.5–7.0 can adsorb phosphate by ligand exchange. These compounds are most stable between pH 5 and 6.5 so that P adsorption in acid soils that have a predominance of phyllosilicates in the clay fraction is usually at a maximum in this pH range (section 10.3).

Above pH 6.5, phosphate forms insoluble salts with Ca, such as *octacalcium phosphate*, Ca$_4$H(PO$_4$)$_3$. 5H$_2$O, which reverts to the more stable *hydroxyapatite*, Ca$_{10}$(PO$_4$)$_6$(OH)$_2$, and phosphate may also be present as an impurity in calcite crystals. In the latter case, the release of phosphate depends on calcite dissolution, which is primarily controlled by the partial pressure of CO$_2$ in the soil air (section 11.3).

7.4 PARTICLE INTERACTION AND SWELLING

Attractive and repulsive forces

The attractive forces between clay particles are of two main kinds:

(a) *Electrostatic*, as between negatively charged cleavage faces and positively charged edges; or K$^+$ ions and the adjacent lamellae in an illite crystal; or partially hydrated Ca^{2+} and Mg^{2+} ions and the lamellae of vermiculite;

(b) *Van der Waals' forces* between individual atoms in two particles. For individual pairs of atoms the force falls off with the seventh power of the interatomic distance, but the total force between two particles is the sum of all the interatomic forces which results not only in an appreciable force between macromolecules, but also in a less rapid decay in force strength with distance.

The origin of the attractive forces is such that they vary little with changes in the solution phase, whereas the repulsive forces that develop due to the interaction of diffuse double layers are markedly dependent on the composition and concentration of the soil solution. The model of the double layer presented so far assumes no impediment to the diffuse distribution of ions at the charged surface—the concentration in the midplane between two clay plates is equal to the concentration in the bulk solution. This is reasonably true of dilute clay suspensions in salt solutions, but in concentrated suspensions and in natural soils the particles are close enough for the diffuse layers to interact, when the balance between forces of attraction and repulsion determines the *net attractive force* between the particles. If the net force is attractive (Figure 7.14a), the particles remain close together and are described as *flocculated*. It also follows that soil aggregates, of which the clay particles form a part, will remain stable. Conversely, if the net force is repulsive (Figure 7.14b),

the particles move farther apart until a new equilibrium position is attained under the constraint of an externally applied force, or the particles move far enough to exist as separate entities in the solution—a state described as *deflocculated*.

The excess concentration of ions at the midplane between two interacting particles gives rise to an osmotic or *swelling pressure* of the solution between the two plates. Swelling pressures of 10–20 bars may develop for particle separations from < 1 nm upwards, depending on the type of clay mineral, the inter-lamellar cations and the salt concentration of the solution.

Flocculation–deflocculation and swelling behaviour

FACE-TO-FACE FLOCCULATION
The stacking of single layers of Ca-montmorillonite in roughly parallel alignment, with much overlapping, to form *quasi-crystals* and of larger crystals of Ca-illite or

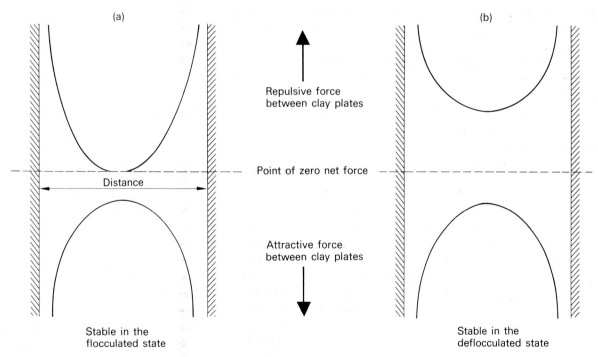

Figure 7.14. (a) and **(b)** The interaction of attractive and repulsive forces between clay plates in salt solutions.

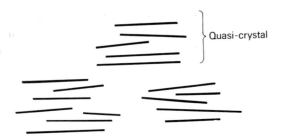

Figure 7.15. Face-to-face flocculation of Ca-saturated mont-morillonite.

Ca-vermiculite to form *domains*, was briefly described in section 4.4. Surface areas for Ca-montmorillonite of $100-150$ m^2 g^{-1} compared with $750-800$ m^2 g^{-1} for the completely dispersed clay suggest that each quasi-crystal consists of 5–8 layers (Figure 7.15). Anions are totally excluded from the interlayer regions and the diffuse double layer develops only at the external surfaces of the quasi-crystal. The interlayer cations appear to occupy a midplane position and provide a strong attractive force that stabilizes the interlayer spacing, irrespective of the external solution concentration.

SWELLING BEHAVIOUR

Except in atmospheres drier than about 50 per cent R.H. (section 6.1), Ca-montmorillonite holds at least two layers of water molecules in the interlayer spaces due to the hydration of the cations. The basal spacing is then 1.5 nm. Additional water is taken up as the soil solution becomes more dilute, but only to the extent of expanding the basal spacing to 1.9 nm which is maintained even in distilled water. This behaviour is typical of *intracrystalline* swelling in Ca-montmorillonite or *intradomain* swelling in Ca, Mg-illite or vermiculite.

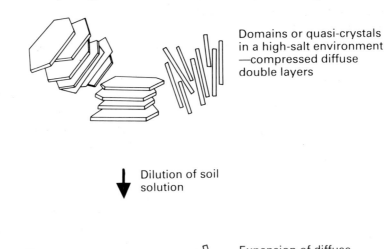

Domains or quasi-crystals in a high-salt environment —compressed diffuse double layers

Dilution of soil solution

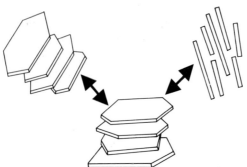

Expansion of diffuse double layers on external surfaces of the domains or quasi-crystals

Figure 7.16. Intercrystalline or domain swelling of a Ca-saturated 2:1 clay.

As observed in section 4.4, however, the quasi-crystals and domains exhibit long-range order over distances up to 5 μm. Diffuse double layers develop at the *outer* surfaces of the quasi-crystals (and domains), leading to swelling and the creation of wedge-shaped spaces where water can be held by surface tension. Swelling of this kind, described as *intercrystalline* or domain swelling, is illustrated in Figure 7.16; it induces particle realignments that release the strains imposed by the crystal packing and bending during the previous drying cycle, and has been suggested as a cause of hysteresis in moisture characteristic curves at high suctions (section 6.5).

The swelling pattern of montmorillonite is entirely changed when divalent cations are replaced by the monovalent cations Na^+ and Li^+ (Table 7.4). As the salt concentration drops below 0.3 M NaCl, repulsive forces prevail and the swelling pressure forces the layers apart to distances that can be predicted from diffuse layer thicknesses calculated from DDL theory.

Table 7.4. Intracrystalline swelling of 2:1 clays saturated with different cations.

Clay mineral	Charge per unit cell	Exchangeable cation	Swelling in dilute solution (nm)
Mica	−2.0	K, Na	none
Vermiculite	−1.3	Ca, Mg, Na	1.4–1.5
		K	1.2
		Cs	1.2
		Li	≫ 4.0
Montmorillonite	−0.67	Ca, Mg	1.9
		K	1.5 and ≫ 4.4
		Cs	1.2
		Na, Li	≫ 4.0

(After Quirk, 1968.)

EDGE-TO-FACE FLOCCULATION
Kaolinite crystals are larger than those of montmorillonite and the hydrous micas and they flocculate face-to-face in concentrated solutions at high pH. Leaching and a reduction in the concentration of the soil solution cause the diffuse layers to expand, and the increased swelling pressure to disrupt the face-to-face arrangement. If there is a concurrent increase in soil acidity, the edge faces become positively charged and the mutual attraction of double layers at cleavage and edge faces encourages *edge-to-face* flocculation (Figure 7.17).

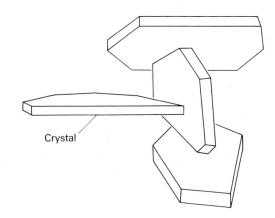

Crystal

Figure 7.17. Edge-to-face flocculation of kaolinite at low pH.

Of the two mechanisms, face-to-face flocculation is the norm for montmorillonitic clays, depending as it does on the predominance of di- and trivalent exchangeable cations or the precipitation of sesquioxidic films that form an 'electrostatic bridge' between opposing surfaces. However, edge-to-face flocculation is more common in kaolinites because of the high positive edge charge. Edge-to-face flocculation is insensitive to changes in the solution concentration but is weakened by the adsorption of anions such as $H_2PO_4^-$ at the edge faces. In the same way, chemisorption of phosphate and organic anions by sesquioxides may actually reverse the surface charge, rendering the oxide negatively charged and ineffective as a clay flocculant.

7.5 CLAY–ORGANIC MATTER INTERACTIONS

Much of the organic matter in soil is complexed in some way with the clay minerals and free oxides. Knowledge of these interactions is therefore of great importance to the understanding of virtually every process—physico-chemical and biological—that occurs in the soil. Broadly speaking, the organic compounds may be divided into cationic, anionic and nonionic compounds of low ($< 1,000$) to high ($> 100,000$) molecular weight. Because of the polarity of charge on clays below pH 7, organic cations are adsorbed mainly on the planar faces and anions at the edges, except that small organic anions can form links with exchangeable cations on the cleavage faces. Uncharged organic molecules interact with the surfaces through polar groups and non-specific van der Waals' forces.

Electrostatic interactions

CATIONS

Many organic cations are formed by protonation of amine groups, as in the alkyl* amines ($R—NH_2$) and amino acids ($RCHNH_2COOH$). Protonation is enhanced by the greater acidity of water molecules solvating cations adsorbed on clay surfaces. The higher the ionic potential of the cation M or its polarizing effect on the $M—OH_2$ bond, the greater is the tendency for a proton to be expelled from one of the solvating water molecules. The resultant protonation of a neutral organic molecule can be represented by:

$$M(H_2O)_n^{m+} + R—NH_2 \rightleftharpoons MOH(H_2O)_{n-1}^{(m-1)+} + R—NH_3^+ \quad (7.18)$$

On dry clay surfaces the polarizing effect of the cations is concentrated on fewer water molecules so the degree of dissociation is increased and the protonation of adsorbed organic molecules correspondingly enhanced. The displacement of the metallic cations by organic cations such as the alkyl amines reduces the affinity of the clay surface for water, making the soil more water-repellant and diminishing the swelling propensity of the clay.

Organic cation adsorption takes place at the internal and external surfaces of 2:1 expanding-lattice clays.

*A chain hydrocarbon group such as $CH_3, C_2H_5 \ldots R$

The adsorption affinity increases with the chain length of the compound due to van der Waals' forces between the adsorbed molecules, so that adsorption in excess of the *CEC* may be achieved. Proteins adsorbed in inter-lamellar regions are less susceptible to microbial attack; conversely, extracellular enzymes, which are functional proteins, often show reduced biochemical activity in the adsorbed state.

ANIONS

Organic compounds are also adsorbed due to the attraction of $—COO^-$ groups to positively charged sites on clay edge faces and oxide surfaces. In addition, the carboxyl groups may undergo ligand exchange with OH and OH_2^+ groups in these surfaces: the latter obviously provides a much stronger form of adsorption than the former.

Organic anions and uncharged polar groups can form *inner-* and *outer-sphere* complexes with adsorbed cations on negatively charged organic or clay mineral surfaces. The former complex involves direct co-ordination of the anion with the metallic cation following the displacement of a solvating water molecule (the *cation-bridge* shown in Figure 7.18). The outer-sphere complex is a weaker bond formed between the anion (or polar molecule) and the metallic cation without the displacement of a water molecule (the *water-bridge* of Figure 7.19). Adsorption of organic molecules by cation-bridging is more likely when the exchangeable cations are monovalent (because these cations dehydrate more easily),

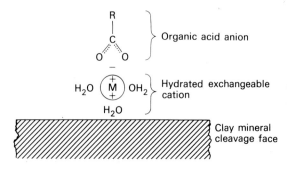

Figure 7.18. Organic acid molecule forming a cation-bridge (after Greenland, 1965a).

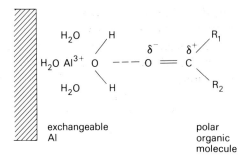

Figure 7.19. Organic molecule forming a water-bridge with exchangeable cation.

whereas water-bridging is favoured when the cations are di- or trivalent.

Phenolic-type compounds such as tannins which are active constituents of the litter of many coniferous trees, form coordination complexes with Fe and Al, and in small concentrations they are very effective *deflocculants* of clays. The compound must contain at least three phenolic groups, of which two coordinate the metal atom and the third dissociates to confer a negative charge on the whole complex. Conversely, long chain polyanions are particularly effective *flocculants* of clays, because the clay particle edges become attached to the anionic groups of the molecule in the manner of a 'string of beads' (Figure 7.20). Such compounds have therefore been applied successfully,

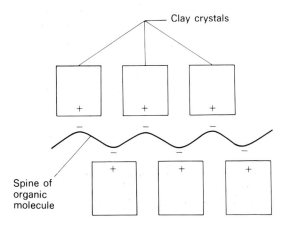

Figure 7.20. 'String of beads' arrangement of organic polyanion and kaolinite crystals.

if expensively, as *soil conditioners* to improve soil structure (section 11.4).

Specific and non-specific van der Waals' forces and entropy stabilization

Low molecular weight alcohols, sugars and oligo-saccharides are not adsorbed in competition with water unless present in high concentration relative to the water. Highly polar compounds can interact with the surface by forming water bridges, as discussed above, or by direct hydrogen-bonding to another organic compound that is already adsorbed (Figure 7.21).

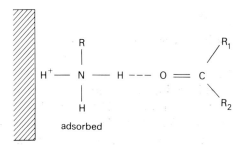

Figure 7.21. Hydrogen-bonding between two organic compounds at a clay surface.

High molecular weight alcohols (e.g. polyvinyl alcohol) and polysaccharides are strongly adsorbed at clay surfaces, even in competition with water. The attraction occurs partly through hydrogen-bonding to O and OH groups in the surface and partly through non-specific van der Waals' forces. The greatest contribution to the adsorption energy probably comes from the large increase in entropy when the organic macromolecule displaces many water molecules from the surface—the entropy increase stabilizes the molecule in the adsorbed state (see Table 4.2).

7.6 SUMMARY

Retention of cations and anions, soil acidity, water adsorption, and the formation and stabilization of structure are but a few of the many soil properties that are determined by reactions at mineral and organic surfaces.

Apart from their large specific areas, these surfaces are important because of the charge they bear: either a *permanent charge* (usually negative) due to isomorphous substitutions in the crystal lattice, or a *pH-dependent charge* due to reversible dissociation of protons from OH and OH_2^+ groups (clay mineral edges and oxides) or COOH and ⟨O⟩–OH groups of organic matter. The preferential attraction of cations and repulsion of anions at planar clay surfaces give rise to an *electrical double layer*, comprising Stern layer cations tightly adsorbed at the surface plus loosely bound cations in the diffuse layer extending into the solution.

The repulsive force developed as diffuse layers of adjacent clay crystals overlap is manifest as a *swelling pressure*, which can disrupt the microstructure of clay domains. Attractive forces between clay crystals include non-specific van der Waals' forces and electrostatic forces emanating from divalent cations, which form *outer-sphere* complexes with the crystal surfaces. Hydroxy–Al polymers can also act as electrostatic 'bridges' between the planar clay surfaces. When such forces exceed the repulsive force, *face-to-face* flocculation of the clay ensues; alternatively, the crystals may flocculate *edge-to-face*, as in the case of kaolinite at pH < 6.

The sum of the moles at cation charge in the double layer, at a reference pH of 7, comprises the *cation exchange capacity (CEC)*. The difference between the *CEC* and Σ(Ca, Mg, K, Na) defines the *exchange acidity*, and the ratio:

$$\frac{\Sigma(\text{Ca, Mg, K, Na})}{CEC} \times 100$$

defines the *per cent base saturation*. Acid clays are primarily Al^{3+}-clays. Hydrolysis of Al^{3+} and polymerization of the hydroxy-Al products accounts for much of the *pH buffering capacity* of a soil over the pH range 4.0–7.5.

Cation exchange between the double layer and the

soil solution may be described by the *Gapon equation*:

$$\frac{\text{K-exchanger}}{\text{Ca-exchanger}} = k_{\text{K–Ca}}\frac{K^+}{\sqrt{Ca^{2+}}}$$

Changes in solution activities that occur on leaching soils, or in soils influenced by saline water, may be explained in terms of the *ratio law*, which predicts that when cations in solution are in equilibrium with a larger number of exchangeable cations, the *activity ratio* $a^+/(a^{n+})^{1/n}$ is a constant over a limited range of salt concentration.

Inorganic and organic anions can be adsorbed at pH-dependent sites when these are positively charged (*non-specific* adsorption). However, oxyanions such as $H_2PO_4^-$, $H_3SiO_4^-$, MoO_4^{2-} and $B(OH)_4^-$ are more likely to be adsorbed by *ligand exchange* in which the O of the anion displaces an OH or OH_2^+ group from the coordination shell of the metal cation in the surface. This results in a very strong bond and the anion is said to be *chemisorbed*. Organic anions also interact with cations adsorbed at negatively charged surfaces by forming *cation-bridges* (displacement of a solvating water molecule) or *water-bridges* (outer-sphere complex formed with the cation). Depending on the sign of the surface charge, organic cations and anions can be important in linking together clays and sesquioxides to form stable microaggregates.

REFERENCES

BECKETT P.H.T. (1971) Potassium potentials—a review. *Potash Review*, Subject 5, pp. 1–41.

GREENLAND D.J. (1965a) Interaction between clays and organic compounds in soils. I. Mechanisms of interaction between clays and defined organic compounds. *Soils and Fertilizers* **28**, 415–425.

PARFITT R.L., RUSSELL J.D. & FARMER V.C. (1976) Confirmation of the surface structures of goethite (a-FeOOH) and phosphated geothite by infrared spectroscopy. *Journal of the Chemical Society, Faraday Transactions*, **72**, 1082–1087.

QUIRK J.P. (1968) Particle interaction and soil swelling. *Israel Journal of Chemistry* **6**, 213–234.

SCHOFIELD R.K. (1939) *The electric charges on clay particles*. International Society of Soil Science, British Empire Section, pp. 1–5, Cambridge.

SCHOFIELD R.K. (1947) A ratio law governing the equilibrium of cations in the soil solution. *Proceedings of the 11th International Congress of Pure and Applied Chemistry (London)* **3**, 257–261.

SPOSITO G. (1984) *The Surface Chemistry of Soils.* Clarendon Press, Oxford.

VAN OLPHEN H. (1977) *An Introduction to Clay Colloid Chemistry*, 2nd Ed. Wiley Interscience, New York.

WHITE R.E. (1969) On the measurement of soil pH. *Journal of the Australian Institute of Agricultural Science* **35**, 3–14.

FURTHER READING

ARNOLD P.W. (1978) Surface–electrolyte interactions, in *The Chemistry of Soil Constituents* (Eds. D.J. Greenland and M.B.H. Hayes). Wiley, Chichester, pp. 335–404.

GREENLAND D.J. (1965b) Interaction between clays and organic compounds in soils. II. Adsorption of soil organic compounds and their effect on soil properties. *Soil and Fertilizers* **28**, 521–532.

MORTLAND M.M. (1970) Clay–organic complexes and interactions. *Advances in Agronomy* **22**, 75–117.

SPOSITO G. (1981) *The Thermodynamics of Soil Solutions.* Clarendon Press, Oxford.

THENG B.K.G. (1979) *Formation and Properties of Clay–Polymer Complexes.* Elsevier, Amsterdam.

THOMAS G.W. (1977) Historical developments in soil chemistry: ion exchange. *Soil Science Society of American Journal* **41**, 230–238.

THOMAS G.W. & HARGROVE W.L. (1984) The chemistry of soil acidity, in *Soil Acidity and Liming* (Ed. F. Adams). Agronomy Monograph, No. 12 (2nd Ed.). American Society of Agronomy, Madison, pp. 3–56.

WHITE R.E. (1980) Retention and release of phosphate by soil and soil constituents, in *Soils and Agriculture* (Ed. P.B. Tinker). Critical Reports on Applied Chemistry, Volume 2. Blackwell Scientific Publications, Oxford, pp. 71–114.

Chapter 8
Soil Aeration

8.1 SOIL RESPIRATION

Respiratory quotient and respiration rate

The importance of soil organisms in promoting the turnover of carbon has been outlined in Chapter 3. *Aerobic respiration* involves the breakdown or dissimilation of complex C molecules, oxygen being consumed and CO_2, water and energy for cellular growth being released. Under such conditions, the *respiratory quotient*, defined as:

$$RQ = \frac{\text{volume of } CO_2 \text{ released}}{\text{volume of } O_2 \text{ consumed}} \qquad (8.1)$$

is equal to 1. When respiration is anaerobic, however, the *RQ* rises to infinity because O_2 is no longer consumed but CO_2 continues to be evolved. The respiratory activity of the soil organisms or *respiration rate*, *R*, is measured as the volume of O_2 consumed (or CO_2 released) per unit soil volume per unit time.

Soil respiration is augmented by the respiration of living plant roots, and, further, the respiration of microorganisms is greatly stimulated by the abundance of carbonaceous material (mucilage, sloughed-off cells and exudates) in the soil immediately around the root.

Table 8.1. Soil respiration as influenced by crop growth and temperature.

	Respiration rate, R(l m^{-2} day^{-1})			
	January		July	
	Fallow	Cropped	Fallow	Cropped
O_2 demand	0.5	1.4	8.1	16.6
CO_2 release	0.6	1.5	8.0	17.4
RQ	1.20	1.07	0.99	1.05

(After Currie, 1970.)

This zone, which is influenced by the root, is called the *rhizosphere*. The stimulus to soil respiration provided by the presence of a crop is demonstrated in Table 8.1, where respiration rates for cropped and fallow soil in the south of England are compared for the months of January (mean soil temperature 3°C at 30-cm depth) and July (mean soil temperature 17°C). This soil had an unusually deep A horizon so that the respiration rates are about twice those normally recorded in the field.

Measurement of respiration rates

An instrument designed to measure respiration rates is called a *respirometer*, and one designed specifically for field measurements is illustrated in Figure 8.1. The soil moisture content is regulated by controlled watering and drainage; temperature is recorded and the circulating air is monitored for O_2, the deficit being made good by electrolytic generators. Carbon dioxide is absorbed in vessels containing soda lime and measured.

Such measurements show that the respiration rate depends on:

(a) soil conditions, such as organic matter content, O_2 supply and moisture;

(b) cultivation and cropping practices;

(c) environmental factors, principally temperature.

The effect of temperature on respiration rate is expressed by the equation:

$$R = R_0 Q^{T/10} \qquad (8.2)$$

where *R* and R_0 are the respiration rates at temperature *T* and 0°C respectively, and *Q* is the magnitude of the increase in *R* for a 10°C rise in temperature, called the *Q-10 factor*, which lies between 2 and 3. Temperature change is the cause of large seasonal fluctuations in soil respiration rate in temperate climates, as illustrated in Figure 8.2.

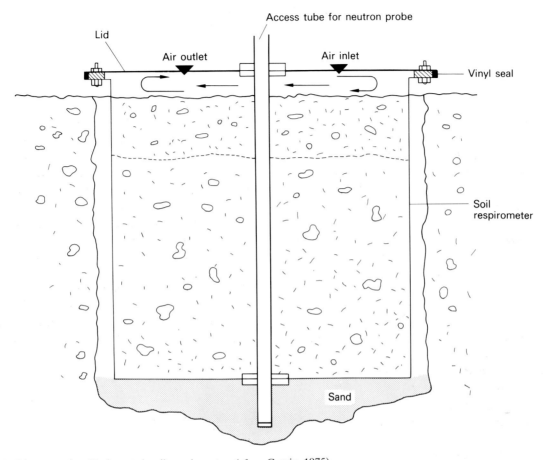

Figure 8.1. Diagram of a Rothamsted soil respirometer (after Currie, 1975).

Cycles of respiratory activity

In southern England the seasonal maxima in R values lag 1–2 months behind those of soil temperature. Superimposed on this seasonal pattern is the effect of weather, for within any month of the year the daily maximum in R may be up to twice the daily minimum, a fluctuation strongly correlated with soil surface or air temperatures. Finally, there are minor variations in R caused by the diurnal rise and fall of soil temperature.

These trends are modified by other factors, for example, at Rothamsted in southern England the respiratory peak is consistently lower during a dry summer than a wet one, and the rate tends to be higher in spring than in autumn for the same average soil temperatures. The latter effect is attributable to two factors, primarily a greater supply of organic residues at the end of the quiescent winter period than after a summer of active decomposition, together with spring cultivations that break up large aggregates and expose fresh organic matter to attack by microorganisms.

8.2   MECHANISMS OF GASEOUS EXCHANGE

The composition of the soil air, as described in section 4.5, is buffered against change by the much larger

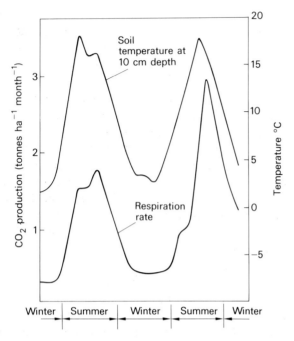

Figure 8.2. Seasonal trends in soil temperature and respiration rate in southern England (after Currie, 1975).

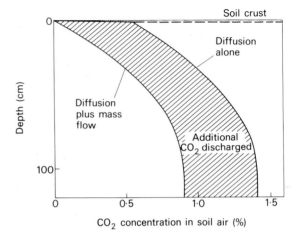

Figure 8.3. Concentration *vs* depth profiles for CO_2 in the soil air under a surface crust (after Currie, 1970).

volume of the Earth's atmosphere. There is also a large reservoir of HCO_3^- in the oceans. Buffering is achieved by the dynamic exchange of gases as O_2 travels from the atmosphere to the soil and CO_2 moves in the reverse direction, a process called *soil aeration*. Three transport processes are involved in soil aeration:

(a) dissolved O_2 is carried into the soil by percolating rainwater; the contribution is small owing to the low solubility of O_2 in water (0.028 ml ml^{-1} at 25°C and 1 bar pressure);

(b) mass flow of gases due to pressure changes of 1–2 mbars created by wind turbulence over the surface;

(c) diffusion of gas molecules through the soil pore space.

Gaseous diffusion is normally far more important than mass flow in either the liquid or gas phase [(a) and (b)]. This is confirmed by the gas concentration profiles shown in Figure 8.3, where diffusion alone reduces the concentration of CO_2 in the soil air at a depth of 1 m to a value < 1.5 per cent. The effect of mass flow, on the

other hand, is merely to reduce the CO_2 concentration by a further 0.5 per cent, virtually all of this decrease occurring across the thin surface crust where mass flow due to barometric pressure changes can occur.

Gas exchange by diffusion

The rate of gas diffusion, or *diffusive flux*, F, is given by the equation:

$$F = -D \, grad \, C \qquad (8.3)$$

where F is the volume of gas diffusing across a unit area perpendicular to the direction of movement in unit time; *grad C* is the change in gas concentration, C (ml ml^{-1} air-space) per unit length in the direction of diffusion, and D is the *gas diffusion coefficient* (compare with equation 6.9 for water flow). Values of D for CO_2 and O_2 in water and air are given in Table 8.2, from which it can be seen that for the same *grad C* the rate of gas diffusion is potentially 10,000 times faster through air-filled pores than water-filled pores. Further, since D_{O_2} is slightly larger than D_{CO_2}, when the soil RQ is 1, O_2 diffuses into the soil faster than the CO_2 that is produced can diffuse out. A pressure difference builds up and mass flow of N_2 occurs to eliminate it. Nevertheless, under steady-state aerobic conditions it is sufficiently accurate to equate the diffusive flux of O_2 into the soil with that of CO_2 out.

More significantly the D values for O_2 and CO_2 through the air-filled pores are always less than the D values in the outside air because of the restricted volume of the pores and the tortuosity of the diffusion pathway.

Table 8.2. Diffusion coefficients of O_2 and CO_2 in air and water ($cm^2\ s^{-1}$)

	Oxygen	Carbon dioxide
Air	2.3×10^{-1}	1.8×10^{-1}
Water	2.6×10^{-5}	2.0×10^{-5}

These effects are expressed in the equation:

$$D_e = \alpha \epsilon D_0 \qquad (8.4)$$

where D_e and D_0 are the diffusion coefficients of the gas in the soil pores, and in the outside air respectively, ϵ is the air-filled porosity (section 4.5) and α is an *impedance factor*, which ranges from 0 to 1 and decreases as the pathway becomes more tortuous.

The value of α varies in a complex way with soil structure, texture and water content, as illustrated in Figure 8.4. The change in α in dry sand (Figure 8.4a) is limited by the extent to which pore-size distribution within the matrix can be altered by changing the packing density and by the addition of finer particles.

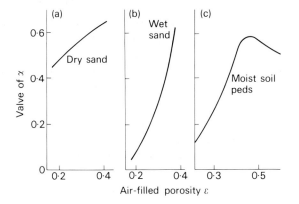

Figure 8.4. The relation between impedance factor, α, and air-filled porosity, ϵ, for dry and wet sand and soil peds (after Currie, 1970).

However, when the sand is wet (Figure 8.4b), water films fill the narrowest necks between pores so that the shape as well as the size distribution of the pores is altered. When water is added to dry soil peds (Figure 8.4c), α initially increases as the *intraped* pores fill and the tortuosity and roughness of the gas diffusion path actually decreases. However, once the larger *interped* pores begin to fill, α declines rapidly as it does in the wet sand, approaching a minimum value at $\epsilon \simeq 0.1$, a value corresponding to the average proportion of discontinuous or 'dead-end pores' in the soil. It can be concluded that in structured soils, the intraped pores make little contribution to gas diffusion and the *major* pathway is through the interped pores, most of which are drained at water contents less than the field capacity. This being so, the value of α will reflect management practices that change the proportion of large pores in the total pore space. Soil structural deterioration through continuous cultivation, for example, often results in reduced total porosity, that is soil compaction, and a coincidental decrease in the macroporosity. Consequently, the value of α, which may be as high as 0.6 in a well-drained, structured soil, may fall to a value as low as 0.16 in an over-worked arable soil.

PROFILES OF O_2 AND CO_2 IN FIELD SOILS
In the field, the concentrations of O_2 and CO_2 in the soil pore space change in space and time due to variations in the air-filled porosity, tortuosity and respiratory activity. Under these conditions, movement of O_2 and CO_2 must be described by combining the diffusive flux equation with a continuity equation that includes a term for soil respiration. For transfer in the vertical direction we may write:

$$\epsilon \frac{\partial C}{\partial t} = -\frac{\partial (F)}{\partial z} \pm R$$

i.e.

$$\epsilon \frac{\partial C}{\partial t} = D_e \frac{\partial^2 C}{\partial z^2} \pm R \qquad (8.5)$$

where D_e is the effective gas diffusion coefficient (from equation 8.4), z is the depth and R the rate of CO_2 production ($+R$) or O_2 consumption ($-R$). The value of R is usually expressed in ml cm^{-3} soil s^{-1}, which accounts for the appearance of ϵ (volume of air/volume of soil) in equation 8.5.

Assuming that steady-state conditions are attained (i.e. $dC/dt = 0$) and that R and D_e do not change substantially with depth, equation 8.5 can be integrated to give an expression relating the change in O_2 concentration (ΔC) in the gas phase to the depth z over which that change occurs, that is:

$$\Delta C = \frac{Rz^2}{2D_e} \qquad (8.6)$$

Substitution of appropriate values for R, z and D_e in this equation gives changes in O_2 concentration generally consistent with field measurements: two contrasting situations serve to illustrate the point.

(a) For a well-drained sandy soil, putting $\epsilon = 0.4$ and $\alpha = 0.6$ in equation 8.4 gives $D_e(O_2) = 5.5 \times 10^{-2}$ cm^2 s^{-1}, and taking $R \simeq 2 \times 10^{-7}$ ml cm^{-3} s^{-1} (Table 8.1), gives:

$\Delta C = 0.004$ ml ml^{-1} gas phase at a depth of 50 cm.

This corresponds to a negligible fall in O_2 partial pressure of 0.004 bars, by Dalton's law of partial pressures (section 4.5).

(b) On the other hand, in a clay soil where poor structure impedes diffusion in the gas phase, D_e may fall to

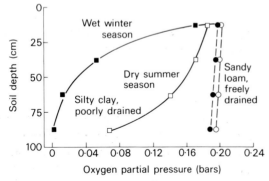

Figure 8.5. Changes in O_2 partial pressure in contrasting soils with depth and season of the year.

1.5×10^{-3} cm^2 s^{-1} and R to 1×10^{-7} ml cm^{-3} s^{-1}, so that

$\Delta C = 0.083$ ml ml^{-1} gas phase at 50 cm,

and the O_2 partial pressure falls to $0.21 - 0.08 = 0.13$ bars.

These calculated results agree well with measured O_2 concentrations for freely and poorly drained soils, shown in Figure 8.5. The restriction on gas diffusion imposed by the predominance of water-filled pores in a poorly drained soil is discussed below.

GAS DIFFUSION THROUGH WATER-FILLED PORES

In the absence of any physical impediment to the supply of O_2, the cytochrome oxidase enzyme system, which catalyses the reduction of O_2 to H_2O in plants and microorganisms, has a Michaelis constant K_m* of 2.5×10^{-8} M. This indicates a very high affinity for O_2. However, in stirred bacterial suspensions it is found that aerobic respiration can only be sustained at dissolved O_2 concentrations above $c.$ 4×10^{-6} M, indicating that diffusion of O_2 to the site of the enzyme in the cell's mitochondria is the rate-limiting step in reduction. The concentration of O_2 in water in equilibrium with the normal partial pressure of O_2 in the atmosphere at 25°C is:

$0.21 \times 0.028 = 0.0059$ ml cm^{-3} water

which is approximately equal to 2.6×10^{-4} M†. Thus, it is likely that soil microorganisms will continue to respire aerobically provided that the O_2 partial pressure in their *immediate* surroundings is one-hundredth or more of that in the atmosphere, that is > 0.002 bars. Although partial pressures above this value may be maintained in the few air spaces remaining in a waterlogged soil, the respiring organisms live in the water-filled pores *within* peds so that O_2 must diffuse through water, a slow process due to the low

---

*K_m is the concentration of substrate (O_2) at which the rate of the reaction is 50 per cent of the maximum.

†The volume of an ideal gas at 0°C and 1 atmosphere's pressure is 22.4 l mol^{-1}.

solubility of oxygen and the low diffusion coefficient for O_2 in water (Table 8.2). In many soils, therefore, anaerobic pockets develop at the centre of peds larger than a certain size when the peds remain wet for any length of time.

To illustrate this effect, equation 8.5 can be solved, assuming the peds to be spherical and that steady-state conditions apply, to give the equation:

$$a = \left(\frac{6 \Delta C D_e}{R} \right)^{1/2} \qquad (8.7)$$

where a is the ped radius, ΔC is the difference in O_2 concentration between the ped surface and its centre, D_e is the effective diffusion coefficient of O_2 *through the water phase* and R the respiration rate inside the ped. If the partial pressure of O_2 in the air space at the ped surface is assumed to be 0.21 bars, while at the centre of the ped it is zero, equation 8.7 can be solved for a range of R values to give the maximum radius of ped, the centre of which will just be anaerobic. If D_e in the water phase is given a value of 1.5×10^{-5} cm^2 s^{-1} (Greenwood, 1975), the results shown in Figure 8.6 can be obtained.

It follows that, provided there is some continuity of air-filled pores ($\epsilon > 0.1$), anaerobic conditions are unlikely to occur unless the sites of active respiration are separated from the gas phase by more than 1 cm of

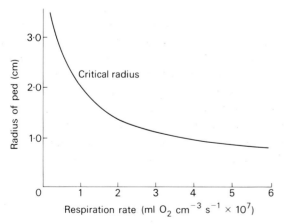

Figure 8.6. Critical radius for the induction of anaerobic conditions at ped centres (after Greenwood, 1975).

water-saturated soil. This is probably a common occurrence in heavy clay soils in Britain during the winter and spring months. Taking the same condition of a wet ped of radius a, just less than the critical radius for the onset of anaerobic conditions, one may calculate the maximum accumulation of CO_2 within the ped by equating the influx of O_2 with the efflux of CO_2. Using equation 8.3 we have:

$$D_{O_2} \frac{\Delta C'}{a} = D_{CO_2} \frac{\Delta C''}{a} \qquad (8.8)$$

i.e.

$$\Delta C'' = \frac{D_{O_2}}{D_{CO_2}} \Delta C' \qquad (8.9)$$

where $\Delta C'$ and $\Delta C''$ are the concentration differences in the water phase for O_2 and CO_2 respectively. Substituting in equation 8.9 for D_{O_2} and D_{CO_2} from Table 8.2, and putting $\Delta C' = 0.0059$, gives:

$$\Delta C'' = 0.0077$$

As the solubility of CO_2 at 25°C and 1 bar pressure equals 0.759, it follows that the CO_2 partial pressure at the ped centre is $c.$ 0.01 bars. An increase in partial pressure of this magnitude has negligible effect on plant and microbial metabolism. Nevertheless, if the O_2-free zones in the soil are extensive, as in a waterlogged soil, the CO_2 partial pressure can rise to 10 times this value and may have detrimental effects on the growth of some plants and microorganisms.

8.3 EFFECTS OF POOR SOIL AERATION ON ROOT AND MICROBIAL ACTIVITY

Plant root activity

The picture emerging from the previous discussion is one of a mosaic of aerobic and anoxic zones in heavy textured soils during wet periods of the year, as illustrated in Figure 8.7. The anoxic zones exert no serious ill effect (other than through the reduction of NO_3^-, discussed later) as most of the roots grow through the

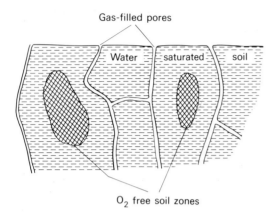

Figure 8.7. Distribution of water and oxygenated zones in a structured soil (after Greenwood, 1969).

larger pores (> 100 μm diameter) where O_2 diffusion is rapid. Oxygen can also diffuse for limited distances through plant tissues from regions of high to low O_2 supply, an attribute best developed in aquatic species and lowland rice (*Oryza sativa*), because of larger intercellular air spaces in which the value of D_{O_2} approaches 1×10^{-1} cm^2 s^{-1}.

However, when heavy rain coincides with temperatures high enough to maintain rapid respiration rates (typically in spring), the *whole* root zone can become anaerobic, a condition, which even if sustained for only a day, has serious effects on the metabolism and growth of sensitive plants. Species intolerant of waterlogging experience an acceleration in the rate of glycolysis and the accumulation of endogenously produced ethanol and acetaldehyde. Cell membranes become leaky, and ion and water uptake is impaired. The characteristic symptoms are wilting, yellowing of leaves and the development of adventitious roots at the base of the stem. Stimulation of the production of endogenous ethylene (C_2H_4) may also cause growth abnormalities, such as leaf epinasty—the downward curvature of a leaf axis.

Soil microbial activity

Soil aeration determines the balance between aerobic and anaerobic metabolism which has profound effects on the energy available for microbial growth and the type of end-products. *Aerobic organisms* are generally beneficial to soil and plants: the *heterotrophs* produce much biomass, water and CO_2, and unlock essential elements, such as N, P and S from organic combination; whereas specialized groups of *autotrophs*, such as the nitrifying bacteria and sulphur-oxidizing bacteria, produce NO_3^- and SO_4^{2-}. Some of the rhizosphere-dwelling species of bacteria produce growth regulators, including gibberellic acid and indolacetic acid, which may stimulate plant growth. *Anaerobic organisms* are usually much less numerous than aerobes, comprising about 10 per cent of the total soil population as measured by dilution plate counting (section 3.2). Because less metabolic energy is generated during anaerobic respiration, cell growth is slower than in aerobic respiration; the products formed are also more varied (see below). In addition, species of facultative and obligate anaerobes catalyse chemical reactions that are often undesirable, such as the reduction of NO_3^-, MnO_2 and SO_4^{2-} (section 8.4).

AEROBIC PROCESSES: NITRIFICATION
Nitrification entails the biological oxidation of inorganic N forms to nitrate (NO_3^-). Nitrification is discussed here because it is an essential prior step to the loss of N by NO_3^- reduction and the evolution of N_2O and N_2—a process called *denitrification*.

The principal nitrifying organisms are chemoautotrophic bacteria of the genera *Nitrosomonas* and *Nitrobacter*. These organisms are unique in deriving energy for growth solely from the oxidation of NH_4^+ produced by saprophytic decay organisms, or accruing to the soil from fertilizers and rain. The oxidation occurs in two steps:

$$(1) \quad NH_4^+ + \tfrac{3}{2}O_2 \rightarrow NO_2^- + 2H^+ + H_2O + \text{energy} \quad (8.10)$$

carried out by the *Nitrosomonas* species, and:

$$(2) \quad NO_2^- + \tfrac{1}{2}O_2 \rightarrow NO_3^- + \text{energy} \quad (8.11)$$

carried out by *Nitrobacter* species. These energy-releasing reactions are coupled to cellular synthesis and to the reduction of CO_2, which is the sole source of C for these organisms. As CO_2 is plentiful, the growth rate and hence bacterial numbers are limited primarily by the supply of oxidizable substrates: for example, the demand for NH_4^+ created by the more numerous soil heterotrophs restricts the rate of NH_4^+ oxidation by the slow-growing nitrifiers, and the absence of detectable NO_2^- in well-aerated soils attests to the limitation that NO_2^- supply imposes on *Nitrobacter* numbers.

Growth of the organisms and rate of nitrification are also influenced by temperature, moisture, pH and especially O_2 supply. The *temperature* optimum lies between 30 and 35°C, but some nitrification proceeds at temperatures down to 0°C. The optimum *moisture content* is approximately 60 per cent of field capacity. *Nitrosomonas* species are more susceptible to dry conditions than *Nitrobacter*, but sufficient bacteria survive short periods of desiccation for increased nitrification to follow the flush of organic decomposition contingent upon the rewetting of an air-dry soil. An alkaline *soil pH* is most favourable, *Nitrosomonas* species attaining peak activity between pH 7 and 9 and

Nitrobacter species having a somewhat lower optimum (section 12.2).

ANAEROBIC PROCESSES: FERMENTATION AND REDUCTION PRODUCTS

The pathway of cellular oxidation of carbohydrates is the same in plants and microorganisms, in the presence and absence of O_2, as far as the key intermediate, pyruvic acid ($CH_3COCOOH$). This is called *glycolysis*. Normally, reduced coenzymes in the cell are reoxidized as electrons are passed along a chain of respiratory enzymes to the terminal acceptor, oxygen. In the absence of O_2, however, there are two possible alternatives:

(a) other organic compounds serve as final electron acceptors and are reduced in turn—the process of *fermentation*;

(b) inorganic compounds such as NO_3^-, $Fe(OH)_3$, MnO_2 and SO_4^{2-} are reduced (section 8.4).

As indicated in Figure 8.8, several organic acids of low molecular weight—the volatile fatty acids (VFA)—are formed as intermediates in carbohydrate fermentation. The more abundant of these VFAs are

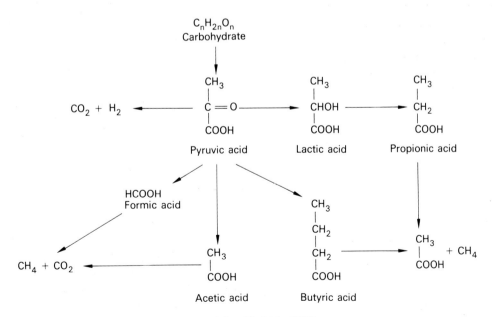

Figure 8.8. Products of the fermentation of carbohydrates (after Yoshida, 1975).

acetic and butyric acids, the concentrations of which can sometimes exceed toxic levels (> 1–10 mM) for several weeks in flooded soils. However, these acids are eventually dissimilated to the simple gases, H_2, CH_4 and CO_2, by the obligate anaerobes which multiply slowly in completely waterlogged soil. Even in normally aerobic soils, when unfavourable conditions develop (through surface compaction and excess water), crop residues may ferment to produce sufficiently high concentrations of soluble phytotoxic substances, usually acetic acid, for seed germination and seedling establishment to be adversely affected.

Less N is immobilized in a given time under anaerobic conditions because decomposition of the substrate is slower and the growth yield of the microorganisms is lower (section 3.3). Consequently, more N appears as NH_4^+, and putrefaction products, such as volatile amines, thiols and mercaptans, may also occur.

ETHYLENE PRODUCTION
One soil organism found to produce the hydrocarbon gas ethylene (C_2H_4) is the early-colonizing sugar fungus *Mucor*, primarily *Mucor hiemalis*, which only grows aerobically. However, as the O_2 partial pressure falls and the growth of the fungus is slowed, the production of C_2H_4 per unit mass of mycelium increases. It appears, therefore, that an anaerobic/ aerobic interface is important for promoting C_2H_4 production in soil. Very low concentrations of ethylene (~ 0.1 ppm) stimulate root elongation, but concentrations of 1 ppm and greater, which may occur in wet, poorly drained soil or beneath the smeared plough layer in a clay, seriously inhibit root growth (Figure 8.9). Some microorganisms also metabolize C_2H_4 —coryneform bacteria consume C_2H_4 as their sole source of C. The ethylene concentration therefore reflects the balance between microbial production, and diffusive losses and decomposition, which explains why very little C_2H_4 accumulates in well-aerated soils where it is destroyed or lost as rapidly as it is formed.

Indices of soil aeration
From the foregoing it can be seen that the aeration

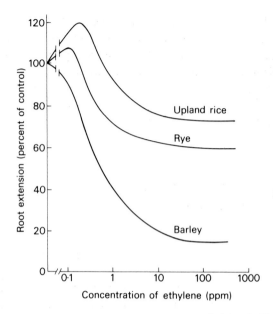

Figure 8.9. The effect of ethylene on root growth (after Smith and Robertson, 1971).

status of a soil may be assessed in one of several ways. (1) Measurement *in situ* of the rate of O_2 diffusion through the soil. A popular method relies on the principle that reduction of O_2 at the surface of a platinum (Pt) electrode inserted into the soil, to which a fixed voltage is applied, causes a current flow proportional to the rate of oxygen diffusion to the electrode. The problems encountered in applying this method have been reviewed by McIntyre (1970). Alternatively, the concentration of O_2 in a volume of air or water extracted from the soil can be measured by gas–liquid chromatography. Obviously, where aerobic and anaerobic zones exist in close proximity (as in Figure 8.7), such a measurement is only a spatial average for a volume of soil in which the exact distribution of O_2 partial pressure remains undefined. Based on such measurements, plant root elongation is found to be inhibited at O_2 partial pressures between 0.06 and 0.01 bars.
(2) The detection of fermentation and putrefaction products by their odour (especially the volatile amines and mercaptans), and, where possible, by quantitative analysis (the VFAs, CH_4 and C_2H_4).

(3) By noting the associated chemical changes, for example, the solution and reprecipitation of Mn as small black specks of MnO_2, or the appearance of a bluish-grey colour of reduced Fe compounds.

Once the O_2 supply in the soil has been depleted, and anaerobic respiration initiated, the *redox potential* is a useful measure of the intensity of the ensuing reducing conditions.

8.4 OXIDATION–REDUCTION REACTIONS IN SOIL

Redox potential

Oxidation–reduction or *redox* reactions are those in which a chemical species (a molecule or ion) goes from a more oxidized to a less oxidized (reduced) state, or *vice versa*, through the transfer of electrons. One such reversible reaction is:

$$H_2(gas) \rightleftharpoons 2H^+ + 2e^- \qquad (8.12)$$

where e^- stands for a free electron. This reaction may be coupled with another reversible reaction, for example:

$$\tfrac{1}{2}O_2(gas) + 2e^- \rightleftharpoons O^{2-} \qquad (8.13)$$

in such a way that electrons flow from the species of lower electron affinity (H_2) to higher electron affinity (O_2). The resultant *coupled* reaction may be written as:

$$\tfrac{1}{2}O_2(gas) + 2H^+ + 2e^- \rightleftharpoons H_2O(liquid) \quad (8.14)$$

The general form of the redox reaction is therefore:

$$Ox + mH^+ + ne^- \rightleftharpoons Red \qquad (8.15)$$

for which (Ox) and (Red) are the activities of the oxidized and reduced species, respectively, and m and n are stoichiometric coefficients. If the free electron is assigned an activity (e), it is possible to write, following reaction 8.15, the expression

$$pe = \frac{1}{n}[\log K_e + \log(Ox) - \log(Red) - mpH] \quad (8.16)$$

where pe is the negative logarithm of the free electron activity and K_e is the thermodynamic equilibrium constant for the reaction. If K_e, (Ox), (Red), pH, m and n are known, pe can be calculated. Large positive values of pe (low electron activity) favour the existence of oxidized species or those that are electron acceptors. However, in heterogeneous soil systems, it is virtually impossible to assign values to all these parameters and variables: instead the oxidation–reduction status of the system is *measured* by the *redox potential*, which is the difference in electrical potential between the two half-reactions of the coupled reaction (8.15).

By combining the equation for the free energy change in a reversible reaction:

$$\Delta G = \Delta G^0 + RT \ln \frac{(Products)}{(Reactants)} \qquad (8.17)$$

with the equation for electrical work done for a given free energy change (ΔG):

$$\Delta G = -nFE \qquad (8.18)$$

where n = number of electrons transferred through an electrical potential difference, E, and F is Faraday's constant (in this analysis the electron has charge but no mass), we may write:

$$E = E^0 - \frac{RT}{nF} \ln \left\{ \frac{(Red)}{(Ox)(H^+)^m} \right\} \qquad (8.19)$$

i.e.

$$E_h = E^0 + 2.3\frac{RT}{nF} \left\{ \log_{10}\frac{(Ox)}{(Red)} - mpH \right\} \ (8.20)$$

In equation 8.20, E_h is the equilibrium redox potential and E^0 is the standard redox potential of the reaction $H^+ + e^- \rightleftharpoons \tfrac{1}{2}H_2(gas)$. E^0 is zero when the pressure of H_2 gas at 25°C is 1 atmosphere and the activity of H^+ ions in solution is one (i.e. pH = 0). It is clear that oxidizing systems will have relatively positive E_h values and reducing systems more negative E_h values. It also follows that for a given ratio of (Ox)/(Red), the lower the pH the higher the E_h value at which reduction occurs.

In practice, it is more realistic to refer the measured E_h to pH 7, rather than pH 0, in which case the standard redox potential is defined by:

$$E_7^0 = E^0 - \left(2.3\frac{RT}{F}\right)\frac{m}{n}\text{pH}$$

$$= E^0 - 0.413\frac{m}{n} \qquad (8.21)$$

Equation 8.20, therefore, becomes:

$$E_h = E_7^0 + \frac{0.059}{n}\log_{10}\frac{(\text{Ox})}{(\text{Red})} \qquad (8.22)$$

To measure the redox potential, a Pt electrode and suitable reference electrode (calomel or Ag/AgCl) are placed in the moist soil and the electrical potential difference recorded. Aerobic soils have E_h values between 0.3 and 0.8 V, but reproducible values are only obtained in anaerobic soils, in which the potential ranges from 0.3 to -0.4 V.

The redox potential will show least change on the addition or removal of electrons when (Ox) = (Red): this is the condition of maximum *poise* and occurs, as can be seen from equation 8.22, when $E_h = E_7^0$. Equation 8.22 also reveals that the change in E_h of the system as the ratio (Ox)/(Red) changes is inversely related to n, the number of electrons transferred.

Sequential reductions in soil deprived of oxygen

Once part or all of the soil has become anaerobic, substances other than O_2 are used as terminal acceptors for the free electrons produced during respiration. A redox reaction will proceed spontaneously in the direction of decreasing free energy $(-\Delta G)$, but the rate at which it proceeds may be very slow if the activation energy is high or if the redox half-reactions are physically not well coupled. In soil, redox reactions are coupled through various microorganisms growing facultatively or obligately under anoxic conditions (section 8.3). In addition to the fermentation of carbohydrates, the following sequence of reactions involving inorganic compounds occurs:

$$2NO_3^-(\text{soln}) + 12H^+(\text{soln}) + 10e^- \rightleftharpoons N_2(\text{gas}) + 6H_2O$$
$$(\text{liquid}) \qquad (8.23)$$

$$MnO_2(\text{solid}) + 4H^+(\text{soln}) + 2e^- \rightleftharpoons Mn^{2+}(\text{soln}) + 2H_2O$$
$$(\text{liquid}) \qquad (8.24)$$

$$Fe(OH)_3(\text{solid}) + 3H^+(\text{soln}) + e^- \rightleftharpoons Fe^{2+}(\text{soln}) + 3H_2O$$
$$(\text{liquid}) \qquad (8.25)$$

$$SO_4^{2-}(\text{soln}) + 10H^+(\text{soln}) + 8e^- \rightleftharpoons H_2S(\text{gas}) + 4H_2O$$
$$(\text{liquid}) \qquad (8.26)$$

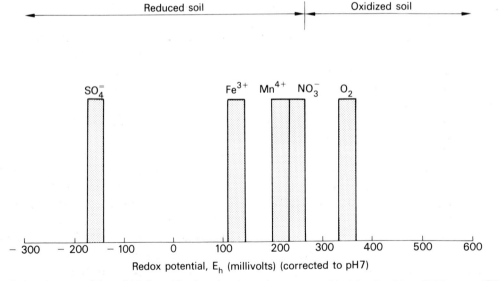

Figure 8.10. Threshold redox potentials at which the oxidized species shown become unstable (after Patrick and Mahapatra, 1968).

The approximate E_h value at which each oxidized form becomes unstable is shown in Figure 8.10. Each reaction poises the soil in a particular E_h range until most of the oxidized form has been consumed and then the E_h drops to a lower value as the reducing power intensifies.

DENITRIFICATION

The first reaction (8.23) summarizes the process of *denitrification* that is carried out by facultative anaerobic bacteria, predominantly *Pseudomonas* and *Achromobacter* species, once the partial pressure of O_2 has fallen to a very low level. The complete pathway for NO_3^- reduction is as follows:

$$NO_3^- \rightleftharpoons NO_2^- \rightleftharpoons N_2O \rightleftharpoons N_2$$
nitrous
oxide

aerobic   anaerobic

Laboratory incubations under anaerobic conditions indicate an unusually high temperature optimum of 65–70°C for denitrification, with the rate being very slow at temperatures $< 10°C$. Temperature and pH affect the ratio of $N_2O:N_2$ evolved during denitrification in closed systems. At low temperatures and pH < 5, the ratio of $N_2O:N_2$ is 1 or greater, but at temperatures of 25°C or greater and pH > 6, most of the N_2O is reduced to N_2. In the field, however, both gases can escape from the site of formation by diffusion and N_2O, which is about as soluble as CO_2, is lost in the drainage water. The variable ratio of $N_2O:N_2$ produced during denitrification makes difficult the measurement of field losses of N by this pathway (section 10.2).

When carbon substrates are abundant under anaerobic conditions, dissimilatory reduction of NO_3^- to NH_4^+ may also occur (the bacterium *Bacillus lichenformis* can reduce NO_3^- to either N_2O and N_2 or NH_4^+). The NH_4^+ produced in this way is difficult to distinguish from that produced by ammonification (section 3.1), and is, of course, susceptible to immobilization by other organisms. Dissimilatory reduction of NO_3^- to NH_4^+ is, however, normally insignificant.

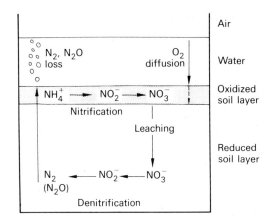

Figure 8.11. Nitrification and denitrification in the oxidized and reduced zones of a flooded soil (after Patrick and Mahapatra, 1968).

Denitrification losses can be serious in agricultural soils, especially those experiencing changes from aerobic to anaerobic conditions (see Figure 8.7), and the reversal of these conditions with time. An extreme situation can arise in padi rice culture, as illustrated in Figure 8.11; the NO_3^- produced aerobically diffuses or is leached into anaerobic zones where it is denitrified. Accordingly, N fertilizer in such systems is best applied in an NH_4^+ form and placed directly in the anoxic zone below the soil surface.

More intensive reducing conditions

Reactions 8.24 and 8.25 are characteristic of *gleying*, a pedogenic process discussed in section 9.2. The former is less effective in poising the soil than either NO_3^- or Fe^{3+} reduction because MnO_2 is very insoluble and relatively few microorganisms use it as a terminal electron acceptor. Nevertheless, in very acid soils waterlogging may lead to high levels of exchangeable Mn^{2+}, which cause manganese toxicity in susceptible plants. Anaerobiosis and Fe^{3+} reduction generally increase the phosphate concentration in the soil solution, especially in acid soils where phosphate is immobilized in very insoluble Fe compounds. The effect is complicated, however, by the inevitable rise in pH on prolonged waterlogging ($3H^+$ ions are consumed for each electron transferred in reaction 8.25). At pH > 6 ferrous compounds become less soluble and an oxide containing Fe in the ferric and ferrous states precipitates, probably hydrated magnetite $Fe_3(OH)_8$, which presents a highly active surface for the readsorption of phosphate. Certainly all the dissolved P is readsorbed when aerobic conditions return and amorphous $Fe(OH)_3$ precipitates.

At very low E_h values (Figure 8.10), SO_4^{2-} is reduced due to the activity of obligate anaerobes of the genus *Desulphovibrio*. As the concentration of weakly dissociated H_2S builds up in the soil solution, ferrous sulphide (FeS) precipitates and slowly reverts to iron pyrites (FeS_2), according to the reaction:

$$FeS(solid)+S(solid) \rightleftharpoons FeS_2(solid) \qquad (8.27)$$

Iron pyrites is a mineral characteristic of marine sedimentary rocks and of sediments currently deposited in river estuaries and mud flats.

8.5 SUMMARY

During respiration by soil organisms and plant roots, complex C molecules are dissimilated to provide energy for cellular growth. When conditions are *aerobic*, one mole of CO_2 is released for each mole of O_2 consumed and the *respiratory quotient (RQ)* is 1. Respiration continues aerobically, at a rate determined by soil moisture, temperature and substrate availability, provided that O_2 from the atmosphere moves into the soil fast enough to satisfy the organisms' requirements, and CO_2 can move out. This dynamic exchange of gases is called *soil aeration*.

Diffusion of gases through the air-filled pore space is the most important mechanism of soil aeration. The rate of diffusion or diffusive flux, F, is given by:

$$F = -D \, grad \, C$$

but under non-steady state conditions (gas concentrations, C, changing in space and time), the time rate of change in concentration in the soil pores is given by:

$$\epsilon \frac{\partial C}{\partial t} = D_e \frac{\partial^2 C}{\partial z^2} \pm R$$

where R, the soil respiration rate, is positive when measured as CO_2 release and negative when measured as O_2 uptake. D_e is the effective gas diffusion coefficient in the vertical (z) direction and ϵ is the fraction of the soil's volume that is air-filled.

Sandy soils with ϵ values > 0.1 maintain O_2 partial pressures close to the atmospheric level of 0.21 bars. However, in clay soils that remain at field capacity for many weeks (English winter conditions), sites of active microbial respiration within peds become *anaerobic* if separated from air-filled pores by more than 1 cm of water-saturated soil. The reason for this is the fact that the D values of O_2 and CO_2 in water are 10,000 less than in air and gas exchange through water-filled pores is very slow.

Although the cytochrome oxidase enzyme system has a very high affinity for O_2, restrictions imposed by the need for O_2 to diffuse to the enzyme site within an organism result in aerobic respiration ceasing at O_2 partial pressures of about 0.002 bars. Root elongation appears to be inhibited at partial pressures below 0.06–0.01 bars. However, once the O_2 supply is exhausted, which may happen within 1–2 days of a soil becoming completely waterlogged, bacteria that can respire either aerobically or anaerobically continue to grow and obligate anaerobes also begin to multiply. Carbohydrates are incompletely oxidized or fermented to *volatile fatty acids*, which can attain localized concentrations (1–10 mM) high enough to be toxic to seedlings, before being dissimilated to CO_2, CH_4 and H_2.

In the absence of O_2, other substances serve as electron acceptors in bacterial respiration, and reduction of NO_3^-, MnO_2, Fe^{3+} compounds and SO_4^{2-} occurs in that sequence as reducing conditions intensify. The reducing power of the soil system is measured by the *redox potential*, E_h which for the coupled reversible reaction:

$$Ox+mH^++ne^- \rightleftharpoons Red$$

is defined by the equation:

$$E_h = E^0 + \frac{2.3RT}{nF}\left\{ \log_{10} \frac{(Ox)}{(Red)} - mpH \right\}$$

REFERENCES

CURRIE J.A. (1970) Movement of gases in soil respiration, in *Sorption and Transport Processes in Soils*. Society of Chemical Industry Monograph **37**, pp 152–169.

CURRIE J.A. (1975) Soil respiration, in *Soil Physical Conditions and Crop Production*. MAFF Bulletin **29**, pp 461–468.

GREENWOOD D.J. (1969) Effect of oxygen distribution in soil on plant growth, in *Root Growth*. (Ed. W.J. Whittington). Butterworths, London, pp 202–221.

GREENWOOD D.J. (1975) Measurement of soil aeration, in *Soil Physical Conditions and Crop Production*. MAFF Bulletin **29**, pp 261–272.

McINTYRE D.S. (1970) The platinum electrode method for soil aeration measurement. *Advances in Agronomy* **22**, 235–283.

PATRICK W.H. & MAHAPATRA I.C. (1968) Transformation and availability to rice of nitrogen and phosphorus in waterlogged soils. *Advances in Agronomy* **20**, 323–359.

SMITH K.A. & ROBERTSON P.D. (1971) Effect of ethylene on root extension of cereals. *Nature New Biology* **234**, 148–149.

YOSHIDA T. (1975) Microbial metabolism of flooded soils, in *Soil Biochemistry*, Volume 3. (Eds. E.A. Paul and A.D. McLaren). Marcel Dekker, New York, pp 83–122.

FURTHER READING

CANNELL R.Q. (1977) Soil aeration and compaction in relation to root growth and soil management, in *Applied Biology*, Volume 2 (Ed. T.H. Coaker). Academic, London, pp 1–86.

PONNAMPERUMA F.N. (1972) The chemistry of submerged soils. *Advances in Agronomy* **24**, 29–96.

ROWELL D.L. (1981) Oxidation and reduction, in *The Chemistry of Soil Processes* (Eds. D.J. Greenland and M.H.B. Hayes). Wiley-Interscience, Chichester.

TIEDJE J.M., SEXTONE A.J., PARKIN T.B., REVSBECH N.P. & SHELTON D.R. (1984) Anaerobic processes in soil. *Plant and Soil* **76**, 197–212.

Chapter 9
Processes in Profile Development

9.1 THE SOIL PROFILE

Soil horizons

In Chapter 1, the course of profile development from bare rock to a *brown forest soil* (Inceptisol) was described in outline. Except in peaty soils, the downward penetration of the weathering front greatly exceeds the upward accumulation of litter on the surface, and the mature profile may have the vertical sequence previously illustrated in Figure 1.4, that is:

Litter layer, L

A horizon (mainly eluvial)

B horizon (mainly illuvial)

C horizon (parent material)

Predictably, this simple model does not encompass the full complexity of profile development. A more comprehensive model with an enlarged glossary of descriptive terms, the *horizon notation*, intended to convey briefly the maximum information about the soil to the observer, is that of the Soil Survey of England and Wales (Hodgson, 1974), which is consistent with current international usage.

There are two main kinds of horizon, *organic* and *mineral*, which are distinguished by their organic matter content. An organic horizon, which usually bears a superficial undecomposed litter layer, L, must have either: (a) greater than 30 per cent organic matter (18 per cent C) when the mineral fraction has 50 per cent or more clay; or (b) greater than 20 per cent organic matter (12 per cent C) when the mineral fraction has no clay.

When an organic horizon remains wet most of the time, it is called a *peaty* or O horizon; otherwise it is simply called an F horizon when the organic matter is only partly decomposed, or an H horizon, when the

organic matter is low in minerals and well decomposed. Wetness favours an increasing thickness of the O horizon, which qualifies as *peat* when more than 40 cm deep and the organic matter content is > 50 per cent. *Sandy peat* can have between 20 and 50 per cent organic matter, but must have 50 per cent or more of sand, whereas a *loamy peat* has a similar organic content, but less than 50 per cent sand.

A *mineral* horizon is low in organic matter and overlies consolidated or unconsolidated rock; it may be designated A, E, B, C or G, as indicated in Table 9.1, depending on its position in the profile, its mode of formation, colour,* structure, degree of weathering and other properties that can be recognized in the field. Three kinds of complex horizon are also recognized: a horizon with features of two horizons, such as a strongly gleyed C horizon, is designated C/G, whereas a relatively homogeneous horizon lying between two other horizons, for instance A and B, is designated AB. Less commonly, a transitional horizon may comprise discrete parts (more than 10 per cent by volume) of two contiguous horizons, then it is called A & B, or E & B for example. Specific attributes of each major horizon are indicated by lower-case suffixes attached to the horizon symbol (see Table 9.1).

Table 9.1. Soil horizon notation.

Horizon	Brief description
O—organic	
Of	Composed mainly of fibrous peat
Om	Composed mainly of semi-fibrous peat
Oh	Mainly amorphous organic matter; uncultivated
Op	Amorphous organic matter mixed by cultivation
A—at or near the surface, with well mixed humified organic matter (formerly designated A_1)	
Ah	Uncultivated with > 1 per cent organic matter

*Standard colour descriptions are obtained from Munsell colour charts.

Table 9.1. *continued*

Horizon	Brief description
Ap	Mixed by cultivation
Ag	Partially gleyed due to intermittent waterlogging; rusty mottles along root channels

E—eluvial, of low organic content, having lost material to lower horizons (formerly designated A$_2$)

Ea	Lacking mottles, bleached particles due to the removal of organic and sesquioxidic coatings
Eb	Brownish colour due to uniformly disseminated iron oxides; often overlies a Bt horizon
Eg	See Ag; often overlies a Bt horizon, and sometimes a thin iron-pan

B—subsurface, showing characteristic structure, texture and colour due to illuviation of material from above and the weathering of the parent material

Bf	Thin iron-pan or placon; often enriched with organic matter and aluminium
Bg	Partially gleyed; blocky or prismatic structure; variable black MnO$_2$ mottles
Bh	Translocated organic matter with some Fe and Al
Bs	Enriched with sesquioxides, usually by illuviation; orange to red colour
Bt	Accumulation of translocated clay as shown by clay coatings on ped faces
Bw	Shows alterations to parent material by leaching, weathering and structural reorganization

G—gleyed, bluey-grey to green colours predominate throughout a structureless soil matrix

C—parent material

Ck	Contains > 1 per cent secondary CaCO$_3$ as concretions or coatings (also Ak and Bk)
Cy	Contains secondary CaSO$_4$ as gypsum crystals
Cm	Continuously cemented, other than by a thin iron-pan
Cr	Weakly consolidated but dense enough to prevent root ingress except along cracks
Cu	Unconsolidated and lacking evidence of gleying, cementation, etc.
Cx	With fragipan properties, as in Pleistocene deposits; dense but uncemented; firm when dry but brittle when moist

(After Hodgson, 1974.)

Soil layers

Unless the record of rock lithology or stratigraphy suggests the contrary, it is assumed that the parent material originally extended unchanged to the soil surface. As indicated in section 5.2, this assumption is often invalid for soils on very old land surfaces subjected to many cycles of weathering, erosion and deposition that have resulted in the layering of parent materials. An example of layering is the *K cycle* of soil formation on weathered and resorted materials in south-east Australia (Figure 9.1).

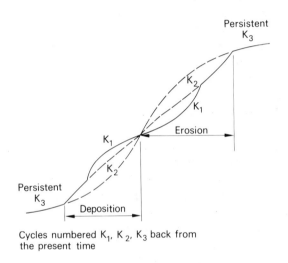

Cycles numbered K$_1$, K$_2$, K$_3$ back from the present time

Figure 9.1. K cycles of soil formation on a hill slope (after Butler, 1959).

Layers of different lithology are identified by numbers prefixed to the horizon symbols, the uppermost layer being 1 (prefix usually omitted) and those below labelled sequentially 2, 3, etc. It is also possible for an old soil to be buried by material of similar lithology, in which case the buried soil horizons are prefixed by the letter *b*. Buried horizons are found in very old cleared areas where man has shifted top soil for farming or construction purposes; similarly, the course of soil formation near old settlements has frequently been altered by man's deliberate deposition of organic residues, such as seaweed and waste. Soils of this kind are called man-made or *plaggen soils*.

9.2 PEDOGENIC PROCESSES

In this section, the more important processes operative in profile development (*pedogenesis*) are discussed. The magnitude of their effect is governed by the inter-action of factors defining the initial state of soil formation (*parent material, relief*), the environment (*climate, organisms*) and the extent of reaction (*time*), as discussed in Chapter 5.

Congruent and incongruent dissolution

HYDROLYSIS
Rock minerals become unstable when exposed to a changed environment. Chemical reactions occur that may be relatively subtle (e.g. oxidation of Fe^{2+} to Fe^{3+}) or drastic, such as the complete dissolution of the mineral in water. Minerals, such as *halite* (NaCl) or *gypsum* ($CaSO_4 . 2H_2O$), dissolve to release their constituent elements in the same molar ratio as they occur in the solid, that is:

$$NaCl(solid) \rightleftarrows Na^+(liquid) + Cl^-(liquid) \quad (9.1)$$

and:

$$CaSO_4.2H_2O(solid) \rightleftarrows Ca^{2+}(liquid) + SO_4^{2-}(liquid) + 2H_2O \quad (9.2)$$

This is called *congruent* dissolution, as opposed to *incongruent* dissolution, where the molar ratio of elements in solution differs from that in the solid. Because of this inequality, a compound different from the original mineral gradually forms in the solid phase. Chemical weathering of the primary silicate minerals is a good example of incongruent dissolution by surface hydrolysis. Water molecules at the mineral surface dissociate into H^+ and OH^- and the small, mobile H^+ ions (actually H_3O^+) penetrate the crystal lattice,

creating a charge imbalance which causes cations such as Ca^{2+}, Mg^{2+}, K^+ and Na^+ to diffuse out. The potash feldspar *orthoclase* hydrolyses to produce a weak acid (silicic), a strong base (KOH) and leaves a residue of the clay mineral *illite*:

$$3K\,Al^{IV}Si_3O_8 + 14H_2O \rightleftarrows K(Al\,Si_3)^{IV}Al_2^{VI}O_{10}(OH)_2 \\ + 6Si(OH)_4 + 2KOH \quad (9.3)$$

Hydrolysis of the calcic feldspar, *anorthite*, on the other hand, leaves a residue of vermiculite-type clay, which is a weak acid, and the strong base $Ca(OH)_2$:

$$3Ca\,Al_2^{IV}\,Si_2O_8 + 6H_2O \rightleftarrows 2H^+2[(Al\,Si_3)^{IV}Al_2^{VI} \\ O_{10}(OH)_2]^- + 3Ca(OH)_2 \quad (9.4)$$

Equations (9.3) and (9.4) are specific examples of the general reaction:

$$[M^+silicate^-] + H_2O \rightleftarrows H^+[silicate\,secondary\,mineral]^- \\ + M^+OH^- \quad (9.5)$$

in which M^+ is an alkali or alkaline earth cation, or sometimes a transition element (Fe, Mn). Owing to the formation of the strong base, MOH, nearly all the primary silicates have an alkaline reaction—the *abrasion* pH—when placed in CO_2-free distilled water (pH $\simeq$ 6). However, reaction 9.5 proceeds more rapidly to the right when more H^+ ions are present to neutralize the OH^- ions of the strong base. The main sources of this extra acidity in natural, well-drained soils are:
(a) CO_2 dissolved in the soil water, at partial pressures up to at least 0.01 bars (section 8.2); and
(b) carboxyl and phenolic groups of humified organic matter (section 3.4).

The hydrolysis, neutralization and cation exchange reactions involving a weathering primary silicate, organic matter and the clay mineral residues are illustrated in Figure 9.2. As basic cations are removed from the soil by leaching, the clay minerals and organic matter gradually become saturated with H^+ ions. However, H-clays are unstable and disintegrate slowly from the crystal edges inwards to release Al and SiO_2: the silica goes into solution as $Si(OH)_4$ and the Al is adsorbed as Al^{3+} by both clay minerals and humified

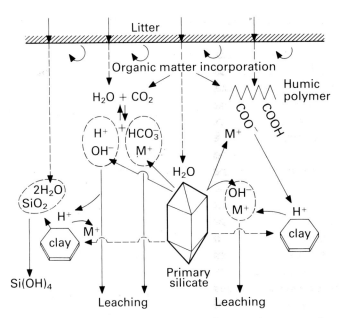

Si(OH)$_4$

Figure 9.2. Scheme for silicate mineral hydrolysis in the presence of CO$_2$-charged rainwater and humified organic residues.

organic matter (section 7.2). The subsequent slow hydrolysis of the Al^{3+} produces more H$^+$ ions, the neutralization of which is achieved by further lattice decomposition so the resultant pH is higher than that of a pure H-clay or organic acid. Ultimately, clay minerals are completely destroyed to yield gibbsite, or goethite/hematite mixtures in soils high in ferro-magnesian minerals. The change in mineralogy and pH

Table 9.2. Mineralogical changes in weathering basalt.

	Depth (cm)	pH (in water)	Kaolinite (%)	Other clay fraction minerals
Soil surface				
	10–20	6.5	15–25	H, Q, G
Decreasing	60–90	6.2	30–40	H, Q, G
intensity	120–150	6.1	40–50	H, G
of	180–210	6.0	45–55	H
weathering	270–300	5.7	55–65	H
	570–600	4.9	65–75	H
Parent rock				

H = hematite; Q = quartz; G = gibbsite.
(After Black and Waring, 1976.)

with depth in a highly weathered soil on basalt (Table 9.2) illustrates this sequence of events.

OXIDATION

The weathering of iron-bearing rock minerals in waters containing dissolved O$_2$ and CO$_2$ is another example of incongruent dissolution. Hydrolysis of the mineral releases the iron, present mainly as Fe^{2+}, which then oxidizes to the Fe^{3+} form and precipitates as an insoluble oxyhydroxide—usually either *ferrihydrite* (Fe$_2$O$_3$.2FeOOH.nH$_2$O), which is a necessary precursor of *hematite* (α-Fe$_2$O$_3$), or the stable mineral *goethite* (α-FeOOH) (section 2.4). The precipitate may form a coating over the mineral surface which slows down the subsequent rate of hydrolysis. Accordingly, in an anaerobic environment where ferric oxide precipitation is inhibited, the rate of weathering of iron-bearing minerals is expected to be faster.

Note that the oxidation of Fe^{2+} to Fe^{3+} according to:

$$Fe^{2+}+2H_2O+\tfrac{1}{2}O_2 \rightleftarrows Fe(OH)_3+H^+ \qquad (9.6)$$

is an acidifying reaction. The localized production of H$^+$ ions will generally accelerate the rate of mineral weathering, as discussed under 'HYDROLYSIS'. Some rock minerals alter directly from one form to another by *in situ* oxidation of Fe^{2+}-iron in the lattice: the weathering of biotite mica to vermiculite is one example.

Leaching

DIFFERENTIAL MOBILITIES

The mobility of an element during weathering reflects its solubility in water and the effect of pH on that solubility. Relative mobilities, represented by the Polynov series in Table 9.3, have been established from a comparison of the composition of river waters with that of igneous rocks in the catchments from which they drain. The low mobility of Al and Fe can be explained in terms of the formation and strong adsorption of hydroxy-Al and hydroxy-Fe ions at pH > 4 and 3, respectively, and the precipitation of the insoluble oxyhydroxides. The least mobile element is titanium, which forms the oxide TiO_2 (anatase), insoluble at pH > 2.5. It is therefore used as a reference material to determine the relative gains or losses of other elements in the profile.

Table 9.3. The relative mobilities of rock constituents.

Constituent elements or compounds	Relative mobility*
Al_2O_3	0.02
Fe_2O_3	0.04
SiO_2	0.20
K	1.25
Mg	1.30
Na	2.40
Ca	3.00
SO_4	57

*Expressed relative to Cl taken as 100.
(After Loughnan, 1969.)

PERCOLATION DEPTH

Leaching of an element depends not only on its mobility, but also on the rate of water percolation through the soil. In arid areas, even the most soluble constituents, mainly NaCl and, to a lesser extent, the chlorides, sulphates and bicarbonates of Ca and Mg, tend to be retained and give rise to *saline soils* (section 9.6). As the climate becomes more humid, losses of salts and silica increase and the soils become more highly leached, except in the low-lying parts of the landscape where the drainage waters and salts accumulate and modify the course of soil formation (section 9.4). The effect of leaching is well illustrated by the steadily increasing depth to the carbonate layer in soils formed on calcareous loess in the USA, along a line drawn from semi-arid Colorado to humid Missouri (Figure 9.3).

Cheluviation

In certain instances, leaching of the immobile Al^{3+} and Fe^{3+} cations is enhanced by the formation of soluble organo–metal complexes or *chelates* (section 3.4). Aluminium forms an electrostatic-type complex with the carboxyl groups of humified organic matter in which the average charge per mole of complexed Al is

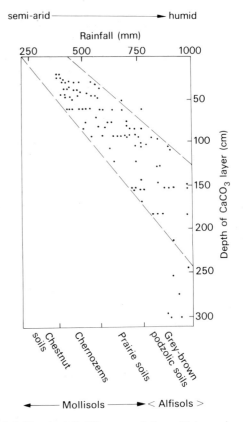

Figure 9.3. Depth of $CaCO_3$ accumulation with increasing rainfall (after Jenny, 1941).

about $+1$; that is:

$$R—COO^-[Al(OH)_2(H_2O_4]^+,$$

where R is the stem of the organic molecule. Ferric iron, on the other hand, is likely to be reduced to ferrous iron at the expense of plant-derived reducing agents, with the concomitant formation of a soluble ferrous–organic complex:

$$\text{Organic reductant } (e^-)+Fe(OH)_3+3H^+ \rightleftharpoons Fe^{2+}\text{-}$$
$$\text{organic ligand}+3H_2O \qquad (9.7)$$

The organic reductant supplies the electrons that drive this reaction to the right. The stability of the Fe^{2+}-organic ligand complex is favoured by anaerobic conditions (low redox potential, E_h, see section 8.4), but because H^+ ions are also involved on the left-hand side of equation 9.7, the complexes will form at higher E_h values (more oxidizing conditions), the lower the local pH. Thus, the most active complexing agents are organic acids, particularly those of the fulvic acid (FA) fraction leached from litter, and strongly reducing polyphenols that are leached from the canopies of conifers and heath plants and their freshly fallen litter.

The formation and subsequent leaching of these complexes is called *cheluviation*. Polyphenols not involved in cheluviation are oxidatively polymerized to help form, together with the 'tanned' proteins and cellulose, the characteristic H-layer of mor humus found under coniferous forest and heath vegetation (section 3.3).

PODZOLIZATION

Podzolization involves the cheluviation of Al and Fe from insoluble oxides in the upper soil zone and their deposition, with organic matter, deeper in the profile. The removal of iron oxide and organic coatings renders the sand grains bleached in appearance, which is a feature of an Ea horizon. As the ferrous–organic complex is leached more deeply it becomes less stable through a combination of factors:
(a) microbial decomposition of the organic ligands;
(b) complexing of more metal ions, which increases the ratio of cation to ligand in the complex;
(c) a rise in pH, which favours the oxidation of Fe^{2+} at a given E_h;
(d) the flocculating effect of higher concentrations of Ca^{2+} and Mg^{2+} ions in the soil solution near the weathering parent material.

The iron precipitates as ferrihydrite, giving a uniform orange-red colour to the B horizon: the precipitated hydroxide adsorbs fresh ferrous–organic complexes, in which the iron subsequently oxidizes, and also acts as a coarse filter for humus and Al–organic complexes eluviated from the A horizon. Hence, the Bs horizon that forms is usually overlain by a Bh horizon.

The combination of Bs and Bh horizons generally satisfies the criteria for a *spodic horizon*, which is the diagnostic horizon of the soil order Spodosol (section 14.3). The end-product of podzolization is therefore a Spodosol, or *podzol*, in the terminology of Table 5.3, the exact form of which is determined by the interplay of vegetation, drainage and parent material. Podzolization is favoured by permeable, siliceous parent material, a P/E ration > 1 for most of the year and slow rates of organic decomposition because of the type of litter or the acidity of the soil, or both. Two types of *podzol* profiles are shown in Figure 9.4.

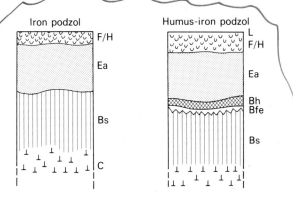

Figure 9.4. Diagrammatic profiles of an *iron podzol* and a *humus-iron podzol* (Spodosols).

Lessivage

Duchaufour (1977) used the term *lessivage* to describe the mechanical eluviation of clay particles from the A horizon without chemical alteration; but in the 1982 translation by T. R. Paton the term 'pervection' is substituted for lessivage. The clay is washed progressively downwards in the percolating water to be deposited in oriented films (argillans) on ped faces and pore walls, gradually forming an horizon of clay accumulation (Bt).

Lessivage is unlikely to occur in calcareous soils, or neutral soils that retain a high proportion of exchangeable Ca^{2+} ions, because the clay remains flocculated. Under acid conditions, if exchangeable Al^{3+} and hydroxy-Al ions are predominant on the clay surfaces, the clay should also remain flocculated (section 7.4). However, when adsorbed Al, and the Fe of oxide films stabilizing microaggregates, are complexed by soluble organic compounds, the clay particles are more likely to deflocculate and be translocated.

The formation of a lessived soil (Duchaufour's *sol lessivé*) predisposes the soil to further pedogenic processes, the course of which depends on the drainage of the profile. When the drainage is free, with increasing lessivage and acidification in the upper profile, podzolization ensues and the soil may pass through successive stages in the maturity sequence:

lessived brown soil → acid lessived soil → podzolized lessived soil → podzol

The first three stages correspond to the *podzolic* or *podzolized* soils of the Great Soil Group terminology (Table 5.3) or the Alfisols of Soil Taxonomy. Essentially, the difference between a *podzol* and lessived soil is that clay is destroyed rather than translocated in the *podzol*, and the intensity of cheluviation of Fe and to a lesser extent Al is greater in the *podzol*. The latter effect is illustrated by the free oxide ratios in Table 9.4. In the lessived soil the ratios SiO_2/Al_2O_3 and Al_2O_3/Fe_2O_3 remain reasonably constant down the profile; but in the *podzol* these ratios are high in the Ea horizon and much lower in the Bh and Bs horizons.

Table 9.4. Free oxide ratios in the clay fractions of a typical *podzol* and lessived soil.*

	Podzol (Spodosol)			Lessived soil (Alfisol)	
Horizon	$\frac{SiO_2}{Al_2O_3}$	$\frac{Al_2O_3}{Fe_2O_3}$	Horizon	$\frac{SiO_2}{Al_2O_3}$	$\frac{Al_2O_3}{Fe_2O_3}$
Ea	3.22	5.72	Eb	3.15	3.98
Bh Bs	1.85	0.84	Bt1	3.05	3.66
Bs	1.83	2.40	Bt2	3.70	3.12
C	2.12	3.71	C	3.21	3.10

*Grey-brown podzolic soil (after Muir, 1961).

Where the drainage of the lessived soil is, or becomes impeded, waterlogging and consequent *gleying* processes may supervene at variable depths.

Gleying

Even in a well-drained soil, if the climate favours the formation of an O horizon or peat, the surface may remain saturated for most of the year. In other situations the whole soil may be waterlogged for all or part of the year due to high groundwater.

Collectively, soils that are influenced by waterlogging are called *hydromorphic* soils (section 9.5). The most important feature of such soils is *gleying* (section 8.4). True gley soils have uniformly blue-grey to blue-green colours and are depleted of iron due to its solution and removal as Fe^{2+} ions. The reduction of iron is primarily biological and requires both organic matter and microorganisms capable of respiring anaerobically. However, most of the iron exists as Fe^{2+}–organic complexes in solution or as a mixed precipitate of ferric and ferrous hydroxides, which is responsible for the characteristic gley colour. This precipitate forms as follows:

$$2Fe(OH)_3+Fe^{2+}+2OH^- \rightleftharpoons Fe_3(OH)_8 \text{ or}$$
$$Fe_3O_4 . 4H_2O \text{ hydrated magnetite} \quad (9.8)$$

Mottling is a secondary effect in gleys resulting from the reoxidation of iron in better aerated zones, especially around plant roots and in the larger pores. The orange-red mottles have a much higher Fe content than the surrounding blue-grey matrix. Thus, a C/G or B/G horizon frequently develops a coarse prismatic structure in which the ped interiors have a uniform gley colour and the exterior faces, pores and root channels have rusty mottling. This is typical of a *groundwater gley*.

By contrast, in soils with impeded surface drainage, gleying is confined to the A, E or top of the B horizon and is subject to seasonal weather conditions. In soils where a summer *SMD* develops (section 6.5), the Fe^{2+}-iron is oxidized and rusty mottles develop against the blue-grey colours of the ped matrices, darkened in the A horizon by the presence of organic matter which does not accumulate to any depth on the surface. This is typical of the *pseudogleys*. Conversely in colder and wetter climates, the surface soil can remain saturated for most of the year, allowing a thick O horizon or even peat to develop, and the blue-grey gley colours persist in the mineral soil. This is typical of a *stagnogley*. Pseudogleys and stagnogleys fall into the category of

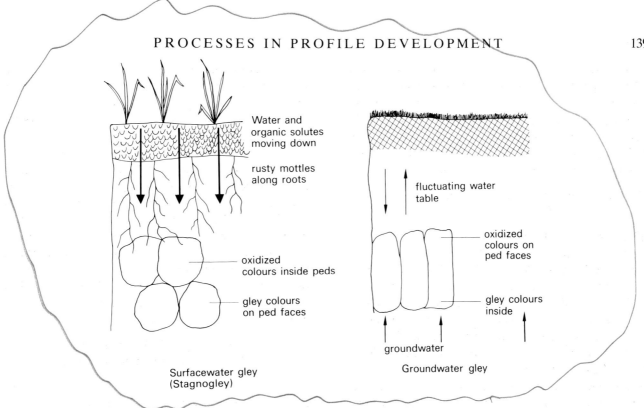

Figure 9.5. Profile features of groundwater gley and surfacewater gley soils (after Crompton, 1952).

surfacewater gleys. Profile features of groundwater and surfacewater gleys are illustrated in Figure 9.5.

Illustrative examples of these pedogenic processes in action in temperate and tropical soils are given in the following sections.

9.3 FREELY DRAINED SOILS OF HUMID TEMPERATE REGIONS

The soils over substantial areas of Europe and North America have formed on comparatively uniform parent materials of glacial or periglacial origin for a similar period of time. On well-drained sites of slight to moderate relief, two soil sequences are common: one on calcareous, the other on siliceous, parent materials.

Soils on calcareous, clayey parent material
The general sequence of soil formation with the passage of time, and increased leaching, is shown in

Figure 9.6. The plant succession advances from grassland to deciduous forest. In eastern Canada under deciduous forest the soil may change from a *brown forest soil* (Inceptisol) to *grey-brown podzolic soil* (Alfisol), but to a *grey wooded soil* (Alfisol) under cooler conditions where conifers flourish. The major difference is that the *grey-brown podzolic soil* has a mull-like Ah horizon and the *grey wooded soil* a mor humus layer. Both soil types degenerate, with further leaching, into *brown podzolic soils* (Spodosols), in which the whole profile becomes acidic and podzolization more obvious. The latter corresponds to the intermediate stages (4 and 5) of Duchaufour's maturity sequence on siliceous parent rock (Figure 9.7).

Soils on permeable, siliceous parent material
The early stages of this soil sequence, illustrated in Figure 9.7, feature clay lessivage, under a deciduous forest cover, leading to the formation of a *lessived brown soil* (Alfisol) and *acid lessived soil* (Alfisol).

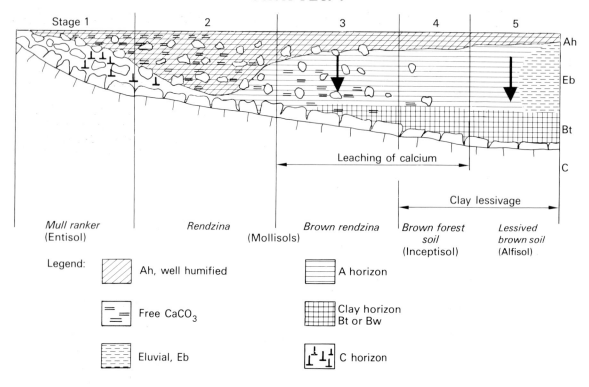

Figure 9.6. Soil maturity sequence on calcareous clayey parent material under a humid temperate climate (after Duchaufour, 1982).

With further time and leaching, the podzolizing influence strengthens and an *iron podzol* or *humus–iron podzol* (Spodosol) forms in the surface of the deep permeable Eb horizon of the lessived soils. Mackney (1961) has described the terminal stages (5 to 7) of this sequence under oakwoods on heterogeneous acid sands and gravels of the Bunter Beds in the English Midlands. On less acid parent material, however, continued degradation of the *podzolized lessived soil* (Alfisol) to the *humus–iron podzol* (Spodosol) is normally associated with a change in the vegetation from climax deciduous forest to coniferous forest or heathland.

Soils on mixed siliceous and calcareous parent material
A sequence similar to that shown in Figure 9.7 occurs under beechwoods in the Chiltern Hills of England,

where the original parent material consisted of some siliceous glacial drift mixed with soliflucted material from the Chalk that has, however, become acidic after prolonged weathering and leaching (Avery, 1958). As shown in Figure 9.8, the soils on the steepest slopes, A, where erosion is active, are dominated by the Chalk rock and a typical *rendzina* (Mollisol) is formed (*cf.* stage 2, Figure 9.6). On the intermediate slopes, B, mantled by thin drift and solifluction deposits, immature *brown forest soils* (Inceptisols) are found. On the deeper drift and acid clay-with-flints of the plateaux and gentle slopes, C, *lessived brown soils* and *acid lessived soils* (Alfisols) occur. Some gleying occurs in the Bt horizons of the lessived soils in the valleys, whereas on the plateaux, mor humus accumulates and occasionally *micro-podzols* (Spodosols) are found in the Eb horizons of the lessived soils.

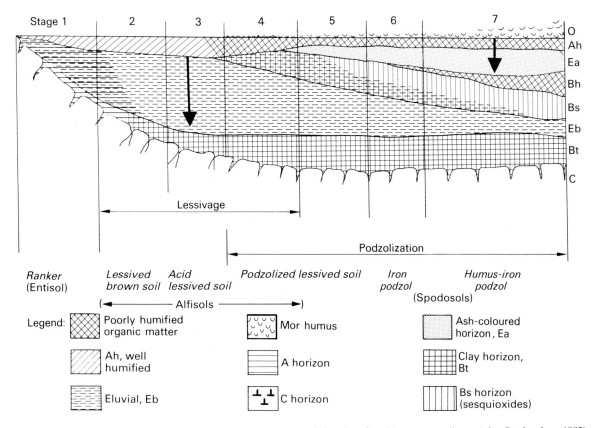

Figure 9.7. Soil maturity sequence on permeable, siliceous parent material under a humid temperate climate (after Duchaufour, 1982).

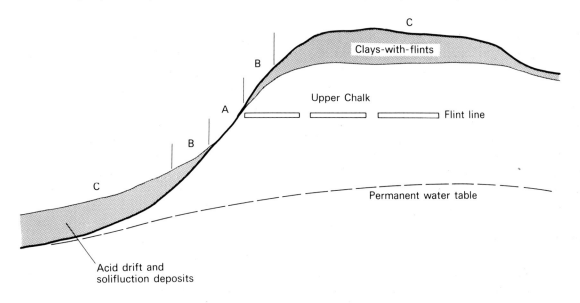

Figure 9.8. Parent material variations with relief in an old Chalk landscape of the Chiltern Hills (after Avery, 1958).

9.4 SOILS OF THE TROPICS AND SUBTROPICS

Tropical and subtropical lands are generally *hot* so that chemical weathering and biochemical reactions, such as organic matter decomposition, proceed more rapidly than in temperate climes. Except for the humid equatorial regions, fluctuations in soil moisture are also more extreme. Age is another factor that sets tropical soils apart from temperate soils. Many tropical land surfaces are pre-Pleistocene and some date back to the early Tertiary and late Mesozoic ($60-100\times10^6$ y BP). There has been little recent folding, apart from orogenic belts along the Andes in South America and in south-east Asia, but the ancient stable shields and tablelands have been gently upwarped or downwarped to form continental swells and basins on a grand scale. Ice sheets have been absent, and erosion on these relatively flat surfaces has not been active enough to remove the products of weathering. Therefore, the soils have developed on deeply weathered materials and the influence of surface organic matter on profile development is much less than in the temperate regions. As mentioned in section 5.3, age and relief are important factors influencing pedogenesis in the tropics and subtropics, although trends associated with differences in parent material are also discernible.

Duchaufour (1982) considers there are three basic kinds of weathering in hot climates, which lead to distinctive soil types. In the humid tropics all three may occur together, the change from one to another being determined by local parent material, relief, etc. so that it is possible to treat the processes as *stages* in the same weathering sequence:

fersiallitization → *ferrugination* → *ferrallitization*

(1) *Fersiallitization*. During this process, primary minerals weather to 2:1 clays, partly inherited and partly neoformed (section 2.3); Fe^{2+} is liberated, but it is rapidly oxidized and precipitated as ferric oxy-hydroxides (section 9.2). The formation of *hematite*, in particular, as a coating on clays gives many of these soils a characteristic red colour—the process of *rubification*. Such soils are readily formed in humid Mediterranean-type climates on a wide range of parent materials. Lessivage leading to the development of a red Bt horizon is a characteristic feature, probably aided by the rapid movement of water down cracks as the soil rewets at the end of the dry summers (section 6.4). These soils are generally of high base saturation (> 35 per cent) and therefore are classed as Alfisols.

In the humid tropics, fersiallitization of relatively young basic rocks may produce *eutrophic brown soils* (Inceptisols), which, if freely drained, will go on to form *ferruginous soils* (Alfisols and Ultisols). On poorly drained sites, however, such soils develop into mature *vertic eutrophic brown soils* (Vertisols) which have > 35 per cent of predominantly 2:1 clays and a pronounced tendency to shrink and swell with changing moisture content.

(2) *Ferrugination*. Weathering is more complete (only resistant primary minerals like orthoclase, muscovite and quartz remain), and 1:1 clays are synthesized. The iron oxides formed may or may not be rubified, and 2:1 clays are translocated to form a Bt horizon. Base saturation is variable, depending on the climate: under the marked wet and dry seasonality of the savanna regions, base saturation is usually > 35 per cent and *tropical ferruginous soils* (Alfisols) form. In humid subtropical regions, however, ferrugination leads to the formation of deeper (> 3 m), more highly leached *ferrisols* (Ultisols and some Oxisols). In the humid tropics and on old landscapes, ferrisols may persist at high altitudes (where they are humus-rich and very acidic) and on steep slopes.

(3) *Ferrallitization*. For a range of parent materials, on reasonably level but well-drained sites, soil formation in the humid tropics eventually proceeds to the end-member of the series:

tropical fersiallitic soil → *tropical ferruginous soil* → *ferrisol* → *ferrallitic soil*

Virtually all the primary minerals have been decomposed leaving quartz, kaolinite (neoformed), iron and aluminium oxides. Cation exchange capacities < 16 cmol charge (+) kg^{-1} clay and the absence of a Bt horizon place most *ferrallitic soils* in the class Oxisol. They are generally acidic (pH 4–5) and of low chemical fertility, but have good physical properties.

Parent material and drainage exert their influence on the degree of ferrallitization in the following way. On acidic materials with imperfect drainage, sufficient silica is retained to allow the neoformation of kaolinite in the lower profile, which can be up to 50 m deep overall (*ferrallitic soil with kaolinite*). However, basic rocks weathering on well-drained sites produce shallower soils containing much less kaolinite and an abundance of gibbsite and goethite or hematite (see Table 9.2). The latter soils are true *ferrallites*, also known as *krasnozems* in Australia.

LATERITE

Where the organic-rich A horizon of an old *ferrallitic soil* is stripped off by erosion, the exposed sesquioxide zone becomes indurated by dehydration at high temperatures to form a geological structure called *laterite*. The profile of a typical *ferrallitic soil with kaolinite* that degenerates to a deep laterite is illustrated in Figure 9.9. It comprises:

(1) an uppermost *concretionary layer* of iron oxides, with secondary gibbsite, which may be massive if cemented, but loosely nodular or pisolitic* if uncemented;

(2) an intermediate *mottled zone*—orange to red iron oxide deposits in a pale-grey matrix of kaolinite;

*Resembling small dried peas in size and shape.

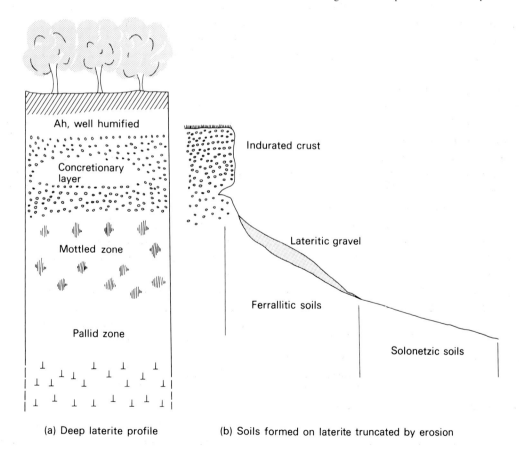

(a) Deep laterite profile   (b) Soils formed on laterite truncated by erosion

Figure 9.9. (a) A deep *laterite* (Oxisol) profile. **(b)** Soil catena on a truncated *laterite* (Oxisol).

(3) lowermost, a *pallid zone* of pale-grey to white kaolinite and bleached quartz grains, sometimes weakly cemented by secondary silica, and usually increasing in salt concentration towards the weathering zone.

Ancient laterites that were much eroded and dissected during the Pleistocene often survive as prominent escarpments and crusts in the landscapes of tropical Africa, Australia and south-east Asia. Truncation by erosion gives rise to a new weathering surface (Figure 9.9b) on which distinctive contemporary soils have formed: impoverished *ferrallitic soils* on the upper two laterite layers, and solonetz-type soils (section 9.6) in the pallid zone where salts had accumulated.

Whereas layers 1 and 2 of the laterite profile represent the residuum of prolonged intense weathering, the pallid zone appears to form due to the reduction and solution of iron under anaerobic conditions in a zone of groundwater influence. The reduced iron moves in solution both laterally and vertically, the vertical movement following seasonal, or longer-term, fluctuations in the height of the watertable. Reoxidation and precipitation of insoluble $Fe(OH)_3$ may in some cases account for the development of the overlying mottled zone. Nevertheless, iron moving laterally in percolating drainage water is eventually precipitated at some point in the landscape where better aeration supervenes, generally because of drier conditions, and the ironstone or *orstein* deposits typical of a *groundwater laterite* (Oxisol) form (Figure 9.10).

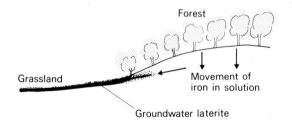

Figure 9.10. *Groundwater laterite* formation in a humid tropical environment.

9.5 HYDROMORPHIC SOILS

Soils showing some hydromorphic features are found in catenas in many parts of the world: the *vertisols* and *groundwater laterites* are two examples that have already been cited. Nevertheless, hydromorphism is especially prevalent in humid temperate regions, such as the British Isles, where groundwater gleys and surfacewater gleys can be found on most parent materials. Two soil sequences found in the north and west of England and Wales serve to illustrate the point.
(1) At altitudes > 300 m, where the annual rainfall is $> 1,250$ mm, the high P/E ratio and low temperatures favour blanket peat formation (see below), even on well-drained parent materials. A typical catena, proceeding downslope, consists of:

Blanket peat $\rightarrow$ *thin iron-pan soil* $\rightarrow$ *podzol or brown*
(Histosol)   *podzolic soil*
                                            (Spodosols)

as illustrated in Figure 9.11.

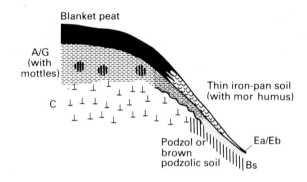

Figure 9.11. Catenary sequence on cool wet uplands in the west of England and Wales.

Gley colours predominate under the peat and to the base of the wet soil zone as the peat attenuates to a thin organic horizon downslope. It is roughly in this mid-slope area that the distinctive iron-pan—shiny black above and rusty-brown beneath—is found, as reduced iron in solution moves down into the better aerated lower horizons and is oxidized. This is an example of a *surfacewater gley* (stagnogley type).

(2) At lower altitudes under an annual rainfall of 750–1,000 mm, a catenary sequence similar to that of Figure 9.12 forms on slopes of gentle relief often overlying relatively impermeable glacial till. The sequence consists of:

Brown podzolic soil → peaty podzol → peaty gley soil
(← Spodosols————————→)   (Histosol)

The better-drained upslope soils show rusty mottling along root channels in an otherwise weakly gleyed Ea or Eb horizon underlying a thin Ah; but as the water-table rises downslope, the whole profile becomes gleyed, apart from weak rusty mottling near the surface, and a *peaty gley soil* (Histosol) forms. This is an example of a *groundwater gley*.

Peat formation
A prerequisite of peat formation is a very slow rate of organic decomposition due to a lack of oxygen under continuously wet conditions. This condition is achieved by various combinations of relief or climate (especially *P/E* ratio) with permeability and base status of the underlying rocks. The two main types of peat are described below.

BASIN OR TOPOGENOUS PEATS
Irrespective of climate, basin peats form where drainage water collects. If the water is neutral to alkaline, even though plant decomposition rates are substantial, the greater growth of sedges, grasses and trees produces an accumulation of well-humified remains in the form of *fen peat*. The ash content of this peat is high (10–50 per cent) due to the accession of colloidal mineral particles in the drainage waters. Large areas of fen peat are found, for example, in Florida, the Fens of England and in Botswana, the latter two being areas of relatively low rainfall.

BLANKET BOG (UPLAND OR CLIMATIC PEAT)
A *P/E* ratio > 1 for all or part of the year and cool temperatures predispose to the formation of *blanket peats*, described as ombrogenous deposits, over much of Ireland, the west of Scotland and the Pennines in England, especially on acidic rocks (Figure 9.11). The vegetation of hardy grasses, Ericaceous plants and *Sphagnum* moss provides an acid litter, and as the only source of water is rain which is low in bases, the rate of plant decomposition is extremely slow and the humification imperfect. The ash content of the peat is usually < 5 per cent.

The formation of basin and fen peats began earlier in the post-Pleistocene period than did the blanket peats;

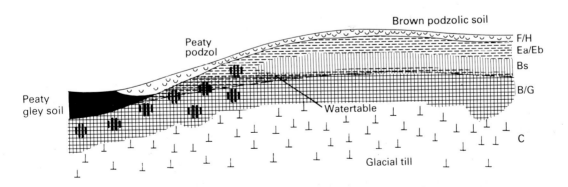

Figure 9.12. Catenary sequence on cool, imperfectly drained lowlands in the west of England and Wales.

many of which originated during the warm, wet Atlantic phase lasting from *c.* 7,500 BP to 4,000 BP. Where present-day climate is favourable (*P/E* ratio > 1), basin and fen peats are often transformed to *raised bog* (Figure 9.13). The build-up of sediment and peat in a wet hollow eventually raises the living vegetation above the level of the groundwater so that continued plant growth becomes dependent solely on rainwater. Slowly, acid-tolerant and hardy species such as heather, cottonwood and *Sphagnum* invade and displace the species typical of the fen or basin peat.

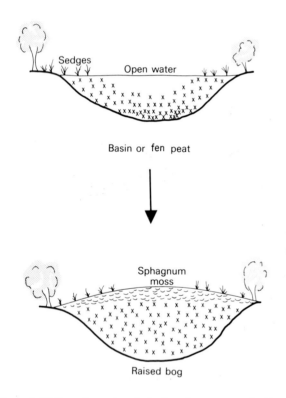

Figure 9.13. Transformation of a basin or fen peat to a raised bog.

9.6 SALT-AFFECTED SOILS

Past cycles of geologic weathering have resulted in the accumulation of salts in the oceans, lakes and groundwater. Sediments laid down under water naturally entrain salts (*connate salts*) which are inherited by the soils that ultimately form on these sediments. During soil formation, salts continue to be released by weathering (of residual primary minerals and secondary silicates), and they also accrue from atmospheric deposition. Wind turbulence over the sea creates aerosols of salt crystals, which may subsequently be deposited on land as *dry deposition*; or they may act as condensation nuclei for raindrops. Such accessions may be as much as 100–200 kg ha^{-1} y^{-1} for distances up to 50–150 km from the sea; they are significant inputs for very old soils containing few weatherable minerals, such as those found in the western part of Australia.

Salt-affected soils also arise through man's activities, especially those that result in the rise of saline groundwater too near the soil surface. Examples of this are discussed in Chapter 13. A soil is *saline* if it has a salt concentration exceeding about 2,500 ppm, corresponding to an electrical conductivity of the saturation extract > 4 mmhos cm^{-1} (section 13.2). Naturally occurring saline soils, common in arid regions, are called *white alkali soils* or *solonchaks* (Aridosols): typically, they have a uniform dark-brown Ah horizon containing superficial salt efflorescences as well as interflorescences within the pores. The salts are brought up by capillary rise of the saline groundwater. Below the Ah horizon the soil is structureless and frequently gleyed. Although there is exchange of Na$^+$ for Ca^{2+} and Mg^{2+} on the clay surfaces, in accordance with the ratio law (section 7.2), the clay remains flocculated and the structure stable provided that a high salt concentration is maintained. Frequently, sufficient secondary CaCO$_3$ precipitates out to form an Ak horizon and the soil pH ranges from 7 to 8.1, depending on the partial pressure of CO$_2$ in the soil air.

Solonchak–solonetz–solod soil sequence
Climatic change with rainfal increasing, or a lowering of the watertable, can have serious consequences for a saline soil, especially if the exchangeable Na$^+$ content exceeds 10–15 per cent of the *CEC* (section 13.3). Dilution of the soil solution as the salts are slowly

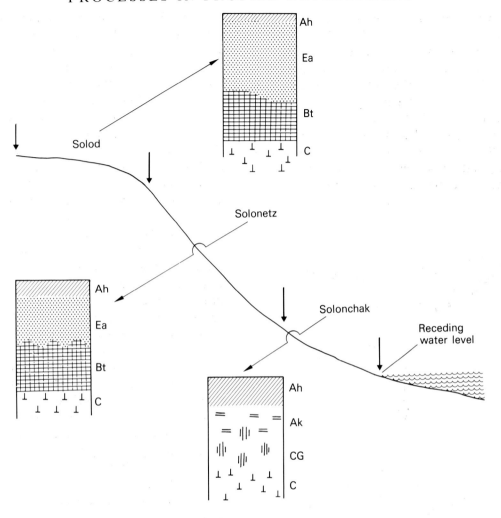

Figure 9.14. *Solonchak–solonetz–solod* maturity sequence resulting from a falling watertable (after Kubiena, 1953).

leached causes the diffuse double layers at clay surfaces to expand, resulting in a pH rise and increased swelling pressure within clay domains. For a soil solution dominated by Ca and Mg, a reduction in the concentration at equilibrium from 10^{-2} to 10^{-4} M would cause the pH to rise by 1 unit (say, from 8 to 9), if the activity ratio $H^+/(Ca^{2+}+Mg^{2+})^{1/2}$ remained constant. As the soil solution becomes more dilute, it is possible that adsorbed Na^+ ions exchange for solution H^+ ions, produced by the dissociation of water molecules, and

the pH rises further according to:

$$Na\text{-clay}+HOH \rightleftharpoons H\text{-clay}+NaOH \quad (9.9)$$

pH values as high as 9–9.5 have been recorded. The NaOH reacts with CO_2 of the soil atmosphere to form $NaHCO_3$ and Na_2CO_3. These salts maintain a reversible hydrolytic equilibrium with NaOH, that is:

$$NaHCO_3+H_2O \rightleftharpoons NaOH+H_2CO_3 \quad (9.10)$$

The soil is now at the *black alkali* stage, where the

high pH causes the organic matter to disperse, further weakening the cohesion of the aggregates. The unstable Na-clay also disperses and is illuviated into the lower profile where it is flocculated by the higher salt concentration. The Bt horizon that develops is very poorly drained, often with signs of gleying, and has massive columnar peds which are capped with organic coatings (see Figure 4.4). The overall process of clay dispersion and illuviation is called *solonization*, and the soil formed is a *solonetz* (Alfisol).

Continued leaching of the *solonetz* leads to the complete removal of soluble salts from the A horizon, which becomes acid, and the soil becomes a *solodized-solonetz* (Alfisol). The H-clay decomposes to release Al^{3+}, Mg^{2+} and SiO_2 (section 7.2). Thus, with the advance of salt leaching and acidification down the profile into the Bt horizon, the H,Na-clay becomes an Al,Mg-clay: the massiveness of that horizon decreases and permeability improves, facilitating further leaching. This, the final step in the degradation of a *solonchak*, is called *solodization*, and the resultant soil with its sandy, bleached E horizon and disintegrating Bt horizon is called a *solod* or *soloth* (sodic Alfisol or Natrustalf). The typical profiles of this maturity sequence are illustrated in Figure 9.14.

9.7 SUMMARY

A *soil profile* develops as the weathering front penetrates more deeply into the parent material and organic residues accumulate at the surface. The principal pedogenic processes operating are *mineral dissolution* (including *oxidation* and *hydrolysis*), *leaching, cheluviation, lessivage, gleying* and *salinization*, variations in the intensity of which with depth give rise to one or more *soil horizons*. Superimposed lithological discontinuities give rise to *soil layers*.

Pedogenesis in temperate climates is often dominated by the influence of surface organic matter, and the soluble complexing agents released from it. Many of the parent materials are also young, having been deposited when the Pleistocene ice sheets retreated. With the passage of time, a soil passes through the stages of a *maturity sequence*, examples of which are given in Figures 9.6 and 9.7. Furthermore, subtle changes in the influence of parent material, climate, organisms and relief can produce comparable profile variations in soils of the same age, as illustrated in Figures 9.11 and 9.12.

In hot climates, chemical weathering and organic matter decomposition proceed more rapidly than in temperate climes, and extremes of moisture are common. Compared with recently glaciated and periglacial areas of the Northern Hemisphere, much of the tropical land surface is very old so that soils have been forming on the same parent material for millions of years. The main trends in pedogenesis can be differentiated by the extent of primary mineral decomposition, the leaching of bases and silica, and the neoformation of clay-fraction minerals (kaolinite and the sesquioxides). In the humid tropics, the most severe weathering environment, the three processes of:

$$\textit{fersiallitization} \rightarrow \textit{ferrugination} \rightarrow \textit{ferrallitization}$$

can occur side by side, the change from one to another being determined by local parent material and relief.

Hydromorphic processes (including gleying) can affect soils under any climate giving rise to *surfacewater gleys* (*pseudogleys* and *stagnogleys*) and *groundwater gleys*. Continuously wet and cold climates produce *peaty gleys* and *blanket peats* (Histosols).

Salinization due to inherited salts or the rise of saline groundwater produces *solonchak* soils (Aridosols) which, on being leached, undergo *solonization* (*solonetz* phase—Alfisols) and finally *solodization* (*solod* or *soloth* phase—Sodic Alfisols).

REFERENCES

AVERY B.W. (1958) A sequence of beechwood soils on the Chiltern Hills, England. *Journal of Soil Science* **9**, 210–224.
BLACK A.S. & WARING S.A. (1976) Nitrate leaching and adsorption in a krasnozem from Redland Bay, Qld. II. Soil factors influencing adsorption. *Australian Journal of Soil Research* **14**, 181–188.

BUTLER B.E. (1959) *Periodic phenomena in landscapes as a basis for soil studies.* CSIRO Soil Publication No. 14.

CROMPTON E. (1952) Some morphological features associated with poor soil drainage. *Journal of Soil Science* **3**, 277–289.

DUCHAUFOUR P. (1977) *Pédologie. I. Pedogenèse et Classification.* Masson, Paris.

DUCHAUFOUR P. (1982) *Pedology, Pedogenesis and Classification* (Translated by T.R. Paton). Allen & Unwin, London.

HODGSON J.M. (1974) (Ed.) *Soil Survey field handbook.* Soil Survey of England and Wales, Technical Monograph No. 5.

JENNY H. (1941) *Factors of Soil Formation.* McGraw-Hill, New York.

KUBIENA W.L. (1953) *The Soils of Europe.* Murby, London.

LOUGHNAN F.C. (1969) *Chemical Weathering of the Silicate Minerals.* Elsevier, New York.

MACKNEY D. (1961) A podzol development sequence in oakwoods and heath in central England. *Journal of Soil Science* **12**, 23–40.

MUIR A. (1961) The podzol and podzolic soils. *Advances in Agronomy* **13**, 1–56.

FURTHER READING

ANDERSON H.A., BERROW M.L., FARMER V.C., HEPBURN A., RUSSELL J.D. & WALKER A.D. (1982) A reassessment of podzol formation processes. *Journal of Soil Science* **33**, 125–136.

AUBERT G. & TAVERNIER R. (1972) Soil survey, in *Soils of the Humid Tropics.* National Academy of Sciences, Washington, D.C.

BRIDGES E.M. (1978) *World Soils*, 2nd Ed. Cambridge University Press, Cambridge.

DAVIES R.I. (1971) Relation of polyphenols to decomposition of organic matter and to pedogenetic processes. *Soil Science* **111**, 80–85.

GOUDIE A. (1973) *Duricrusts in Tropical and Subtropical Landscapes.* Clarendon Press, Oxford.

D'HOORE J.L. (1964) *Explanatory monograph to soil map of Africa.* Commission for Technical Cooperation in Africa, Publication **93**, pp 1–126.

NEWBOULD P.J. (1958) Peat bogs. *New Biology* **26**, 89–104.

SANCHEZ P.A. & ISBELL R.F. (1979) A comparison of the soils of tropical Latin America and tropical Australia, in *Pasture Production in Acid Soils of the Tropics* (Eds. P.A. Sanchez and L.E. Tergas). Centro Internacional de Agricultura Tropical, Colombia, pp 25–53.

Part 3
Utilization of Soil

'And earth is so surely the food of all plants that
with the proper share of the elements, which each
species requires, I do not find but that any
common earth will nourish any plant.'
Jethro Tull (1733) in
Horse-hoeing Husbandry,
republished by William Cobbett in 1829

Chapter 10
Nutrient Cycling

10.1 NUTRIENTS FOR PLANT GROWTH

The essential elements

According to present knowledge, there are sixteen elements without which green plants cannot grow normally and reproduce. On the basis of their concentration in plants, these *essential elements* are subdivided into the *macronutrients*:

C, H, O, N, P, S, Ca, Mg, K and Cl

which occur at concentrations greater than 1,000 ppm, and the *micronutrients*:

Fe, Mn, Zn, Cu, B and Mo

which are generally less than 100 ppm in the plant. Carbon, H and O are only of passing interest, because they are supplied as CO_2 and H_2O, which are abundant in the atmosphere and hydrosphere; likewise Cl, which is abundant and very mobile as the Cl^- ion. Of the others, with the exception of N which comprises 79 per cent of the atmosphere, the major source is weathering minerals in the soil and parent material. The nutrient-supplying power of a soil is a measure of its *fertility*, the full expression of which depends on diverse properties of the plant, its environment, and on management (section 11.2).

Plants absorb other elements, some of which are beneficial though not essential, such as Si (for sugar cane), Na (for sugar beet and mangolds) and Al (for many ferns). Cobalt is essential for symbiotic N_2 'fixation' in legumes and blue-green algae, while Se and I, although of no benefit to plants, are vital for the health and normal reproduction of animals. V may substitute for Mo to some extent in the N_2 fixation process of bacteria.

Nutrient cycling

The flow of nutrients in the biosphere is continuous between three main compartments:

(a) an *inorganic store*, chiefly the soil;

(b) a *biomass store*, comprising living organisms above and below ground; and

(c) an inanimate *organic store*, on and in the soil, formed by the residues and excreta of living organisms.

The general cycle is illustrated in Figure 10.1. The importance of the various inputs and outputs, and the transformations undergone *en route*, differs for each element, as does the partitioning between soil and non-soil. For example, contrast the behaviour of P where more than 90 per cent of the total element may be in the top 30 cm of soil, with that of K where the proportion is usually less than 50 per cent.

There are also differences in the way the elements are distributed within the soil. Elements that are released at depth by mineral weathering are absorbed by plant roots and transported to the shoots, to be deposited finally on the soil surface in litter and animal excreta. Accessions in rainfall add to the surface store. Nitrogen always accumulates in the organic-rich A horizon, the content declining gradually with depth; P is similar, but the decline with depth is more abrupt because of the immobility of phosphate ions in the soil (Figure 10.2a). Sulphur, like N, accumulates in the surface of temperate soils, but not necessarily in tropical soils, due to the mineralization of organic-S and the downward displacement of SO_4^{2-}. Many of the latter soils show a bimodal S distribution, with organic S in the surface and retention of sulphate ions in the subsoil at sites of positive charge (Figure 10.2b). The distribution of most cations—Fe^{3+}, Ca^{2+}, Mg^{2+}, Mn^{2+}, Cu^{2+} and Zn^{2+} for example—correlates with clay accumulation, except in peaty soils, where Cu in particular tends to be associated with the organic matter.

THE INORGANIC STORE
For elements such as P, K and Ca, plant roots tap only a

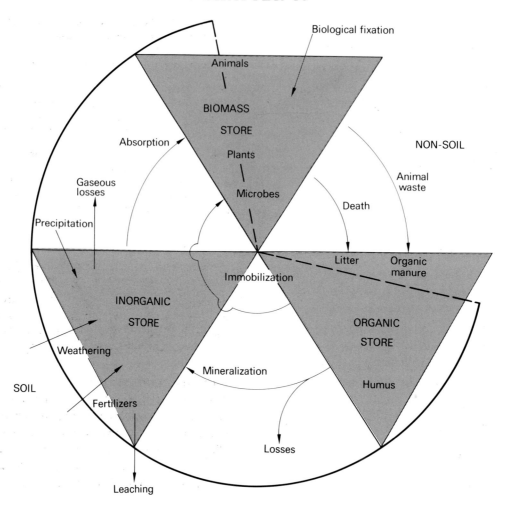

Figure 10.1. Fundamentals of a nutrient cycle.

small fraction of the soil's inorganic store during a single growth season. This fraction is withdrawn from the *available pool*, which is made up of:

(a) ions in the *soil solution;*

(b) *exchangeable ions* adsorbed by clay minerals and organic matter.

Less readily available ions are held in sparingly soluble compounds like gypsum, calcite and insoluble phosphates, or 'fixed' in non-exchangeable forms in clay lattices (K^+ and NH_4^+).

Inputs to the available pool occur by:

(1) weathering of soil and rock minerals (section 5.2);

(2) precipitation and dust fallout;

(3) mineralization of organic matter (section 3.1 *et seq.*);

(4) the application of organic manures, organic and inorganic fertilizers (section 11.1 and Chapter 12).

THE BIOMASS STORE

The storage of nutrients in the plant biomass depends on the mass of vegetation produced. As Table 10.1 shows, the store of K, Ca and Mg (less so for N) in

(a)

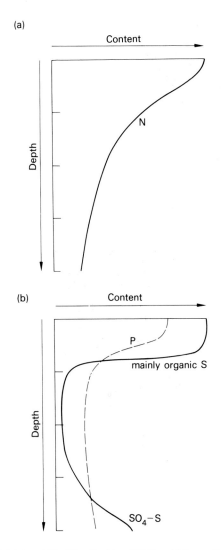

Figure 10.2 (a) and **(b).** Distribution of N, P and S in a soil profile.

Table 10.1. The distribution of selected elements between the plant biomass and the soil for different vegetation types in West Africa (kg ha^{-1}).

Plant biomass (including roots and litter)				Soil (0–30 cm)			
N	K	Ca	Mg	N	K	Ca	Mg
Mature forest (40 y old)							
2040	910	2670	350	4590	650	2580	370
Imperata cylindrica grassland							
46	105	13	24	1790	190	2910	380

(After Nye and Greenland, 1960.)

mature forest under high rainfall in the tropics is comparable to that in the top 30 cm of soil; but the store of these elements in rather poor grassland is much less, except for K which cycles rapidly through the soil–plant system. All biomass is important in nutrient cycling, irrespective of the size of the store, because the remnants of living organisms and their excreta are the substrate for soil microorganisms.

Although small (0.5–2 t C ha^{-1}), the soil microbial biomass has been identified as the 'eye of the needle' through which all the C returned to the soil must eventually pass. Virtually the same is true of N; 98 per cent or so of the soil N is in organic combination, and the metabolism of this N by the soil microorganisms is intimately associated with that of C.

THE ORGANIC STORE

The cycle shown in Figure 10.1 is completed by the decomposition of litter and animal excreta by the soil mesofauna and microorganisms, a topic discussed in Chapter 3. Specific aspects of mineralization, as they affect the availability to plants of individual nutrients, are discussed in the following sections.

10.2   THE PATHWAY OF NITROGEN

Mineralization and immobilization

Depending on the soil and environmental conditions, the quantity of soil N to maximum rooting depth (*c.* 1 m) ranges from 1 to 10 t ha^{-1}, much of which occurs in the top 15 cm or so. Thus, for fertile agricultural soils in Britain, the average figure for soil N is 4,000 kg ha^{-1} 15 cm^{-1} (Anon., 1983). Organic N is transformed during microbial decomposition of organic matter according to the reaction:

$$\text{Organic N} \rightarrow NH_4^+ + OH^- \qquad (10.1)$$
$$\text{(proteins, nucleic acids)}$$

This step is often referred to as *ammonification*, and it is obvious this is an alkalizing reaction. Subject to any limitations imposed by temperature, moisture, pH and aeration, NH_4^+ is oxidized NO_3^- according to:

$$NH_4^+ + 2O_2 \rightarrow NO_3^- + H_2O + 2H^+ \qquad (10.2)$$

which is called *nitrification* and is an acidifying reaction. The whole process is described as mineralization and both NH_4^+ and NO_3^- comprise the pool of *mineral N* on which plants feed. Ammonium is held as an exchangeable cation that is readily displaced into the soil solution and, with NO_3^-, it moves to the roots in the water absorbed by the transpiring plant, a process of *mass flow*. Both ions can also move from regions of higher to lower concentration by *diffusion*. Thus, all the mineral N (except for NH_4^+ 'fixed' in micaceous clay lattices) is available to the plant.

The balance between mineralization and immobilization during decomposition is dependent on the C:N ratio of the substrate (section 3.3), but the *net* mineralization rate can be described approximately by the equation:

$$\frac{dN}{dt} = -kN \qquad (10.3)$$

where N is the amount of organic N per unit soil volume and k is a rate constant varying between 0.01 and 0.06 y^{-1}. During an interval of 1 year, the amount of N mineralized is given by kN, the value of which generally ranges between 10 and 250 kg ha^{-1}. Whether this is enough to meet the needs of plants depends on their uptake of N, and the gains and losses in the N cycle (see below).

Table 10.2. N contents of selected agricultural crops.

	N content (kg ha^{-1})
Wheat, 6 t ha^{-1} (grain and straw)	120
Potatoes, 50 t ha^{-1}	180
Grass, 10 t ha^{-1} (dry matter)	250
Maize, 13 t ha^{-1} (grain and stover)	360
Sugar cane, 120 t ha^{-1}	450

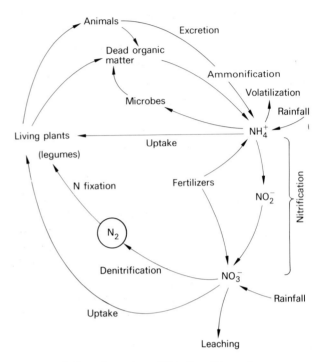

Figure 10.3. Transformations of N in the soil ecosystem.

PLANT UPTAKE

In natural ecosystems, plant growth rates are low and the annual uptake of N is relatively small: Cole and others (1967) quote figures of 25–78 kg N ha^{-1} y^{-1} for a coniferous forest in North America. One of the highest figures is 248 kg N ha^{-1} y^{-1} estimated by Nye and Greenland (1960) for mature secondary rainforest in Ghana. Cultivated crops are much more demanding; N uptakes range from 100 to 450 kg N ha^{-1} (Table 10.2), so that the mineralizing capacity of the soil is often insufficient to maintain optimum growth. In this case, the magnitude of the gains and losses in the N cycle (Figure 10.3) are of great significance.

Gains of soil N

RAINFALL AND DRY DEPOSITION

Apart from N released as N_2 and N_2O during denitrification (section 8.4), N enters the atmosphere as the oxides NO and NO_2, which are formed during the combustion of fossil fuels (the primary source) and by lightning discharges. It is also released as NH_3 gas by volatilization from manures, rotting vegetation and soil, especially calcareous soil following the application of ammonium fertilizers (section 12.2). Dust swept into the air may carry both NH_4^+ and NO_3^- ions, as well as organic N.

Oxides of N (NO_x) react with hydroxyl free radicals to form nitric acid, which contributes to 'acid rain' (about 30 per cent of the total rainfall acidity in the UK). Some NH_3 is absorbed through plant leaves while the remainder is dissolved in rain or forms salts, such as $(NH_4)_2SO_4$ (see equation 10.7c), which come down in the rain or are deposited on soil and plant surfaces (dry deposition). The total N input from the atmosphere ranges from 5 to 30 kg ha^{-1} y^{-1}. Within the total it is difficult to identify precisely the relative contributions of rainfall and dry deposition, but a ratio of 1:1 seems reasonable. The average atmospheric N contribution in the UK (excluding biological N_2 fixation) is about 12 kg ha^{-1} y^{-1} (Anon., 1983), which is small compared with other inputs to the N cycle for agricultural soils.

FERTILIZERS AND MANURES

Under natural vegetation, plant litter and organic manures provide the bulk of the nitrogen returning to the soil. In agriculture, however, where much of the crop is removed from the land, the return of N in chemical fertilizers or through the growth of legumes is necessary to bolster the soil N supplies for subsequent crops.

BIOLOGICAL N FIXATION

A small minority of microorganisms, either living free or in symbiotic association with plants, is exceptional in being able to reduce molecular N_2 to NH_3 and incorporate it into amino acids. This process, called *nitrogen fixation*, enables them to grow independently of mineral N. The annual turnover of biologically fixed N_2 in the biosphere is $\sim 10^8$ tonnes.

The splitting of the dinitrogen molecule (N≡N) and the reduction of each moiety to NH_3 are catalysed by the *nitrogenase* enzyme, which consists of two proteins—an Fe protein and a Mo—Fe protein, roughly in the ratio of 2:1. Reduction proceeds according to the reaction:

$$N_2 + 8H^+ + 6e + n\text{Mg}.\text{ATP} \rightarrow 2NH_4^+ + n\text{Mg}.\text{ADP} + n\text{PO}_4 \qquad (10.4)$$

for which there are three prerequisites:
(a) a reductant of low redox potential (usually ferredoxin);
(b) an ATP-generating system;
(c) a low partial pressure of O_2 at the site of nitrogenase activity.

FREE-LIVING N_2 FIXERS

Regulation of N_2 fixation in aerobic organisms is complicated because the reduced nitrogenase must be protected from O_2, yet O_2 is required for the cell's respiration and oxidative phosphorylation. It is not surprising, therefore, that the ability to fix N_2 is not widespread among non-photosynthetic, aerobic organisms. The more important of the free-living or non-symbiotic N_2 fixing organisms are listed in Table 10.3.

Table 10.3. The major free-living N_2 fixing organisms.

Bacteria	Aerobes	*Azotobacter, Beijerinckia* spp.
	Microaerobic	*Azospirillum brasilense* *Thiobacillus ferrooxidans*
	Facultative anaerobes	*Klebsiella* spp. *Bacillus* spp.
	Obligate anaerobes	*Clostridium pasteuranium* *Rhodospirillum,* *Chlorobium* and *Desulphovibrio* spp.
Blue-green algae	Photosynthetic aerobes	*Nostoc, Anabaena, Gloecapsa* spp.

Apart from the oxygen status, factors such as pH, supply of organic substrates, mineral N concentration and the availability of trace elements, especially Cu, Mo and Co, influence the distribution and activity of N_2 fixing organisms in soil. *Azotobacter* and the blue-green algae are restricted to neutral and calcareous soils, but spore-forming anaerobes like *Clostridium* are tolerant of a wide range of pH. At high levels of mineral N, the N_2 fixers succumb to competition from more aggressive heterotrophic species. Thus, the rhizosphere of plant roots with its abundance of exudates of high C:N ratio is a favoured habitat for N_2 fixing organisms. In recent years, nitrogenase activity has been identified in the rhizosphere of many crops, such as wheat, maize, pasture grasses and sugar cane, a discovery made possible by the use of the *acetylene reduction technique*. When nitrogen is excluded from the nitrogenase site by an excess of acetylene, the latter is reduced to ethylene according to the reaction:

$$C_2H_2 + 2H^+ + 2e^- \rightarrow C_2H_4 \qquad (10.5)$$

and the ethylene detected by gas chromatography. However, the identification of a functional nitrogenase enzyme does not necessarily mean that N_2 is being fixed.

Because of the high energy consumption of N_2 fixation (12–15 moles ATP per mole N_2 reduced), photosynthetic organisms tend to make the greatest contribution to non-symbiotic N_2 fixation: for example, up to 50 kg ha^{-1} y^{-1} due to blue-green algae and photosynthetic bacteria in padi rice culture. Estimates for other farming systems lie in the range 0–10 kg ha^{-1} y^{-1}. There is evidence that contributions from free-living organisms may be higher in temperate deciduous woodlands (20–30 kg N ha^{-1} y^{-1}), and exceptionally up to 100 kg ha^{-1} y^{-1} in some tropical forests.

SYMBIOTIC N_2 FIXATION

Symbiosis denotes the cohabitation of two unrelated organisms that benefit mutually from the close association. In the case of N_2 fixation, the invasion of roots of the host plant by a microorganism (the endophyte) culminates in the formation of a *nodule* in which carbohydrate is supplied to the endophyte and amino acids formed from the reduced N are made available to the host.

There are three main symbiotic associations:
(1) nodulating species of the family Leguminosae, in which the endophyte is a bacterium, *Rhizobium*;
(2) non-legumes, including the genera *Alnus* (alder), *Myrica* (bog-myrtle), *Elaeagnus* and *Casuarina*, in which the endophyte is usually an actinomycete (*Frankia*);
(3) lichens, which are an association of a fungus and a blue-green alga, and the aquatic fern, *Azolla*, living in association with the blue-green alga, *Nostoc*.

In terms of their numbers and widespread distribution, the legumes are by far the most important symbiotic N_2 fixers. A pictorial summary of the processes of root infection, nodule initiation and N_2 fixation in a legume root is presented in Figure 10.4.

The site of fixation is the bacteroid surface to which electrons are transported by a reduced co-enzyme and donated to N≡N bound to nitrogenase. Synthesis of the nitrogenase is induced in the microorganism under the favourable conditions existing within the host tissue. As in the aerobic free-living bacteria, regulation of the O_2 partial pressure on a microscale is essential for N_2 fixation to occur: this is achieved through the pigment *leghaemoglobin* (giving a characteristic pinkish-red colour to nodular tissue that is fixing N_2), which, because of its very high affinity for O_2, provides a means of transferring O_2 to the respiring bacteroids while keeping the free O_2 concentration in the membrane-bound sacs very low.

Other factors affecting N_2 fixation operate at any

Table 10.4. Estimation of N_2 fixation by legumes as measured in kg ha^{-1} y^{-1}.

Temperate species		Tropical and subtropical species	
Clovers (*Trifolium* spp.)	55–600	Grazed grass–legume pastures:	
Lucerne (*Medicago sativa*)	55–400	*Stylosanthes humilis*	10–30
Soyabeans (*Glycine max*)	90–200	*Macroptilium atropurpureum*	44–129
Beans (*Vicia faba*)	200	Grain and forage legumes:	
Peas (*Pisum* spp.)	50–100	Beans (*Phaseolus vulgaris*)	64
		Pigeon pea (*Cajanus cajan*)	97–152
Median	c. 200	Median	c. 100

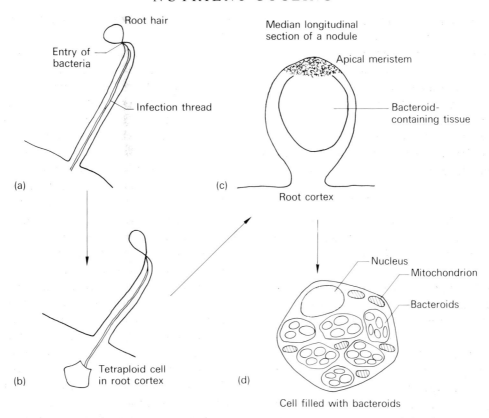

Figure 10.4. Sequence in nodule formation on a legume root. **(a)** Bacteria invade a root hair through an invagination of the cell wall—the infection thread. **(b)** Tetraploid cell penetrated—this cell and adjacent cells induced to divide and differentiate to form a nodule. **(c)** Bacteria released, divide once or twice and enlarge to form bacteroids which respire but cannot reproduce. **(d)** Groups of bacteroids enclosed with a membrane and N_2 fixation begins (after Nutman, 1965).

one of several points in the complex sequence of events: low pH and low Ca inhibit root hair infection and nodule initiation, at which stage also the competition between effective and non-effective bacterial strains is crucial. A non-effective strain produces many nodules which do not fix N_2, yet their presence inhibits nodulation by an effective strain. High mineral N, especially NO_3^-, suppresses nodulation, whereas inadequate photosynthate and moisture stress reduce a nodulated root's fixation capacity. Nutritionally, nodulated legumes need more P, Mo and Cu than unnodulated legumes, and have a unique requirement for Co. Further details are given in recent reviews, see, for example, Subba Rao (1984).

Because of environmental constraints, and inherent differences in host–bacterial strain performance, the quantity of N fixed is variable. Estimates for several temperate and tropical legumes are given in Table 10.4, which shows that under favourable conditions, the contribution of legumes to soil N can be substantial. The legume N is made available by the sloughing-off of nodules, through the excreta of grazing animals, or by decomposition of the plant when it is ploughed into the soil.

Losses of soil N

CROP REMOVAL

Much of the crop N is removed in the harvested material, except under grazing conditions and then approximately 85 per cent of the N is returned in animal excreta. The fate of cereal crop residues is important, because the straw from a 4 t ha^{-1} wheat crop contains 20–25 kg N. The straw is baled and used as animal bedding, eventually forming farmyard manure, or burnt, when most of the N is volatilized, or occasionally ploughed back into the soil.

LEACHING

Nitrogen in the NO_3^- form is very vulnerable to leaching. Soil nitrate levels fluctuate markedly due to the interaction of temperature, pH, moisture, aeration and species effects on nitrification. The following trends should be noted.

(a) *Temperature*. In temperate regions, low winter temperatures ($< 5°C$) retard nitrification, which then attains a peak in spring and early summer when NO_3^- concentrations of 40–60 ppm N may be found in surface soils under bare fallow. Higher values are recorded in warmer, drier soils because of capillary rise and minimal leaching.

(b) *Moisture*. In warm climates of little seasonal temperature variation, soil mineral N fluctuates with the cycle of wet and dry seasons, as has been observed in East Africa. Following partial sterilization of the soil during the long dry season, at the onset of the rains the surviving organisms multiply rapidly and produce a flush of mineralization (Figure 10.5). High NO_3^- levels are the result, but these gradually decline as the readily available organic N is consumed and nitrate is removed by plant uptake and leaching.

(c) *Natural grassland and forest communities*. The mineral N found in grassland and forest soils is predominantly NH_4^+, which suggests that nitrification is suppressed, possibly due to the secretion of inhibitory compounds by the roots of certain species. Normally the input of N in rain roughly balances the loss by leaching in natural plant communities; clear-felling a forest, however, accelerates the rate of nitrification and in the absence of plants to absorb the NO_3^- leads to large N losses in the runoff (section 10.4).

(d) *Soil physical properties*. Soil texture and structure influence NO_3^- leaching. Sandy soils, which generally have high hydraulic conductivities and low water-holding capacities, lose NO_3^- rapidly by leaching when rainfall exceeds evaporation. For example, in a soil of $\theta = 0.2$ at field capacity, 100 mm of rain may displace NO_3^- downwards to a depth of $100/0.2 = 500$ mm. Conversely, in a clay soil for which $\theta \simeq 0.5$ at *FC*, 100 mm of rain is expected to displace nitrate to only 200 mm by 'piston displacement' (Figure 10.6a). But if the soil is well structured, some of the nitrate will move to

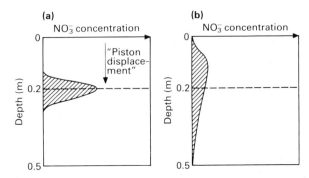

Figure 10.6. Effects of soil structure on depth of leaching of NO_3^- originally concentrated at the surface, following 100 mm of rain. **(a)** Structureless clay soil—the symmetrical 'bulge' in NO_3^- concentration reflects the effect of hydrodynamic dispersion (section 6.4). **(b)** Well-structured clay soil showing redistribution of nitrate between 0 and 50 cm depth.

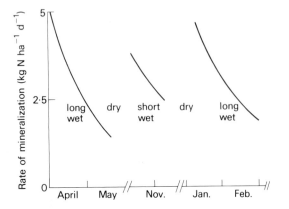

Figure 10.5. Seasonal changes in mineral N in an East African soil (after Birch, 1958).

depths much greater than 200 mm due to fast flow down large continuous interped pores, while the remainder moves more slowly by displacement through the fine pores of the soil matrix (section 6.4). The net result is that nitrate originally concentrated near the surface becomes distributed through a considerable depth of soil (Figure 10.6b).

GASEOUS LOSSES

Animal excreta containing urea or uric acid (in the case of poultry) are susceptible to volatile losses of N. For example, heavily fertilized grassland (receiving up to 400 kg N ha^{-1} y^{-1}) that is extensively grazed can lose much more N (up to 50 per cent of that applied) by volatilization of NH_3 (and some denitrification) than comparable grassland which is cut for hay only. Catalysed by the enzyme urease, which is widespread in soil, urea is hydrolysed to ammonium carbonate which then hydrolyses to ammonium hydroxide, that is:

$$(NH_2)_2CO + 2H_2O \rightleftarrows (NH_4)_2CO_3 \quad (10.6a)$$

$$(NH_4)_2CO_3 + 2H_2O \rightleftarrows 2NH_4OH + H_2CO_3 \quad (10.6b)$$

The alkaline NH_4OH solution is unstable, particularly at high temperatures, and NH_3 gas volatilizes. Loss of NH_3 also occurs from ammonium and urea fertilizers under favourable conditions (section 12.2).

There are two possible mechanisms of nitrogen loss by *denitrification*: chemo-denitrification and biological denitrification, of which the latter is by far the more important.

(a) *Chemo-denitrification.* Losses of N have been observed, even from well-aerated soils, when NO_2^- has accumulated following heavy applications of ammonium or urea fertilizers. The losses are greatest in acid and neutral soils and appear to involve the reaction of nitrous acid (HNO_2) with the soil organic matter to release gaseous N compounds. The most likely active organic components are phenolic compounds, which are known to react with HNO_2 to form nitrosophenols; these are decomposed to N_2O and N_2 under mildly acid conditions in the presence of excess HNO_2.

(b) *Biological denitrification.* As discussed in section 8.4, biological denitrification occurs in anaerobic soil culminating in the release of N_2O and N_2. Quantification of the N loss is very difficult, because although N_2O concentrations in the soil air can be measured accurately by gas chromatography, N appearing as N_2 gas can only be distinguished from background N_2 gas if the original N substrate is labelled with ^{15}N. Short-term rates of denitrification can be measured in the field by injecting sufficient C_2H_2 into the soil to block the reduction of N_2O to N_2, and then determining the flux of N_2O emitted across a limited area of surface. Rates measured in this way may reach 1–2 kg N ha^{-1} day^{-1}. The rate of denitrification is also very variable, and is dependent on organic substrate availability, temperature, pH and the distribution of anoxic zones in the soil. It is generally insignificant at pH < 6 and temperatures < 10°C. Deep leaching of NO_3^- to a poorly aerated subsoil during winter does not necessarily predispose to denitrification if
(i) the temperature is low, and
(ii) organic substrate is insufficient for rapid microbial growth.

10.3 PHOSPHORUS AND SULPHUR

Plant requirements
Crop analyses indicate that comparable amounts of P and S are removed in harvested products each year (Table 10.5); nevertheless, the two elements differ markedly in the balance of inputs and outputs in the soil–plant cycle, and to a lesser extent in the transformations they undergo in the soil.

Table 10.5. Amounts of P and S in crop products (kg ha^{-1}).

	P	S
Barley, 4 t ha^{-1} (grain and straw)	12	15
Potatoes, 50 t ha^{-1}	25	20
Grass, 10 t ha^{-1} (dry matter)	30	15
Maize, 13 t ha^{-1} (grain and stover)	50	50
Kale, 50 t ha^{-1} (fresh weight)	25	100

(After Cooke, 1975, and others.)

The soil reserve

RAINFALL INPUT

P concentrations in rain are very low, the amount deposited ranging from 0.2 to 0.05 kg ha^{-1} y^{-1}. Amounts of S deposited from the atmosphere are much higher, ranging from 7 to 70 kg ha^{-1} y^{-1}. The major source is SO_2 from the burning of fossil fuels, plus small contributions in the form of alkyl sulphides and H_2S released from marine sediments. Several steps are possible in the deposition of atmospheric S on soil and vegetation:

(i) direct absorption of SO_2 through the stomata of leaves;

(ii) oxidation of SO_2 to SO_3:

$$SO_2 + \tfrac{1}{2}O_2 \rightleftarrows SO_3 \qquad (10.7a)$$

Reaction 10.7a is normally slow, but it is accelerated in the presence of O_3 or H_2O_2. Sulphuric acid also forms according to:

$$SO_3 + H_2O \rightleftarrows H_2SO_4 \qquad (10.7b)$$

and is brought down in the rain (hence the term *acid rain*) or neutralized by reaction with NH_3 gas (section 10.2); that is:

$$2NH_3 + H_2SO_4 \rightleftarrows (NH_4)_2SO_4 \qquad (10.7c)$$

Some ammonium sulphate is deposited directly on vegetation, which together with SO_2 absorption comprises *dry deposition* as distinct from S in rain. Dry deposition can provide one-quarter to one-third of a plant's S needs.

S emissions account for about two-thirds of the acidity in rain over Britain. The acidifying *effect* of this rain is not necessarily eliminated by reaction 10.7c, because the H^+ ions are released again when NH_4^+ is oxidized in the soil (equation 10.2). Even the NH_4^+ ions taken up by plant roots can have an acidifying effect on the rhizosphere because H^+ ions are released to balance any net surplus of cations over anions absorbed when N is supplied primarily as NH_4^+.

FORMS OF SULPHUR

Soil S is derived originally from sulphidic minerals in rocks that are oxidized to sulphate on weathering. Clay and shales of marine origin are frequently high in sulphides, and recently formed soils on marine muds may have SO_4-S contents up to 50,000 kg ha^{-1}. Normally, soil S is predominantly in an organic form and ranges between 200 and 2,000 kg ha^{-1}.

Organic S is released by microbial decomposition, depending on the C:S ratio of the residues, which lies between 50 and 150 for well-humified organic matter. Sulphur appears to be stabilized in humus in much the same way as N, and indeed the N:S ratio (between 6 and 10) is much less variable than the C:S ratio. The sulphate released is only weakly adsorbed at positively charged sites, in competition with phosphate and organic anions, and so it is prone to leaching.

FORMS OF PHOSPHORUS

Soil P contents range from 500 to 2,500 kg ha^{-1}, of which 15–70 per cent may exist in the strongly adsorbed or insoluble inorganic forms described in section 7.3; the remainder occurs as organic P which is the major source of P for the soil microorganisms and mesofauna. The phosphate of nucleic acids and nucleotides is rapidly mineralized, but when the C:P ratio rises above *c.* 100, the P is immobilized by the microorganisms, especially bacteria, which have a relatively high P requirement (1.5–2.5 per cent P by dry weight compared to 0.05–0.5 for plants). Bacterial phosphate residues comprise mainly the insoluble Ca, Fe and Al salts of inositol hexaphosphate, which are called *phytates*. Thus, in closed ecosystems with insignificant P inputs, the soil organisms are highly competitive with higher plants for P, as illustrated by the P cycle within a natural grassland (Figure 10.7).

Not only do microorganisms mineralize organic P, but some groups secrete organic acids, such as α-ketogluconic acid, which attack insoluble Ca phosphates and release the phosphate. Some species of *Aspergillus*, *Arthrobacter*, *Pseudomonas* and *Achromobacter*, abundant in the rhizosphere of plants, have this ability. However, most of the P is absorbed by microorganisms themselves in the intensely competitive environment of the rhizosphere.

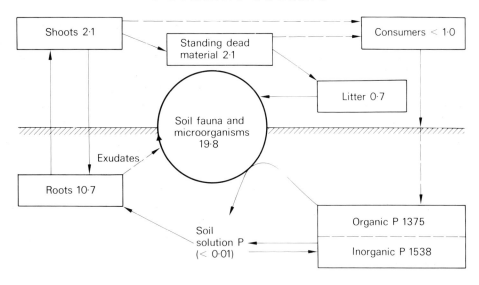

(Figures in kg ha^{-1} 30 cm depth)

Figure 10.7. P cycles in a native prairie grassland in Western Canada (after Halm and others, 1972).

PHOSPHATE AVAILABILITY

Orthophosphate ($H_2PO_4^-$ and HPO_4^{2-}) that is released by mineralization is rapidly adsorbed by the soil particles where its availability steadily declines with time, a process called *phosphate 'fixation'*. Phosphate on surfaces that can be readily desorbed, plus phosphate in solution, is called *labile P* to distinguish it from the P held in insoluble compounds or organic matter, which is *non-labile*. Transformations between the different forms of P in soil are represented in Figure 10.8.

Soil pH, clay and sesquioxide content and exchangeable Al^{3+} all influence P availability (see section 7.3).

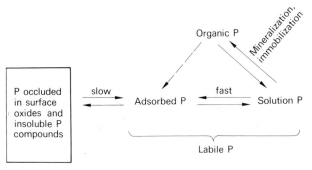

Figure 10.8. Phosphate transformations in soil.

Where the effects of the free Fe and Al oxides are predominant, or in fertilized acid soils where compounds of composition $FePO_4 . nH_2O$ and $AlPO_4 . nH_2O$ exist and dissolve incongruently to release more P than Fe or Al, P in the soil solution increases with pH rise. By contrast, in P-fertilized alkaline and calcareous soils, insoluble Ca phosphates such as hydroxyapatite and octacalcium phosphate comprise the bulk of the non-labile P, and the solubility of these compounds increases as the pH falls. Consequently, the availability of P in such soils is greatest between pH 6 and 7 (Figure 10.9).

However, acid soils of low P status that retain a surplus of clay minerals relative to sesquioxides show a contrary trend in P availability with pH change. Between pH 4.5 and 7, exchangeable Al^{3+} hydrolyses to form hydroxyaluminium ions which act as sites for P adsorption because of their residual positive charge. These ions also polymerize readily so that a hydroxy-aluminium complex with $H_2PO_4^-$ ions *occluded* in its structure forms on the surface of the clay mineral, rendering a *minimum* in P concentration over the pH range 5–6.5. Above pH 7, Al begins to dissolve as the aluminate anion and P is released.

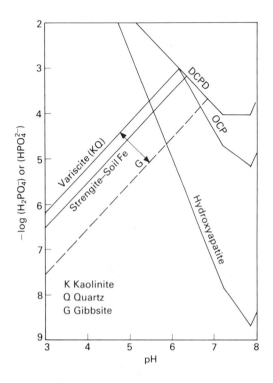

Figure 10.9. pH–P solubility relations at 25°C for Ca phosphates (DCPD, OCP and hydroxyapatite) and Fe and Al phosphates (strengite and variscite, respectively). The position of these lines depends on the presence of other compounds (Fe(OH)$_3$, kaolinite, gibbsite and calcite), which also affect the activities of Ca^{2+}, Fe^{3+} and Al^{3+} ions in the soil solution (after Lindsay, 1979).

MYCORRHIZAS

Because of low P concentrations in soil solutions and competition from microorganisms in the rhizosphere, many plants have evolved mechanisms to enhance the absorption of P when it is in short supply; one of the most important of these is the *mycorrhiza*, a symbiotic association of fungus and root. There are two types of mycorrhiza.

(a) *The ectotrophic form.* This is primarily associated with trees: members of the Pinaceae—pines; Fagaceae—oak, beech; Betulaceae—birch, alder; also *Eucalyptus* and poplar. The fungal mycelium grows externally and forms a sheath some 0.05 mm thick around the root cylinder. The external hyphae afford a

small increase in the soil volume exploited by the root, but the major benefit derives from the fungus's ability to store P in its tissues to be released to the host in times of P deficiency, and the greater longevity of infected roots which continue to absorb P long after non-mycorrhizal roots have ceased active uptake. In return, the fungus obtains carbon substrates and other growth requirements from the host.

(b) *The endotrophic form.* This form, in which the endophyte is a Phycomycete of the genus *Endogone*, is commonly called a *vesicular–arbuscular* (V–A) mycorrhiza. The fungal spores are present in most soils, but only germinate when a suitable host root is near: if the young mycelium does not penetrate a host root, it dies. However, once established in the root (and the host–fungal relationship is not very strain-specific), the hyphae ramify widely outside the host tissue.

Two structures are characteristic of the internal mycelium: first, the *vesicles*, which are temporary storage bodies in the intercellular spaces; second, the *arbuscules*, which are branched structures growing intracellularly that probably release solutes into the cell (Figure 10.10). V–A mycorrhizas are found in nearly all the angiosperms and in many conifers, ferns and lower plants. The incidence of infection is especially high in impoverished soils, where the mycorrhizal state confers a distinct advantage in P uptake, because the external hyphae add up to 50 per cent to the length of absorbing root and provide a pathway for the rapid transport of phosphate across the depletion zone around the root. The benefit is most notable in plants such as onion and citrus which have no or very few root hairs.

V–A mycorrhizas also enhance plant uptake of Cu and possibly Zn from the soil. They are of considerable importance in the cycling of P in natural ecosystems, and for the establishment of pioneer species in harsh habitats such as on sand dunes and mine spoil. The incidence of infection declines, however, with the addition of P fertilizer, therefore little benefit is gained by cultivated crops in general because fertilizer has a more rapid effect on growth than does the mycorrhizal

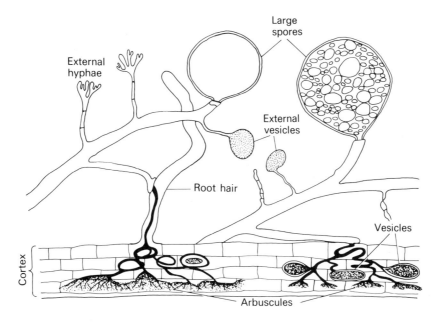

Figure 10.10. Diagram of a vesicular–arbuscular mycorrhiza (after Nicholson, 1967).

infection. This suppression of V–A mycorrhizas by P fertilizer may explain the appearance of Cu and Zn deficiencies in some P-fertilized crops, especially on high pH soils (section 10.5). Members of the Cruciferae (brassicas) and Chenopodiaceae (buckwheat) are non-mycorrhizal and have evolved with other mechanisms for enhancing P uptake from P-deficient soils.

LEACHING LOSSES

With the exception of very sandy soils, and other situations described in section 12.3, phosphate losses by leaching are very small, usually < 1 kg ha^{-1} y^{-1}.

Sulphate forms sparingly soluble gypsum in arid region soils, but is leached from soils of humid regions, except those high in sesquioxides and of low pH, where it is retained at depth (see Figure 10.2b). Data obtained from three sites on neutral and calcareous soils in southern England illustrate the contrast between leaching losses of sulphate and phosphate (Table 10.6).

Table 10.6. P and S losses in drainage from neutral and calcareous soils in southern England (kg ha^{-1} y^{-1}).

	PO$_4$-P	SO$_4$-S
Woburn—neutral, sandy soil over Oxford Clay	0.05	100
Saxmundham—sandy, calcareous boulder clay	0.12	86
Wytham—brown calcareous soil over Oxford Clay	0.13	78

(After Williams 1975 and others.)

10.4 POTASSIUM, CALCIUM AND MAGNESIUM

Plant requirements

Annually agricultural crops remove between 5 and 25 kg Mg ha^{-1}, and exceptionally up to 50–60 kg ha^{-1} for high yields of maize or oil palm; Ca requirements range from 10 to 100 kg ha^{-1} and the K requirement is even higher at 100–300 kg ha^{-1}.

There are few estimates of the rate of cation uptake by perennial vegetation, especially forests, because it is difficult to measure the amount and composition of the annual growth increment. By assuming that in a mature forest the rate of nutrient return balances the rate of uptake from the soil, Nye and Greenland (1960) obtained the estimates of cation uptake shown in Table 10.7a. Much lower figures for K and Ca uptake were obtained by Cole and others (1967) for a 36-year-old Douglas fir plantation in Washington, USA (Table 10.7b).

Table 10.7(a). Annual return (= uptake) of K, Ca and Mg by a 40-year-old forest in West Africa (kg ha^{-1} y^{-1}).

	K	Ca	Mg
Litter fall	68	206	45
Timber fall	5	82	8
Leaf wash	220	29	18
Total	293	317	71

(After Nye and Greenland, 1960.)

Table 10.7(b). Annual return (= uptake) of K and Ca by 36-year-old Douglas fir (kg ha^{-1} y^{-1}).

	K	Ca
Litter fall	3	11
Stemflow	1	1
Leaf wash	11	4
Total	15	16

(After Cole and others, 1967.)

Soil reserves

The pool of exchangeable cations Ca^{2+}, Mg^{2+} and K^+ provides the immediate source of these elements for plants. Insoluble reserves may occur as calcium and magnesium carbonates, potash feldspars and inter-layer K^+ in micaceous clays. Total reserves are therefore very variable, reflecting the conditions of soil formation, but exchangeable Ca^{2+} values usually lie between 1,000 and 5,000 kg ha^{-1}, Mg^{2+} between 500 and 2,000 kg ha^{-1} and $K^+ \sim$ 1,000 kg ha^{-1}.

RAINFALL INPUT

Amounts of K and Ca in all forms of precipitation are comparable and lie between 1 and 20 kg ha^{-1} y^{-1}; Mg can be much higher, especially on sea coasts (for ocean water is rich in Mg), amounting to as much as 150 kg ha^{-1} y^{-1}). Thus, Mg inputs roughly balance combined losses through plant removal and leaching, but not so for K and Ca. A striking feature of K input to the soil is the magnitude of its leaching from leaves, amply demonstrated by the figures for the tropical species and Douglas fir in Table 10.7. This process is one of the main reasons for the rapid cycling of K in natural ecosystems in which plants and their residues can accumulate from 40 to 60 per cent of the total labile K within the root zone.

MINERAL WEATHERING

Apart from some recent glacial materials, the reserves of weatherable minerals are usually small in the top 30 cm of soil where 80–90 per cent of the plant roots are located. Nevertheless, the slow release of nutrients from the 'geologic reserve' at greater depth is often crucial for the stability of natural plant communities wherein elemental losses by leaching and erosion occasionally exceed atmospheric inputs. For example, data from West Africa suggested that 58, 64 and 14 kg ha^{-1} of K, Ca and Mg, respectively, were pumped up annually from the subsoil by deep tree roots.

LEACHING LOSSES

Cation leaching is accelerated in cultivated soils, where nitrification occurs, for two reasons (a) the H^+ ions produced can displace exchangeable Ca^{2+}, Mg^{2+} and K^+ from exchange surfaces; (b) the increase in NO_3^- concentration must be balanced by an increased cation concentration in the water percolating through the soil. For agricultural soils in Britain, Ca leaching normally ranges between 160 and 320 kg ha^{-1} y^{-1}.

Conversely, suppression of nitrification in soils under natural grassland or forest tends to reduce cation losses by leaching. For example, in a deciduous forest catchment in New Hampshire, the net leaching loss of K, Ca and Mg from an acid podzolic soil was 2, 9 and 3 kg ha^{-1} y^{-1}, respectively, and there was a small gain of 2 kg N ha^{-1} y^{-1} (Likens et al., 1970). However, in two years after the forest was felled, but not burnt nor the timber removed, about 120 kg N ha^{-1} was lost annually (mainly as NO_3^-) and K, Ca, Mg losses rose to 36, 90 and 18 kg ha^{-1} y^{-1}, respectively. Most of this increase was due to the mineralization of litter and soil organic matter exposed after the forest clearance, and accelerated nitrification which depressed the pH of the drainage water from 5.1 to 4.3.

10.5 TRACE ELEMENTS

Soil reserves

Those elements whose total concentration in the soil is invariably < 1,000 ppm are called *trace elements*: they fall into two categories as follows:
(a) the essential *micronutrients* Cu, Zn, Mn, B and Mo which are beneficial to plants at normal concentrations in the plant (ranging from 0.1 ppm for Mo to 100 ppm for Mn), but may become toxic at higher concentrations;
(b) other elements, notably Li, Be, Cd, Cr, Pb and Ni which are of no benefit and are toxic at levels greater than a few ppm.

Iron is the one micronutrient that is not strictly a trace element.

INFLUENCE OF PARENT MATERIAL
Trace elements are largely bound in mineral lattices, to be released only by weathering, so that the type of parent material determines to a large extent the abundance of the individual elements in the soil. Iron, Cu, Mn, Zn, Co, Mo, Ni and Pb occur in ferro-magnesian minerals, common in ultrabasic and basic igneous rocks; Fe, Mn, and Mo also occur as insoluble

oxides (sometimes with Co coprecipitated in the MnO_2), and Zn, Fe and Cu as equally insoluble sulphides in sedimentary rocks. Boron occurs as the resistant mineral *tourmaline* in acid igneous rocks and to some extent may substitute for Al in the tetrahedral sheet of 2:1 clay minerals. Trace element contents of soils formed on parent rocks of different basicity are presented in Table 10.8.

Table 10.8. Trace element contents of soils on parent materials of different rock type (kg ha^{-1}).

	Serpentine	Andesite	Granite	Sandstone
	←————————Increasing basicity————————→			
Co	160	16	< 4	< 4
Ni	1600	20	20	30
Cr	6000	120	10	60
Mo	2	< 2	< 2	< 2
Cu	40	20	< 20	< 20
Mn	6000	1600	1400	400

(After Mitchell, 1970.)

ATMOSPHERIC INPUTS
Trace elements enter the air as gases, aerosols and particulates and return to the soil–plant system mainly by dry deposition. Metals that are used extensively in industry, Cd, Zn, Cu and Pb, show the greatest enrichment in the air of industrialized regions, relative to metals like Fe or Ti that are naturally abundant. The amounts deposited are therefore likely to be greater in industrial regions of the Northern Hemisphere than in remote rural areas of the continents. This is shown in Table 10.9 where estimated annual rates of deposition of several trace elements are given. The minimum values are representative of rural areas, whereas the median values are biased towards urban areas because of the greater density of observations there.

Table 10.9. Estimated rates of trace element deposition from the atmosphere over Europe and North America (g ha^{-1} y^{-1}).

Element	Europe		North America	
	Minimum	Median	Minimum	Median
Mo	< 0.3	—	0.2	—
Co	0.3	—	0.2	4.7
Cd	0.8	—	< 0.2	6.5
Ni	6.3	39	< 16	142
Mn	14	68	0.1	236
Cu	13	536	7.9	441
Zn	20	19	—	788
Pb	87	189	71	4257

(After Sposito and Page, 1984.)

Availability to plants

FREE IONS AND COMPLEX FORMS IN SOLUTION

The processes involved in the turnover of a trace element in the soil–plant system are summarized in Figure 10.11.

Molybdenum, As and Se occur as anions in the soil solution and B as the weak acid $B(OH)_3$ which dissociates to the borate anion $B(OH)_4^-$ at pH > 8. Fe, Cu, Mn, Zn, Co, Cd, Ni and Pb occur as hydrated cations forming complexes of varying stability with inorganic and organic ligands. The principles of complex formation are as follows:

$$aM^{n+} + bL^{l-} \rightleftarrows \left\{ M_a L_b \right\}^q \qquad (10.8a)$$

where charge balance requires that:

$$an - bl = q \qquad (10.8b)$$

and q, the valency of the complex, may be +, 0 or −. The stability of the complex is given by the conditional *stability constant* ^cK_s defined by the expression:

$$^cK_s = \frac{[M_a L_b^q]}{[M^{n+}]^a [L^{l-}]^b} \qquad (10.9)$$

The constant is called *conditional* because concentrations (mol L^{-1}) and not activities are used in equation 10.9. For the divalent cations, the most stable

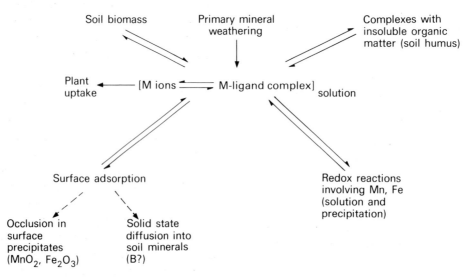

Figure 10.11. Trace element (M) equilibria in the soil (after Hodgson, 1963).

complexes are formed with organic ligands (generally $Cu \gg Pb > Ni > Mn \simeq Zn$), and for the trivalent cations, organic complexes formed by Fe are more stable than Co. The complexes exist in solution as anions or cations of low charge:mass ratio, but they can also be adsorbed (see below).

The cations Fe, Cu, Mn and Zn also hydrolyse at pH values between 2 and 10 and precipitate as insoluble hydroxides. This is a special case of equation 10.8a, for example:

$$Fe^{3+}(soln) + 3OH^-(soln) \rightleftarrows Fe(OH)_3^0(solid) \quad (10.10)$$

However, equation 10.10 is usually considered as a dissolution–precipitation reaction for which the thermodynamic solubility constant K_{sp} is defined by:

$$K_{sp} = \frac{(Fe^{3+})(OH^-)^3}{(Fe(OH)_3)} \quad (10.11)$$

Taking the activity of pure solid $Fe(OH)_3$ as 1, equation 10.11 becomes:

$$K_{sp} = (Fe^{3+})(OH^-)^3 \quad (10.12)$$

Whether the insoluble hydroxide or a soluble organic complex (FeL_b^q) is thermodynamically more stable at a given pH, depends on which chemical form maintains the lower activity of free Fe^{3+} ions in solution. This fact underlies the choice of organic compounds (chelating agents) to form soluble complexes with micro-nutrients, such as Fe, that become unavailable to plants at high pH. Deficiency of Fe is manifest in plants as 'lime-induced' chlorosis*.

Ethylene tetraacetic acid (EDTA) is a synthetic chelating agent which forms stable complexes with most trace elements. Figure 10.12 shows the effect of pH on the balance between the stability of $Fe(OH)_3$ and $FeEDTA^-$. The change from the chelate to the hydroxide as the more stable phase occurs at pH 9 normally, but in calcareous soils it occurs at around pH

---

*Interveinal yellowing of leaves due to the lack of chlorophyll.

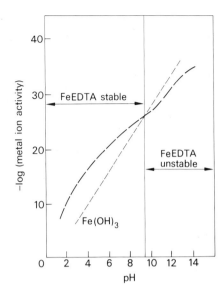

Figure 10.12. pH stability range for Fe EDTA (after Martell, 1957).

8 because of the increasing stability of the $CaEDTA^{2-}$ complex at high pH values. Thus, chelating agents, such as EDDHA (ethylene diamine di(o-hydroxy-phenylacetic acid)) which have a higher affinity for Fe than for Ca, tend to be used on calcareous soils. However, there is evidence that plants take up more Fe from complexes of lower stability, suggesting that the complex may have to be broken down before the Fe can be absorbed into the cell.

TRACE METALS IN THE SOLID PHASE
The surface adsorption component shown in Figure 10.11 accounts for about 10 per cent of the total of any trace element in soil. The remainder is held in insoluble precipitates, primary and secondary minerals or is complexed with soil organic matter. Nevertheless, the adsorbed pool responds rapidly to changes in the solution concentration and for many elements, there-fore, it is more important in determining bioavail-ability than the other less-reactive solid-phase forms. Because complex formation dominates the surface chemistry of these elements, it is better to measure their bioavailability by extracting the soil with a

reagent such as 0.05 M EDTA rather than the conventional extracting solution used for exchangeable cations (section 7.2). Acetic acid at pH 2.5 is also used (see Table 10.10).

Table 10.10. The effect of drainage on trace element solubility in soil.

	Extractable* element content (ppm soil)			
	Co	Ni	Mo	Cu
Freely drained	1.3	1.3	0.06	2.6
Poorly drained	1.9	3.4	0.19	6.6

*Extracted in CH_3COOH or EDTA.

Anions such as MoO_4^{2-} are most strongly adsorbed at low pH on sesquioxide surfaces by a ligand-exchange mechanism analogous to that described for phosphate (section 7.3). B is also adsorbed on oxide surfaces as $B(OH)_4^-$, but as this species only appears at pH > 8, B solubility decreases with increase in pH. Mo is therefore the only micronutrient to increase in availability as the soil pH rises.

The interaction of trace metal cations with negatively charged surfaces usually involves the formation of outer- and inner-sphere complexes (sections 7.1 and 7.5). The latter are more stable than the former because no water molecules separate the ion from the surface functional groups and an element of ionic or covalent bonding is involved. Hydrated Cu^{2+} ions, for example, form outer-sphere complexes with the siloxane ditrigonal cavities of 2:1 clay surfaces from which they are displaced by Ca^{2+}. Cu^{2+} ions forming inner-sphere complexes with COOH groups on organic matter, on the other hand, are not exchangeable to Ca^{2+} ions. Although stable complexes are formed between trace metal ions and organic ligands both in solution and on surfaces, because the organic C in solution (at 50–150 ppm) is very much less than the C in the solid phase, complex formation tends to *immobilize* metals rather than to solubilize them.

Sewage sludge, particularly that derived from industrial wastes, often contains relatively high concentrations of trace metals, such as Cu, Zn, and Ni in fairly labile organic combination which may pose problems when large applications are made to agricultural soils (section 11.1).

THE EFFECT OF POOR DRAINAGE

The manganese and iron of insoluble compounds are mobilized in waterlogged soils as Mn^{2+} and Fe^{2+} ions, as discussed in section 8.4. In addition, the solubility of Co, Ni, Mo and Cu (elements not reduced at the redox potentials attained in soils) may be increased under waterlogged conditions due to the accelerated weathering of ferromagnesian minerals (Table 10.10). The dissolution of manganese oxides can also release coprecipitated Co into the soil solution.

CROP REMOVAL AND LEACHING LOSSES

Estimates of the removal of micronutrients in harvested products vary widely but the range (in g ha^{-1} y^{-1}) is approximately as follows:

Mn	34–500
Mo	< 0.2–500
B	27–160
Zn	31–80
Cu	6–80

A comparison with the minimum figures for atmospheric inputs in Table 10.9 suggests that soils in remote rural areas may suffer a net depletion of these elements by crops. On the other hand, these elements occur as trace contaminants in fertilizers and liming materials (e.g. basic slag) and unknown amounts are released through mineral weathering each year.

Because of precipitation in insoluble compounds and strong retention by mineral and organic surfaces, micronutrient losses by leaching are very small, except for B, and Fe and Mn in some gleyed soils. Nevertheless, without significant atmospheric inputs, cumulative losses from very old soils eventually lead to widespread micronutrient deficiencies, as occur in the coastal areas of eastern and southern Australia.

10.6 SUMMARY

Gains and losses of nutrients in natural ecosystems, comprising the soil–plants and animals–atmosphere, are roughly in balance so that continued growth depends on the cycling of nutrients between the *biomass*, *organic* and *inorganic* stores. Removals from agricultural systems, especially of crop products, generally exceed natural inputs unless these are augmented by fertilizers and organic manures.

Nutrients in the biomass store pass to the organic store by excretion from or on the death of living organisms. The greater part of the organic and inorganic stores is in the soil. Additions to the inorganic store occur through mineralization of organic residues, weathering, rainfall and *dry deposition*, and from fertilizers in agricultural systems. That part of the inorganic store on which plants feed is called the *available* or *labile* pool, consisting of ions in solution and adsorbed by clays, sesquioxides and organic matter, which is distinct from nutrients held in insoluble precipitates or strongly adsorbed complexes, which are considered *non-labile*.

The ability of a minority of plants and micro-organisms to reduce N_2 gas to NH_3 within the cell (*nitrogen fixation*) provides a unique input of N to the biomass and thence organic stores. But substantial losses of N can occur following *nitrification* due to the leaching of NO_3^- in the soil solution and NO_3^- reduction (*denitrification*) to N_2O and N_2 under anoxic conditions. Volatilization of NH_3 gas also occurs following the hydrolysis and decomposition of urea in animal excreta.

Phosphate is rapidly immobilized by soil micro-organisms, by adsorption and by precipitation of insoluble Ca, Fe or Al phosphates. Many plants have evolved fungus–root associations or *mycorrhizas* which enhance P absorption from deficient soils. In contrast to sulphate, which is steadily leached (except from acid sesquioxidic soils), P losses by leaching are negligible because the P concentration in solution is very low.

Calcium, Mg and K are held mainly as exchangeable cations, the supply of which buffers the soil solution against depletion. Cation losses are accelerated when anions such as NO_3^-, SO_4^{2-} and HCO_3^- are plentiful in the percolating water, and also under acid conditions when H^+ and Al^{3-} ions occupy exchange sites. The micronutrient cations, Fe, Mn, Cu, Zn and Co, form complexes with negatively charged surfaces, those formed with organic compounds being the more stable, especially in the case of Cu and Fe. The ability to form stable complexes with soluble organic ligands (*chelates*) can be used to prevent these elements precipitating as insoluble hydroxides at alkaline soil pHs. Molybdenum and boron occur as anions in solution.

REFERENCES

ANON. (1983) *The Nitrogen Cycle of the United Kingdom*. Royal Society Study Group Report.

BIRCH H.F. (1958) The effect of soil drying on humus decomposition and nitrogen availability. *Plant and Soil* **10**, 9–31.

COLE D.W., GESSEL S.P. & DICE S.F. (1967) Distribution and cycling of nitrogen, phosphorus, potassium and calcium in a second-growth Douglas fir ecosystem, in *Primary Productivity and Mineral Cycling in Natural Ecosystems* (Ed. H.E. Young). Ecological Society of America, New York.

COOKE G.W. (1975) *Fertilizing for Maximum Yield*, 2nd Ed. Crosby Lockwood, London.

HALM B.J., STEWART J.W.B. & HALSTEAD R.L. (1972) The phosphorus cycle in a native grassland ecosystem, in *Isotopes and Radiation in Soil–Plant Relationships including Forestry*. International Atomic Energy Agency, Vienna.

HODGSON J.F. (1963) Chemistry of the micronutrient elements in soils. *Advances in Agronomy* **15**, 119–159.

LIKENS G.E., BORNMANN F.H., JOHNSON N.M. & PIERCE R.S. (1967) The calcium, magnesium, potassium and sodium budgets for a small forested ecosystem. *Ecology* **48**, 772–785.

LINDSAY W.L. (1979) *Chemical Equilibria in Soils*. Wiley, New York.

MARTELL A.E. (1957) The chemistry of metal chelates in plant nutrition. *Soil Science* **84**, 13–41.

MITCHELL R.L. (1970) Trace elements in soils and factors that affect their availability. *Geological Society of America, Special paper* **140**, 9–16.

NICHOLSON T.H. (1967) Vesicular–arbuscular mycorrhiza—a universal plant symbiosis. *Science Progress, Oxford* **55**, 561–581.

NUTMAN P.S. (1965) Symbiotic nitrogen fixation, in *Soil Nitrogen* (Eds. W.V. Bartholomew and F.E. Clark). Agronomy No. 10, American Society of Agronomy, Madison.

NYE P.H. & GREENLAND D.J. (1960) *The soil under shifting cultivation*. Commonwealth Bureau of Soils, Technical Communication No. 51, Harpenden, England.

SPOSITO G. & PAGE A.L. (1984) Cycling of metal ions in the soil environment, in *Metal Ions in Biological Systems*, Volume 18 (Ed. H. Siegel). Marcel Dekker, New York.

SUBBA RAO N.S. (1984) *Current developments in biological nitrogen fixation*. Edward Arnold, London.

WILLIAMS R.J.B. (1975) *The chemical composition of water from land drainage at Saxmundham and Woburn (1970–1975)*. Rothamsted Experimental Station Report for 1975, Part 2, pp 37–62.

CLARK F.E. & ROSSWALL T. (1981) (Eds.) *Terrestrial Nitrogen Cycles. Processes, Ecosystem Strategies and Management Impacts*. Ecological Bulletin No. 33, Stockholm.

HARLEY J.L. & SMITH S.E. (1983) *Mycorrhizal Symbiosis*. Academic Press, London.

HAYMAN D.S. (1975) Phosphorus cycling by soil micro-organisms and plant roots, in *Soil Microbiology* (Ed. N. Walker). Butterworths, London.

NICHOLAS D.J.D. & EGAN A.R. (1975) (Eds.) *Trace Elements in Soil–Plant–Animal Systems*. Academic Press, New York.

TINSLEY J. & DERBYSHIRE J.F. (1984) (Eds.) *Biological Processes and Soil Fertility*. Martinus Nijhoff/W. Junk, The Hague.

FURTHER READING

ALLAWAY W.H. (1968) Agronomic controls over the environmental cycling of trace elements. *Advances in Agronomy* **20**, 235–274.

Chapter 11
Maintenance of Soil Productivity

11.1 TRADITIONAL METHODS

Bush fallowing

Shifting cultivation is widespread in the tropics, being practised on approximately 800–1,400 million ha, or a little more than one-half of the potential arable land of these regions. The cycle begins with the clearing and burning of the natural vegetation, which in the case of the rainforest releases a great store of available nutrients for the growth of crops such as maize, cowpeas, cassava and groundnuts. However, loss of nutrients from the ash by leaching can be high and the encroachment of weeds rapid, so that after two or three crops the land is abandoned to the regenerating forest and the farmer moves to a new site, where the cycle is repeated.

Regrowth of the native vegetation, especially the forest, is vital to the stability of such an agricultural system. During this period, which has been called the *bush fallow*, the soil's fertility is restored by the deep-rooting trees or perennial grasses, drawing up nutrients from deep in the subsoil and returning large quantities of litter to the soil surface. There is also an input of N fixed symbiotically in the soil of humid forest regions. Experience in the savanna regions suggests that a cycle of 2–4 years' cropping and 6–12 years' fallow can maintain fertility in the long term, but in the wet forest zones a 1–2-year cropping period and 10–20 years' fallow are preferred.

Rotational cropping

In the more closely settled temperate regions, competition for farming land forced the adoption of intensive cropping systems much sooner than in the tropics. One of the earliest was the simple 3-year rotation of autumn cereal–spring cereal–fallow, practised in England since Celtic times. This wholly arable system was largely displaced in the 18th century by rotations, such as the Norfolk four-course, in which grass–clover pastures and root crops were alternated with cereals, thus permitting the close integration of livestock and arable farming. Added advantages were the nitrogen fixed by the legume and the abundant residues left in the soil by the grass.

However, depressed cereal prices in the 1880s and a growing belief in the value of grass as a 'soil conditioner' led to the temporary pasture or *ley* being extended for 3 or more years: gradually the arable–ley rotation evolved into the practice of *mixed farming*. The essence of this system is the rearing of livestock on farm-produced grass and cereals, in the course of which much of the nutrient gathered from the soil is returned in organic manures.

Organic manures

Farmyard manure (FYM), deep litter (from stall-fed animals), poultry (broiler) manure, sludge and compost are examples of organic manures (Table 11.1). Their beneficial effects lie as much in the improvement of structure accruing from large additions of organic matter, as in the contribution made to the soil's nutrient supply. Organic manures also encourage flourishing populations of small animals, expecially earthworms, as well as microorganisms that may produce growth stimulants (indoleacetic acid, cytokinins) in the rhizosphere. The following categories of organic manure are important.

(a) *FYM* consists of cattle dung and urine mixed with straw. It is very variable in composition and although it is low in macronutrients, at normal application rates of 40–50 t ha^{-1}, the quantities of micronutrients supplied are roughly equivalent to those removed in a succession of 4–5 crops (Table 11.2).

(b) *Slurry* is a suspension of dung in the urine and

Table 11.1. Range and median element contents of organic manures (% fresh weight).

	Dry matter	N	P	K
FYM	11–92 (23)	0.2–3.5 (0.6)	0.04–1.3 (0.13)	0.08–3.6 (0.6)
Poultry manure	5–96 (29)	0.1–6.8 (1.7)	0.04–3.4 (0.6)	0.03–3.6 (0.6)
Liquid cattle manure and slurry	1–60 (4)	0.005–4.8 (0.3)	0.0002–1.1 (0.04)	0.0008–3.5 (0.25)
Pig slurry	1–69 (4)	0.01–4.8 (0.4)	0.004–2.1 (0.09)	0.017–2.7 (0.17)
Compost	24–95 (28)	0.6–1.1 (0.9)	0.18–0.35 (0.22)	0.17–0.75 (0.33)
Sewage sludge (liquid digested)	4	0.2	0.1	0.01
(dewatered digested)	50	1.5	0.5	—

(After MAFF Bulletin 210 and Booklet 2409.)

Table 11.2. Micronutrients in FYM compared with those removed in crops (kg ha^{-1}).

	Mn	Zn	Cu	Mo
FYM (45 t ha^{-1})	3.36	1.12	0.56	0.01
Total in four arable crops in succession	2.50	1.80	0.30	0.01

(After Cooke, 1982.)

washing water coming from animal houses and milking parlours. Compositions of some representative slurries, liquid manures and poultry are given in Table 11.1.

(c) *Compost* is made by accelerating the rate of humification of plant and animal residues in well-aerated, moist heaps. Ideally, a compost of low C:N ratio and acceptable nutrient content can be made in 6–8 weeks (Table 11.1).

(d) *Sewage sludge.* A number of types of sludge may be produced from raw sewage. Liquid *raw sludge* from the primary sedimentation tanks is sometimes offered to farmers, but, more commonly, sludge produced after secondary treatment of sewage is used. Anaerobic fermentation reduces pathogen numbers and smell, and produces a black *liquid digested sludge* containing 2–4 per cent dry matter (Table 11.1). *Dewatered sludges* (raw or digested) contain 40–50 per cent DM and are often called 'sludge cake'. When a flocculant, such as lime or FeSO$_4$ has been used prior to dewatering, the sludge has better 'condition' and is of value in improving the structure of reconstituted soil on mine spoil or derelict land. However, much of the soluble N and P is lost during the dewatering treatment such that the increase in total N and P through dewatering is not in proportion to the increase in dry matter (Table 11.1).

Sludge can be applied at the rate of 50–80 t DM ha^{-1}, provided that the acceptable levels of phytotoxic heavy metals are not exceeded. Currently in the UK, Cu and Ni are considered to be 2 and 8 times as toxic as Zn, and the total *Zn equivalent* of these metals added to agricultural land should not exceed 560 kg ha^{-1} over an extended period. Possibly twice as much could be added to calcareous soils because of the insolubility of the metal hydroxides at high pH (section 10.5).

Green manure

The practice of ploughing in a quick-growing leafy crop before maturity is called *green manuring*. A major benefit is the avoidance of NO$_3^-$ leaching in susceptible soils, for the nitrate is taken up by the green manure crop and released to a subsequent cash crop as the low C:N ratio residues decompose. If the green manure crop is a legume, there can be an additional boost to the soil N supply from symbiotically fixed nitrogen.

The long-term effect on soil organic matter is minimal, as the succulent resides are rapidly decomposed and contribute little to the soil humus. Similarly, effects on structure through the stimulation of soil microbial activity are ephemeral.

Recent developments

Population pressure and a need to grow cash crops for export earnings have induced a *swing away* from traditional methods of agriculture in many parts of the world, especially in the tropics. For example, although shifting cultivation is conservative of natural resources when properly managed, it supports relatively few people (less than one-tenth of the world's population). Land and scarce soil water reserves can rarely be diverted to grow green manure crops on a large scale, and in those areas where freedom from debilitating diseases allows cattle to be kept, the dung has traditionally been burnt as domestic fuel.

With the intensification of agriculture in temperate regions, especially after the Second World War, livestock and arable farming have become increasingly divorced, so that animal manures are not plentiful in areas where they are most needed. Inevitably, there has been a change-over to cereal monoculture—'the one-course rotation'—and increasing reliance on chemical fertilizers in place of traditional manuring and fallowing.

In general, agriculture has been very successful over the last two decades in producing enough food to satisfy a growing population (> 4,000 million, section 1.2). Production has increased at about 1 per cent *per annum* in developed countries and 3 per cent *per annum* in developing countries, although there are obvious problem areas such as North Africa, the Middle East and Africa south of the Sahara. Whereas in developing countries higher yields per ha and more land under cultivation have contributed to the rise in production, in the developed world increased yields per ha, primarily the result of greater fertilizer use and better crop varieties, have been responsible for this rise (Figure 11.1). However, this achievement has not been without cost in terms of more acid soils (section 11.3), deterioration of soil structure (section 11.4) and accelerated erosion (section 11.5). Before discussing such problems, it is appropriate to review the means of assessing soil fertility.

11.2 PRODUCTIVITY AND SOIL FERTILITY

Soil, climate, pests, disease, genetic potential of the crop and man's management are the main factors governing *productivity*, as measured by the *yield* of crop or animal produce per ha. This book is primarily concerned with soil properties, how they interact with soil management, and their effect on productivity. The *fertility* of the soil, and its manipulation by the proper use of fertilizers (Chapter 12), is probably the most important single property affecting productivity. Physical properties, notably structure, are normally secondary although these become of prime importance when soils are susceptible to degradation, or once the fertility problems have been solved.

Soil nutrient supplying power

The assessment of soil fertility has two aspects:
(a) *qualitative*—the aim is to identify which nutrients are deficient (or in excess);
(b) *quantitative*—the aim is to estimate how much of a particular limiting nutrient is required to achieve optimum growth under the prevailing environmental conditions.

Various signs of disorder (chlorotic leaves, stunted

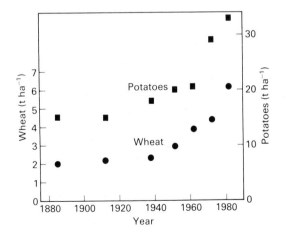

Figure 11.1. Yields (3-year averages) of wheat and potatoes in the UK since 1885–87 (after Cooke, 1984).

growth or withered petioles) are indicative of the deficiency of one or more essential elements. However, visual symptoms appear only after the plant has suffered a check in growth due to 'hidden hunger' for the deficient element. Growth and the concentration of the element in the plant's tissues are closely linked, as shown in Figure 11.2, so that *plant analysis* or *tissue testing* is useful in diagnosing deficiency.

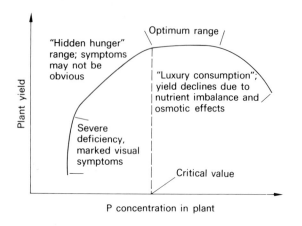

Figure 11.2. The relationship between yield and plant tissue concentration of phosphorus.

DIAGNOSIS OF NUTRIENT DEFICIENCY

Note that the *critical value* is the element concentration in the plant below which an improvement in the supply of the element leads to increased growth. To be of use in deficiency diagnosis, such values must be determined for individual crop varieties, preferably when no other nutrient limits growth. Under these conditions the critical value should be independent of soil type.

Alternatively, the soil can be analysed to estimate the 'availability' of a nutrient, a procedure called *soil testing*. This may involve measuring the concentration of an element in the soil solution, from which the plant draws its immediate supply. A concentration measurement is independent of the size of the system studied and is therefore a measure of the *intensity* (I) of the nutrient's supply—one example is the *phosphate potential* (Schofield, 1955). Another approach is to

measure the amount that can be extracted with chemical solvents, or with an ion exchange resin. Radioactive isotopes, such as ^{32}P, may also be used to label the available fraction, or *labile pool*. The amount of isotopically exchangeable P (called the E value or sometimes the L value, see Figure 12.8) is calculated from the quantity of tracer added (^{32}P soil) and the specific activity of the soil solution after isotopic exchange between the solution and surfaces has taken place, using the following equation.

$$E \text{ value} = {}^{32}P \text{ soil} \times \frac{{}^{31}P \text{ soln}}{{}^{32}P \text{ soln}} \qquad (11.1)$$

The amount extracted or exchanged depends on the size of the system and is referred to as the *quantity* (Q) factor of the nutrient's supply. The change in I with change in Q depends on the slope of the Q/I relation, which defines the *buffering capacity* of the soil for a particular nutrient (section 7.2). Figure 11.3 illustrates the contrasting *phosphate buffering capacities* of two soils. Irrespective of whether Q or I (or both) of a nutrient's supply is measured, the soil test must be calibrated against crop yields, usually on a variety of

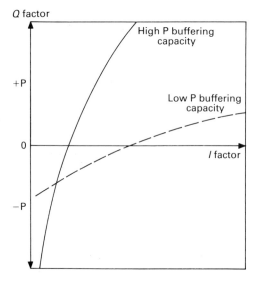

Figure 11.3. Q/I relations for two soils showing contrasting phosphate buffering capacities.

soils with different rates of fertilizer, and a *critical value* for the test established (Figure 11.4). In many countries a solution of 0.5 M NaHCO₃ at pH 8.5 is the standard extractant for soil P. In the UK, no soil test is considered satisfactory for N, but in other European countries, estimates of N fertilizer needs are based on measurements of soil mineral N (exchangeable NH_4^+ and solution NO_3^-) made in late winter.

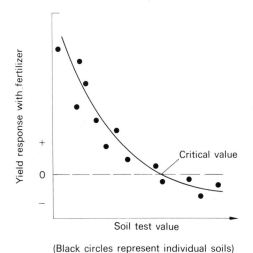

Figure 11.4. Yield response to fertilizer in relation to soil test value (points represent individual soils tested).

QUANTITATIVE ASSESSMENT OF FERTILIZER NEEDS

The amount of fertilizer needed to correct a specific nutrient deficiency can be estimated in the course of calibrating a soil test by means of a *fertilizer rate trial*, in which the yields at different rates of fertilizer (amounts per ha) are measured.

The response to fertilizer frequently follows a *law of diminishing returns*, which means that the yield increase for successive equal increments of fertilizer becomes steadily less. If y is the yield and x the amount of fertilizer element (e.g. N), this law can be expressed in the form:

$$y = A - B \exp(-Cx) \qquad (11.2)$$

As shown in Figure 11.5a, A is the *maximum* (asymptotic) *yield* attainable, (A–B) is the yield without fertilizer and C is the rate of change of y with x.

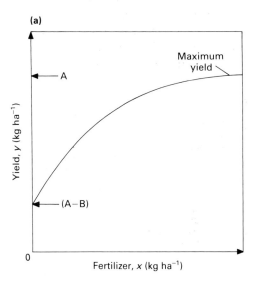

Figure 11.5. (a) A typical fertilizer–response curve obeying equation 11.2

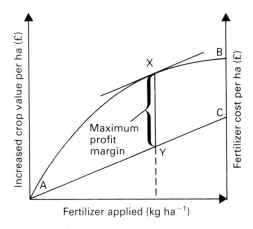

Figure 11.5. (b) Maximizing profit per hectare from fertilizer investment.

Particularly with N, a decline in yield at high-fertilizer levels is sometimes observed which has led to the fitting of quadratic or linear intersecting models to such fertilizer–response curves. However, whichever model is chosen, the precision in predicting the amount of fertilizer needed to attain maximum yield is poor because of factors other than N, such as climate, affecting the yield. More important is the prediction of fertilizer required to produce an *optimum yield*, which may be defined as the yield at which profit per ha is greatest (Figure 11.5b).

For example, when the value of the extra yield due to fertilizer (curve AB) is compared with the fertilizer cost line (AC), the vertical XY indicates the fertilizer rate giving the biggest value–cost differential per ha. If the crop's value, or the fertilizer's cost changes, XY slides to the right or left to a new point of maximum profit.

11.3   SOIL ACIDITY AND LIMING

Origin of acidity

A soil becomes acid if Ca^{2+}, Mg^{2+}, K^+ and Na^+ ions are leached from the profile faster than they are released by mineral weathering, and H^+ and Al^{3+} ions become predominant on the exchange surfaces (section 7.2). The intensity of the acidity, measured by the soil pH, is the result of an interaction of mineral type, climate and vegetation. The main causes of acidification may be summarized as follows.

(a) Microbial oxidation of organic matter, producing humic residues with acidic carboxyl and phenolic groups, and leading to a higher CO_2 partial pressure in the soil air which displaces the carbonic acid equilibria to the right:

$$CO_2+H_2O \rightleftharpoons H^++HCO_3 \rightleftharpoons H^++CO_3^{2-}   (11.3)$$

(b) Nitrification of NH_4^+ ions, producing H^+ ions, and NO_3^-, which is susceptible to leaching (equation 10.2).
(c) Inputs of H_2SO_4, HNO_3 and $(NH_4)_2SO_4$ in dilute solution in rain or by dry deposition from the atmosphere (sections 10.2 and 10.3).
(d) In soils formed on marine muds, or coal-bearing sedimentary rocks, the oxidation of iron pyrites

(formed under intense reducing conditions in the original deposit) gives rise to *acid sulphate soils* or 'cat clays', according to the reactions:

$$2FeS_2+7O_2+2H_2O \rightleftharpoons 4SO^{2-}+4H^++2Fe^{2+}   (11.4)$$

$$2Fe^{2+}+\tfrac{1}{2}O_2+5H_2O \rightleftharpoons 2Fe(OH)_3+4H^+   (11.5)$$

PLANT RESPONSE
Under natural conditions, plants have evolved and adapted to the prevailing soil pH. The *calcifuges* are tolerant of acidity when compared with the *calcicoles*, which are intolerant of acidity or 'lime-loving'. Cultivated plants show a similar diversity in their response to acidity, as the selection of crop and pasture species listed in Table 11.3 shows. Within the legume family, the tropical species are more tolerant of acidity than the medics and clovers of temperate regions.

Table 11.3. Sensitivity of cultivated plants to soil acidity.

Species	pH below which growth is restricted
Rye	4.9
Potato	4.9
Ryegrass	5.1
Wheat	5.4
Maize	5.5
Oats	5.5
White clover	5.6
Barley	5.9
Sugar beet	5.9
Trefoil	6.1
Lucerne	6.1

(After MAFF Bulletin 209.)

Liming

The problems associated with soil acidity—slow turnover of organic matter, poor nodulation of some legumes, Ca and Mo deficiencies, Al and Mn toxicities—can be remedied by liming. Ground limestone, chalk, marl and basic slag are used as liming materials, the active constituent being primarily $CaCO_3$, with some burnt lime (CaO) and hydrated lime $(Ca(OH)_2)$.

When lime is added to moist soil, it hydrolyses slowly to produce an alkaline pH, according to the reaction:

$$CaCO_3 + 2H_2O \rightleftharpoons Ca(OH)_2 + H_2CO_3 \quad (11.6)$$

However, $Ca(OH)_2$ itself reacts with CO_2 from the atmosphere to form $Ca(HCO_3)_2$ so that the net reaction is:

$$CaCO_3 + CO_2 + H_2O \rightleftharpoons Ca(HCO_3)_2 \quad (11.7)$$

The final pH attained can be predicted from the expression:

$$pH = K - \tfrac{1}{2} \log(P_{CO_2}) - \tfrac{1}{2} \log(Ca) \quad (11.8)$$

where P_{CO_2} = partial pressure of CO_2, (Ca) = activity of Ca^{2+} ions and K = 5.05 if the carbonate has the solubility of pure calcite. For soil carbonates, K $\simeq$ 5.2 and the predicted pH for a soil in equilibrium with the partial pressure of CO_2 in the atmosphere (0.00033 bars) is about 8.1.

LIME REQUIREMENT

The *lime requirement* depends on the pH buffering capacity of the soil (section 7.2), and is expressed as the quantity of $CaCO_3$ (t ha^{-1} to a depth of 15 cm) required to raise the soil pH to a desired value. The standard method of measurement is described by Hooper (1973).

A pH of 6.5 is recommended for temperate crops, but liming of grassland to pH 6 is preferred, because grasses are more tolerant of acidity and the possibility of inducing deficiencies of Mn, Cu and Zn is minimized. Tropical species are more acid tolerant and, furthermore, liming to pH 6–6.5 may reduce P availability in soils high in exchangeable Al^{3+} (section 10.3) and destabilize the structure of kaolinitic clay soils (section 7.4). In such soils, liming to hydrolyse the exchangeable Al^{3+}, normally achieved at pH 5.5, is preferable (section 7.2).

Finely ground $CaCO_3$ made to adhere to legume seeds—a process called *lime pelleting*—greatly improves nodulation of sensitive species in acid soils, especially when an effective strain of *Rhizobium* bacteria is included in the coating.

11.4 THE IMPORTANCE OF SOIL STRUCTURE

Structure dependent properties

Good soil management should aim to create optimum physical conditions for plant growth, as manifest by:
(1) adequate aeration for roots and microorganisms;
(2) adequate available water;
(3) ease of root penetration, permitting thorough exploitation of the soil for water and nutrients;
(4) rapid and uniform seed germination;
(5) resistance of the soil to slaking, surface-sealing and accelerated erosion by wind and water.

Two of the most useful indices of structure are *bulk density*, which is inversely related to total porosity (section 4.5), and *aggregate stability*, which is assessed by the coherence of soil aggregates in water. The Emerson stability test (1967) is based on the extent to which aggregates disperse, slake or remain intact in distilled water. The more traditional wet-sieving technique (Loveday, 1974) involves placing a mass of dry soil on a bank of sieves of different mesh sizes which is then immersed in water. After a number of standard vibrations, the proportion of the soil remaining on each of the sieves is determined. Bulk density measurements are more relevant to conditions (1), (2) and (3) above, whereas aggregate stability is more relevant to conditions (4) and (5).

Organic matter, texture and land use modify these properties. For example, it has been found that bulk density increases as organic matter content decreases, and it is highest in the sandy textural classes. Aggregate stability, on the other hand, decreases with declining organic matter and it is least for soils in the silty and fine sandy loam classes.

Aeration and water supply depend on the soil's pore-size distribution, the key indices being the *air capacity* $C_a(50)$ (section 4.5) and the *available water capacity* (*AWC*) (section 6.5). Using these two properties, classes of soil droughtiness at one extreme and susceptibility to waterlogging at the other, may be separated, as illustrated for surface soils in England and Wales which regularly experience a soil moisture deficit $SMD > 100$ mm (Figure 11.6). Soils with > 10 per cent AWC are droughty, and those with > 5 per cent air space at field capacity are likely to be anaerobic. The limiting AWC value is lowered slightly and the $C_a(50)$ value raised to 10 per cent for soils of wetter regions. Irrespective of the individual values of $C_a(50)$ and AWC, a *storage pore space* value (the sum of AWC and $C_a(50)$) less than 23 per cent is undesirable, hence the truncation of the curve separating the 'poor' and 'moderate' classes in Figure 11.6.

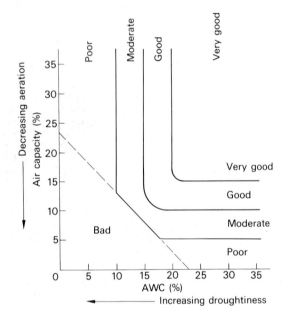

Figure 11.6. Classification of the structural quality of topsoils (after Hall *et al.*, 1977).

The effect of land use

GRASS VS ARABLE

Soil physical conditions are usually best under permanent grass (or forest) and deteriorate at a rate dependent on climate, soil texture and management as the soil is cultivated.

Disruption of peds exposes previously inaccessible organic matter to attack by microorganisms; populations of structure-stabilizing fungi decline and earthworm numbers are markedly depressed. Removal of the protection of a permanent canopy of vegetation exposes the soil to the direct impact of rain, and the loss of surface litter reduces the detention time of water before surface runoff and concomitant erosion begins. Shearing forces generated by tractor wheels and ploughs can rupture macro- and microaggregates (section 4.4). Structure is most vulnerable to damage when the soil is wet ($> FC$) because the distance between clay particles in the domains and quasi-crystals is greatest then. The external force can cause deflocculation and re-orientation of the deflocculated clay particles to form dense layers of low permeability called *plough pans*.

Measurements of water-stable aggregates show that the deterioration in structure is most rapid in the first year out of grass (Figure 11.7). Total porosity and AWC also decline as cultivation is prolonged. These trends are reversed, however, by the introduction of a pasture phase (a ley) into the cropping cycle. The restorative effect of the ley depends on its duration, as illustrated by the comparison of measurements made on a continuously cultivated (fallow–wheat) soil put

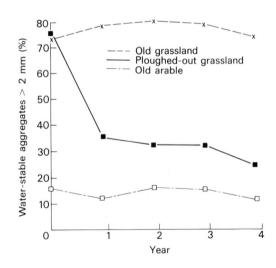

Figure 11.7. Aggregation of the fine earth fraction of a soil under different land use (after Low, 1972).

down to pasture for various lengths of time (Figure 11.8). Note that the improvement in aggregate stability is largely confined to the surface 5 cm or so where the grass roots are concentrated. Improved aggregation following short leys (< 4 y) is primarily due to the effect of polysaccharides (see Figure 4.13). The resulting aggregation is sometimes called 'soft' because it does not survive wet sieving. Longer leys, with an increase in the contribution from humidified organic matter, are needed to produce the more stable 'hard' aggregation. For British soils, 4 per cent organic matter ($\simeq$ 1.6 per cent C) has been suggested as the critical lower limit for aggregate stability under cultivation.

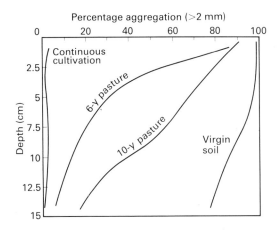

Figure 11.8. The effect of leys of different duration on the water-stable aggregation of a soil under cultivation in southern Australia (after Graecen, 1958).

Figure 11.9. Inverted furrow slices following mouldboard ploughing (courtesy of D.E. Patterson).

Tillage methods

The traditional method of cultivation or tillage in Western agriculture has involved thorough disturbance of the top 20–25 cm of soil, usually with inversion of the furrow slice by mouldboard ploughing (Figure 11.9). Residues of the previous crop and volunteer weeds are buried, and after repeated passes with discs and harrows the soil is reduced to a fine tilth suitable for seeding. This method was exported to North America, Africa and Australasia, often with disastrous results in terms of soil structural deterioration and erosion.

In contrast, traditional methods of cultivation in the tropics were very different. Following the clearing of the natural vegetation (with or without a burn), the first crop was sown with the minimum of soil disturbance. Plant debris left on the surface protected the soil from erosion until the crop was well established. Intercropping with tall and short crops was common so that little bare ground was exposed to the erosive tropical rain (section 11.5).

In developed countries over the past few decades, however, various factors have led to a reappraisal of traditional ploughing methods: first, it was shown in the 1930s at Rothamsted that a fine seedbed tilth did not produce higher yields than less elaborately prepared seedbeds, provided that weeds were controlled; second, manpower and fuel became very costly; and third, the adverse effects of soil erosion, not only on agricultural productivity, but also in polluting the environment, gave increasing cause for concern.

New tillage methods were devised which were a blend of modern and ancient practices. They combine minimum disruption of the natural structure of the soil with the use of new herbicides to control weeds. These practices range from *zero tillage, no-till* or *direct drilling*, where the only mechanical operation is seed sowing, to *reduced* or *minimum tillage*, which involves fewer and shallower passes with tines or discs than does conventional ploughing.

DIRECT DRILLING OR NO-TILL

This practice has gained widespread popularity in the USA, particularly in the southern corn (maize) belt, and also in tropical Africa, South America and Australia. For row crops, such as maize, the seed is almost always sown directly into a grass sod that has been killed with a herbicide, such as paraquat, or into the residues of a previous crop. Soyabeans, for example, are frequently sown into the stubble of a winter cereal (Figure 11.10). In the countries cited, the preservation of plant residues on the soil surface is a major benefit of no-till agriculture because: (i) the mulch formed shades the soil and reduces water loss by evaporation; (ii) the temperature of the surface is kept at tolerable levels; and (iii) surface runoff and erosion are greatly reduced. The last effect is now very well documented, as illustrated by the data in Table 11.4. There is also an energy saving of 7–18 per cent for no-till when compared with conventional ploughing, the higher figure occurring with direct-drilled legumes: slightly higher amounts of N fertilizer are required for

Figure 11.10. Direct-drilling into wheat stubble.

Table 11.4. Average soil losses under different soil management in Illinois, USA (annual precipitation *c.* 1300 mm).

	Average annual soil loss (t ha^{-1})	
	5% slope	9% slope
Conventional tillage, maize and wheat	7.6	21.5
No-till, maize and wheat	0.9	1.3
No-till, continuous maize	0.6	0.9

(After Gard and McKibben, 1973.)

many direct-drilled non-legumes, which offsets some of the energy savings on fuel.

It has been predicted (Phillips *et al.*, 1980) that about 65 per cent of the seven major annual crops in the USA will be under no-till by the year 2000 AD, whereas in Britain only about 30 per cent of the cereal land is considered suitable for this system (Greenland, 1981). The main reasons for this difference lie in the *climate* and the *soil*, and it is appropriate to examine some of the changes in soil properties that occur when no-till replaces ploughing.

Organic matter gradually increases in the top 5 cm and the water stability of the surface peds is enhanced.

On many soils, bulk density is increased down to the old plough depth so that total porosity decreases, mainly at the expense of the macropores. Nevertheless, water infiltration and percolation do not necessarily suffer because earthworms multiply two- or threefold, especially the deep burrowing *Lumbricus terrestris*, and the predominantly vertical orientation and continuity of their holes makes for rapid water flow. Furthermore, the higher bulk density improves the 'trafficability', i.e. load-bearing capacity, of the soil when wet.

Changes also occur in the chemical properties of the soil. Phosphorus and K become concentrated in the 0–5 cm layer, which is no disadvantage provided the soil stays moist (under a mulch for example), but these elements may become less available to the plant if the soil dries out. Total N follows the trend in organic C, but nitrate levels are often lower, possibly because of localized denitrification and greater downward leaching. Consistent with greater NO_3^- leaching are the lower exchangeable Ca and pH values in the 0–5 cm depth of some direct drilled soils, which may therefore require additional lime. Other possible disadvantages are:

(a) slow warming of the wet soil in spring, which retards crop growth (particularly in Britain and northern USA);

(b) the production of volatile fatty acids around fermenting crop residues in the drill silt, which can inhibit seed germination (section 8.3);

(c) an increase in perennial weeds, such as couch and brome grass, which are difficult to control with herbicides alone.

An illustration of the surface mulch on a silt loam soil after several years of no-till maize in Kentucky, USA, is shown in Figure 11.11. Overall, the advantages of no-till agriculture can be summarized as follows:

(i) improved surface soil structure;

(ii) erosion control;

(iii) conservation of soil moisture;

(iv) lower surface soil temperatures (warm and hot climates only);

(v) lower energy requirements.

Figure 11.11. Wheat stubble residues under no-till maize in Kentucky.

Soil conditioners

Because plant and microbial polysaccharides are known to improve aggregation (section 4.4), synthetic macromolecules, called *soil conditioners*, have been used for the same purpose. The main types of soil conditioners are shown in Table 11.5. Application is at the rate of 0.2 g (soluble polymer) or 10 g (emulsifiable polymer) per kg of soil. Moist conditions are necessary to enable the soluble polymers to diffuse into the larger pores (> 30 μm diameter) where they are most effective. With the bitumen emulsions, the soil should be allowed to dry to achieve maximum aggregate stability. A problem with all soil conditioners is that only a shallow depth of soil can be easily treated: for this reason, and for reasons of cost, they tend to be used in

Table 11.5. Summary of the main types of soil conditioners.

Soluble polymers (hydrophilic)	Emulsifiable polymers (hydrophobic)
Polyvinyl alcohol (PVA)	Bitumen
Polyacrylamide (PAM)	Polyvinylacetate (PVAc)
Polyethyleneglycol (PEG)	Polyurethane

special situations, such as for stabilizing sand dunes against wind erosion, or protecting the fine surface tilth of seed beds, especially that of vegetables and sugar beet in Europe (Figure 11.12).

Figure 11.12. Stabilization of surface soil structure—the darker plots have been treated with PVA solution (courtesy of J.M. Oades).

11.5 SOIL EROSION

The term *erosion* describes the transport of soil constituents by natural forces, primarily water and wind. As indicated in section 5.2, past phases of geologic erosion led to the formation of sedimentary deposits which are the parent materials of many present-day soils; but in the short term erosion due to man's activities, especially agriculture, is of greater concern. In the USA, for example, one-fifth of the 175 million ha of cropland loses more than 20 t ha^{-1} of soil annually, one-half loses between 7.5 and 20 t and the remainder less than 7.5 t. Erosion is very damaging to soil fertility because it is mainly the nutrient-rich surface soil that is removed, and of that, predominantly the fine and light fractions—clay and organic matter—leaving behind the inert sand and gravel.

Erosion by water

A global study of stream sediment loads suggests the general relation between erosion by water and rainfall shown by the solid line in Figure 11.13. In arid areas, nearly all the rain falling is absorbed by the parched soil and scanty vegetation, leaving little to run off, whereas

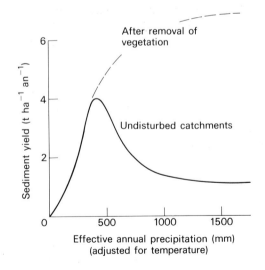

Figure 11.13. Sediment loads in rivers in relation to annual precipitation (adjusted for differences in temperature) (after Langbein and Schumm, 1958).

in humid climates abundant vegetation protects the soil for all of the year. Most vulnerable are the regions of intermediate rainfall, where the incidence of rain is seasonal and the vegetative cover is therefore less effective, especially at the end of each dry season. Nevertheless, the erosive power of high rainfall is attested by the broken curve of Figure 11.13, which applies to humid regions when the natural vegetation is removed.

Soil loss by erosion is dependent on: (a) the potential of rain to erode—the rainfall *erosivity*; and (b) the susceptibility of soil to erosion—the soil *erodibility*.

RAINFALL EROSIVITY

To erode soil, work must be done to disrupt aggregates and move soil particles: the energy is provided by the kinetic energy (*KE*) of falling raindrops and flowing water. The principal effect of raindrops is to detach soil particles whereas that of surface flow is to transport the detached particles. Calculations show that the *KE* of rain is ~ 200 times greater than the *KE* of runoff, and field experiments confirm that erosion from a protected surface can be < 1/100th of that from a bare surface under the same rainfall (Hudson, 1981).

Raindrop impact initiates *splash* erosion (Figure 11.14). On a sloping surface the downhill momentum transmitted to the soil particles is greater than the uphill component, consequently there is a net downslope splash erosion. Slope also accelerates runoff velocity and the two effects combine to produce *wash* or *interrill* erosion. Where the water concentrates, for example in cultivation furrows or tractor wheel marks, *rill* erosion is initiated (Figure 11.15). Rills can be eliminated by normal cultivation methods, but they may deepen to form *gullies*, which cannot be crossed by farm machinery.

Although raindrops vary in size up to 5 mm diameter, the *median drop diameter*, that is the diameter such that half the rain in a storm falls in drops of a smaller size, does not change much with rate of rainfall or *intensity** above 25–50 mm h^{-1}. Raindrop *KE* is

*mm h^{-1} is m^3×10^3 of rain per m^2 of surface perpendicular to the rain's direction per hour.

(a)

(b)

Figure 11.14. (a) Soil surface just before raindrop impact. **(b)** Soil surface immediately after raindrop impact showing splash erosion (courtesy of D. Payne).

given by the equation:

$$KE = 1/2 \, \text{mass} \times (\text{terminal velocity})^2 \quad (11.9)$$

Terminal velocity increases with drop size (mass) so the *KE* of the rain increases sharply up to moderate intensities and then more gradually, as illustrated in Figure 11.16.

Another important characteristic of rainfall is the frequency of high-intensity storms, for which there is a striking difference between tropical and temperate

Figure 11.15. Rill erosion caused by surface runoff on bare soil (courtesy of A.J. Low).

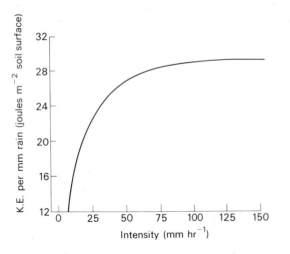

Figure 11.16. Kinetic energy per mm of rain in relation to rainfall intensity (after Hudson, 1981).

regions. Approximately 95 per cent of temperate rain, but only 60 per cent of tropical rain, falls at intensities < 25 mm h^{-1}, which are considered non-erosive. Furthermore, the maximum intensity of tropical rainfall may exceed 150 mm h^{-1}, which is twice that of temperate rainfall. The high average KE and generally greater amounts of tropical rainfall make erosion by water *potentially* a very serious problem in these regions, whereas it is of less importance in humid temperate lands.

SOIL ERODIBILITY

The susceptibility of a soil to erosion is expressed by the equation:

$$\frac{A}{R} = K \qquad (11.10)$$

where A is the annual soil loss in t ha^{-1}, R is the dimensionless rainfall erosivity index, and K is the *soil erodibility* factor. Several indices have been proposed for R, but the most widely used is the EI value, determined as the product of the KE of a storm and its 30-minute intensity. The latter is the maximum rainfall in any 30-minute period, expressed in mm h^{-1}.

The way in which equation 11.10 is usually applied is to set A equal to the soil loss that can be tolerated without productivity seriously declining. This figure varies according to the rate of soil formation (Chapter 5) and ranges between 2 and 11 t ha^{-1} y^{-1}. Once A and R are known, K is then defined by equation 11.10 as the value for a soil which, when multiplied by R, gives the value of A under standard conditions. The standard conditions are a slope length of 22.6 m and gradient of 9 per cent for bare soil that is ploughed up and down the slope. If the actual conditions of slope, surface and management of the land differ from the standards, the consequent effect on A is incorporated into equation 11.10 by rewriting it in the form:

$$A = R \times K \times L \times S \times P \times C \qquad (11.11)$$

The symbols L, S, P and C represent, respectively, a length factor, slope factor, conservation practice factor and crop management factor. Equation 11.11 is called the *Universal Soil Loss Equation* (*USLE*). The values of L, S, P and C are unity when the standard conditions apply: if not, they change in a way that reflects the effect of the changed conditions in either *increasing* or *decreasing* erosion. The effects of these factors on the value of A, the annual soil loss in t ha^{-1}, are briefly examined in the following sections.

(a) *Soil factor, K.* K values depend primarily on texture, organic matter, structural stability and permeability of the soil. Texture is not amenable to change, but the effect of leys or permanent pasture on the other variables affecting K has already been noted (section 11.4). Experiments in the USA on different soil types, but under the standard conditions of L, S, P and C, produced a range of K values between 0.03 and 0.69. Unfortunately, such detailed information is not commonly available in other countries.

(b) *Slope factor, LS.* Both steepness (S) and length (L) of slope affect erodibility: the former because of the greater potential energy of soil and water high on a slope, which can be converted to kinetic energy; the latter because the longer the slope, the greater the surface runoff and its downhill velocity. When soil conservation measures, such as contour terracing or strip cropping, are introduced, the distance, L, of vulnerable soil between structures is adjusted according to the slope, S, so that a composite factor, LS, is used in the *USLE*.

(c) *Conservation practice, P.* Ploughing straight up and down a slope is the worst soil management practice ($P = 1$). Ploughing along contours can reduce P by 10–50 per cent, the greatest effect being achieved on slopes between 2 and 7 per cent. More elaborate procedures range from *contour listing*, small temporary ridges thrown up by cultivation, to *contour ridges* or *banks*, permanent banks with graded channels above for water flow, to *bench terraces*, which convert a steep slope into a series of steps. Generally these structures are designed: (a) to reduce the slope length from which runoff is generated; (b) to hold water on the slope longer so that more of it infiltrates the soil; and (c) to divert surplus water into safe spillways down which it may escape with minimum destructive effect, such as the grassed waterways shown in Figure 11.17. Much earth movement and grading may be required, so that the final design is a compromise between the cost of the work, including the loss of cropping flexibility incurred, and the value of the land and its produce.

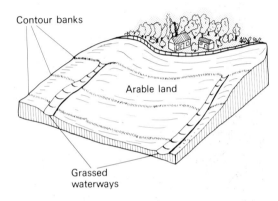

Figure 11.17. Perspective of crop land with contour banks and waterways.

(d) *Crop management, C.* Manipulation of this factor offers the greatest scope for reducing erosion loss because C can vary from 0.001 for well-kept woodland, to 0.05 for continuous forage cropping, to 1.0 for continuous bare fallow. Practices aimed at reducing C must focus on:
(1) not leaving the soil bare when the rainfall erosivity is high; and
(2) growing high-density, healthy crops so that the soil surface is protected and the binding of the soil by the roots is strong.

Leys, permanent pasture, minimum tillage and direct drilling all have merit for erosion control. To these must be added *stubble mulching*, whereby the residues of a previous crop are chopped up and scattered over the soil surface, and *strip cropping* where strips of arable land and pasture alternate down a slope, thereby reducing the length of slope most vulnerable to erosion.

Erosion by wind

Soil movement by wind is prominent in arid areas and on coastlines, where the wind is strong and consistent and the vegetation is adversely affected by salinity. By analogy with water erosion, the severity of wind erosion depends on the *potential* of the wind to erode and the *susceptibility* of the soil to erosion. Although not as widespread as water erosion, the effects of wind erosion are often more dramatic in that the quantities of soil moved per hour across the edge of one hectare can be of the same order as that moved by water erosion in one year.

WIND ACTION

For given soil conditions (moisture, particle density ρ_p), erosion by wind depends on the wind speed and the roughness of the surface. Roughness alters the frictional drag on the wind and the likelihood of turbulence, which can impart an upward velocity component to loose particles or small aggregates (Figure 11.18a). The air-borne particle gains a forward velocity, and because of the increase in wind velocity away from the soil surface (Figure 11.18b), the higher the jump, the greater the particle's momentum. However, surface roughness, through its drag effect, also decreases the gradient in forward wind velocity over the soil and this affects its ability to erode.

Particle or aggregate *size* influences surface roughness and the *density* obviously determines how much energy must be expended to raise a particle of a particular size off the ground. The interaction of these factors determines in a complex way the *threshold wind velocity*, i.e. the minimum velocity required to move a particle, as illustrated in Figure 11.19. This diagram shows that particles or aggregates around 0.1–0.15 mm diameter are the most vulnerable, irrespective of their density. Once a particle has been moved, there are three possible consequences.

(a) Particles < 0.05 mm in diameter that rise more than approximately 30 cm into the air may stay suspended for a long time. The resulting *aerial dispersion*, if sufficiently concentrated (> 0.2 g m^{-3}), becomes a dust storm. In times past, especially during the

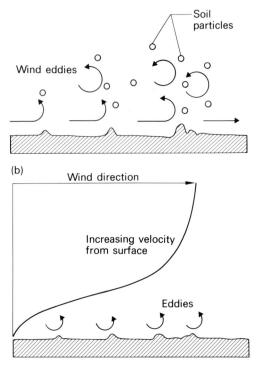

(a)

(b)

Figure 11.18. (a) Turbulent wind eddies over a roughened soil surface. **(b)** Vertical profile of wind velocity over a soil surface (after Hudson, 1981).

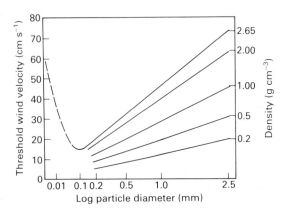

Figure 11.19. The effect of particle size and density ρ_p on the threshold wind velocity for particle movement (after Marshall, 1972).

Pleistocene glaciations, the larger wind-dispersed particles (0.02–0.05 mm) settled to form the extensive *loess* deposits described in section 5.2.

(b) Particles and aggregates between 0.05 and 0.5 mm in diameter do not usually rise more than 30 cm and their jump is followed by a long flat trajectory to earth. On striking the surface, the particle either bounces off or comes to rest, transmitting its *KE* to other particles, which are knocked upwards (Figure 11.20). This self-sustaining process is called *saltation*.

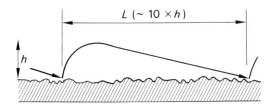

Figure 11.20. Jump and glide path of saltating soil particles (after Hudson, 1981).

(c) On landing, the saltating particle may strike larger, heavier particles which do not rise up but merely roll or *creep* along the surface. Particles and aggregates of a diameter between 0.5 and 2 mm move in this way.

Movement by saltation accounts for more than 50 per cent of wind erosion, with 3–40 per cent of movement occurring in suspension and 5–25 per cent by surface creep. Little can be done to lessen the erosive potential of wind other than by reducing its velocity by means of *wind breaks*, and by shortening its *fetch*, that is the distance of its unhindered travel over erodible land.

SOIL FACTORS

Moist soil supporting healthy vegetation is the most effective deterrent to wind erosion. Cultivated soil is vulnerable when the surface is dry and the structure weak, especially if most of the aggregates are < 1 mm in diameter. Often coarse and fine-textured soils are more erodible than loams, which have the right proportions of clay, silt and sand to form stable aggregates of a non-erodible size. The fen-peat soils of eastern England are particularly vulnerable to wind erosion once drained and cultivated because of the lightness of the organic-rich surface horizon.

With respect to surface roughness, although slight irregularities initiate saltation as the wind speed rises, very rough or cloddy surfaces protect against erosion by reducing the average wind velocity at ground level and trapping saltating particles as they land. *Vegetation*, even if dead, has a similar effect, which is directly proportional to its density and height above the soil. It follows that stubble mulching, pioneered on the wind-swept High Plains of the USA, *trash farming*, whereby large one-way disc ploughs cut through the stubble but do not bury it, and direct drilling through old crop residues are all valuable in controlling wind erosion.

11.6 SUMMARY

Clearing a soil of natural vegetation interrupts the cycling of nutrients and may lead to a decline in fertility. Traditional methods of farming in the tropics (*shifting cultivation* with *bush fallowing* and *green manuring*), and in temperate regions (*rotational cropping* and *mixed farming*) generally maintain fertility, but they have gradually been displaced by intensive mono-cultural systems, such as continuous cereal cropping, in response to the pressure for land, the cost of labour and the demand for food.

Modern intensive farming generally requires inputs of fertilizers and/or organic manures (including sewage sludge) to maintain soil fertility. Qualitatively, soil nutrient deficiencies can be assessed by *soil testing* or *plant analysis*, with the identification of *critical values* for a nutrient element below which plants respond to an increase in that nutrient's supply. For quantitative estimates of fertilizer need, the soil test must be calibrated against fertilizer-rate trials carried out on a representative range of soils. Soil acidity and the consequent problems of Al and Mn toxicity, Ca and Mo deficiency, are remedied by adding lime, the amount needed (the *lime requirement*) depending on the pH buffering capacity of the soil (section 7.2).

Deterioration of soil structure is a more insidious problem that is only revealed through inadequate soil aeration, reduced *AWC*, impedance to root penetration and surface sealing, which inhibits seedling emergence and predisposes the soil to erosion by water and wind. A grass ley following arable cropping is very effective in restoring the water stability of aggregates, especially in the surface 5 cm or so. The improvement is time dependent, about 4 years being required in temperate regions to achieve an improvement in 'hard' aggregation. However, the effect of the ley is relatively short lived once it is ploughed out. A similar improvement in aggregation can be obtained with synthetic organic polymers called *soil conditioners*.

New tillage methods, which are a blend of *modern* (use of rapid-kill, non-persistent herbicides) and *ancient* (sowing seed with minimum soil disturbance) agricultural practices, are gaining widespread popularity, particularly on soils where high surface temperatures, high evapotranspiration, surface sealing and water and wind erosion pose problems. These methods range from *reduced* or *minimum tillage* (fewer passes with implements) to *direct drilling* or *no-till*, where the only disturbance is the sowing of the seed.

Erosion is very damaging to soil fertility because it is mainly the nutrient-rich surface soil that is removed. Water erosion occurs in almost any environment, if the soil is bare and the structure unstable, the magnitude of the loss depending on the rainfall *erosivity* and soil *erodibility*. This relationship is quantified by assigning values to the variables in the *Universal Soil Loss Equation*:

$$A = R \times (K, LS, P, C)$$
(Soil loss) (Erosivity) (Erodibility)

Conservation measures aim at reducing the slope steepness, *S*, and length, *L*, the soil management factor, *P* (through contour banks, terraces, strip cropping) and the crop management factor, *C* (by direct drilling and stubble mulching).

Wind erosion is confined to arid regions or to coastal areas where wind velocities are consistently high. The susceptibility of soil to wind erosion depends on the moisture content of the surface, the diameter and density of the surface particles (those of diameter 0.1–0.15 mm are the most vulnerable) and the degree of protection afforded the surface by vegetation.

REFERENCES

Cooke G.W. (1982) *Fertilizing for Maximum Yield*, 3rd Ed. Granada, London.

Cooke G.W. (1984) Constraints on crop production—opportunities for chemical industry. *Chemistry and Industry* No. 20, 730–737.

Emerson W.W. (1967) A classification of soil aggregates based on their coherence in water. *Australian Journal Soil Research* 5, 47–57.

Gard L.E. & McKibben G.E. (1973) No-till crop production proving a most promising conservation measure. *Outlook on Agriculture* 7, 149–154.

Graecen E.L. (1958) The soil structure profile under pastures. *Australian Journal Agricultural Research* 9, 129–137.

Greenland D.J. (1981) Soil management and soil degradation. *Journal of Soil Science* 32, 301–322.

Hall D.G.M., Reeve M.J., Thomasson A.J. & Wrights V.F. (1977) *Water retention, porosity and density of field soils*. Soil Survey of England and Wales, Technical Monograph No. 9.

Hooper L.J. (1973) Fertilizer recommendations. MAFF Bulletin 209. HMSO, London.

Hudson N.W. (1981) *Soil Conservation*, 2nd Ed. Batsford, London.

Langbein W.B. & Schumm S.A. (1958) Yield of sediment in relation to mean annual precipitation. *Transactions of the American Geophysical Union* 39, 1076–1084.

Loveday J. (1974) (Ed.) *Methods for analysis of irrigated soils*. Commonwealth Bureau of Soils, Technical Communication No. 54.

Low A.J. (1972) The effect of cultivation on the structure and other physical characteristics of grassland and arable soils (1945–1970). *Journal of Soil Science* 23, 363–380.

Marshall J.K. (1972) Principles of soil erosion and its prevention, in *The Use of Trees and Shrubs in the Dry Country of Australia*. Forestry and Timber Bureau, Canberra.

Ministry of Agriculture, Fisheries and Food (1982) *The use of sewage sludge on agricultural land*. Booklet 2409. MAFF Publications, Northumberland.

Ministry of Agriculture, Fisheries and Food (1976) *Organic manures*. Bulletin 210, HMSO, London.

Phillips R.E., Blevins R.L., Thomas G.W., Frye W.W. & Phillips S.H. (1980) No-tillage agriculture. *Science* 208, 1108–1113.

Schofield R.K. (1955) Can a precise meaning be given to 'available' soil phosphorus? *Soils and Fertilizers* 18, 373–375.

FURTHER READING

DE BOODT M. (1979) Soil conditioning for better management. *Outlook on Agriculture* **10**, 63–70.

JONES U.S. (1979) *Fertilizers and Soil Fertility*. Reston Publishing Co., Reston, Virginia.

RUSSELL R.S. (1977) *Plant Root Systems*. McGraw-Hill, London.

STRUTT N. (1970) *Modern farming and the soil*. Report of the Agricultural Advisory Council on Soil Structure and Soil Fertility. HMSO, London.

TISDALE S.L. & NELSON W.L. (1975) *Soil Fertility and Fertilizers*, 3rd Ed. Macmillan, London.

Chapter 12
Agricultural Chemicals and the Soil

12.1 WHAT ARE AGRICULTURAL CHEMICALS?

Agricultural chemicals may be subdivided into two broad categories:
(a) *fertilizers*, briefly introduced in section 11.1, are used to correct nutrient deficiencies and imbalances in plants and animals;
(b) *pesticides* (sometimes called plant-protection chemicals), are used to protect crops from pests and disease or to eliminate competition from aggressive weeds.

Fertilizers and pesticides may be applied as solutions, suspensions, emulsions or solids to soils and plants. When applied as a foliar spray, the properties of the chemical (such as its volatility), the nature of the plant surfaces and the atmospheric conditions determine the efficacy of the application and the subsequent fate of the chemical in the environment. In other cases, the type of chemical, its method of application and the reactions it undergoes in the soil are of prime importance. In this chapter, emphasis is placed on chemical forms and soil reactivity, starting with the N, P, K and S fertilizers.

12.2 NITROGEN FERTILIZERS

Forms of N fertilizer

SOLUBLE FORMS

The compositions of the common water-soluble N fertilizers are given in Table 12.1. Some of these, such as NH_4NO_3 and urea, supply a single macronutrient element (N); others, such as $(NH_4)_2SO_4$, MAP and DAP, provide more than one macronutrient and are classed as *multinutrient* or *mixed* fertilizers. Mixed

Table 12.1. Forms of soluble N fertilizer.

Compound	Formula	N content (%)
Solids		
Sodium nitrate	$NaNO_3$	16
Calcium nitrate	$Ca(NO_3)_2$	17
Ammonium sulphate	$(NH_4)_2SO_4$	21
Calcium cyanamide	$CaCN_2$	21
Ammonium nitrate	NH_4NO_3	34
Urea	$(NH_2)_2CO$	46
Monoammonium phosphate, MAP	$NH_4H_2PO_4$	11–12
Diammonium phosphate, DAP	$(NH_4)_2HPO_4$	18–21
Liquids		
Aqua ammonia	NH_3 in water	25
Nitrogen solutions	NH_4NO_3, $(NH_2)_2CO$ in water	30
Anhydrous ammonia	liquefied NH_3 gas	82

fertilizers may be solid or liquid, an example of the latter being the solutions made by dissolving NH_4NO_3, urea and ammonium polyphosphate in water (e.g. NPK, 19–8–0*), or these constituents plus KCl to give a 5–5–8 mixture.

Highly soluble solid fertilizers can pose handling and storage problems because of their hygroscopicity. This difficulty is largely overcome by *granulation*, which is intended to reduce the contact area between individual particles and also the area available for water absorption. Ideally, a granulated (or granular) fertilizer has uniform granules, roughly spherical in shape and *c.* 3–4 mm in diameter: the ultimate in granulation is the spherical *prills* of NH_4NO_3 and urea. Solid mixed

*In the trade, fertilizer analyses are generally given as the ratio of percentages of $N : P_2O_5 : K_2O$. For simplicity in this book, analyses are given as percentages of the elements N, P, K and S, unless otherwise stated.

fertilizers, made by mixing the ingredients in a slurry before granulation, are homogeneous. Other fertilizers made by mixing individual granular fertilizers, a process of *bulk-blending*, may not be so homogeneous because granules of different size and density may segregate in the mixture. The chemical incompatibility of fertilizers such as KNO_3 and urea also precludes their bulk-blending.

The choice of the most suitable N fertilizer depends on a balance of factors: the cost per kg of N (including transport and application cost), the effects on plant growth (both beneficial and detrimental) and the magnitude of N loss through leaching and denitrification. A high N content and its ease of handling as prills have made urea the most popular form of solid N fertilizer in the world. However, in the USA more N is applied as anhydrous NH_3 than urea and NH_4NO_3 combined because of its very high analysis and the precision with which its application can be controlled. The advantage of a very high N content per unit weight afforded by anhydrous NH_3 is partly offset by higher costs of application, since it must be kept under pressure and injected at least 10 cm below the soil surface. Other effects are discussed below.

SLOW-RELEASE FERTILIZERS

To obviate problems arising from the high solubility of many N fertilizers, and the susceptibility of NO_3-N to leaching, various sparingly soluble N compounds including natural organic materials have been developed for use (Table 12.2). The response to these slow-release fertilizers is usually inferior, per kg of N, when compared with soluble forms, but they are especially useful for the establishment of vegetation on difficult sites, such as reclaimed mine workings, where a slow but assured release of N over an extended period is required. An alternative solution to the leaching problem is to use NH_4-N fertilizers only, together with the specific inhibitor of nitrification 'N-serve' (2-chloro-6 trichloromethyl pyridine) at the rate of 1–2 per cent of the N applied.

Table 12.2. Forms of slow-release N fertilizer.

Fertilizer	Composition	N content (%)
Shoddy	Wool waste	2–15
Dried blood	{ Byproducts of	~ 13
Hoof and horn meal	{ meat processing	7–16
Ureaform	Ureaformaldehyde polymers	21–38
IBDU	Isobutylidene diurea	32
SCU	Sulphur-coated urea	37–40

Reactions in the soil

SCORCH AND RETARDED GERMINATION

There is a possibility of retarded germination, injury to young roots and leaf scorch* because of the concentrated salt solution around dissolving fertilizer granules, particularly if they are combine-drilled into the soil with the seed (Figure 12.1). Generally, fertilizers containing NO_3-N are more harmful than NH_4-N fertilizers; calcium nitrate is satisfactory below 75 kg N ha^{-1} while $(NH_4)_2SO_4$ is safe up to 125 kg N ha^{-1}. Urea is exceptional in being injurious at rates as low as 35 kg N ha^{-1}. The best method of applying urea is to place it in a band, slightly below and to one side of the seed (Figure 12.1), where damage due to high salt concentration and losses by volatilization of NH_3 are minimized. Mixed fertilizers containing N are less injurious to plants than comparable rates of straight N fertilizers.

PH EFFECTS

Ammonia solutions and anhydrous NH_3 produce an alkaline reaction in soil. Urea, whether from animal excreta or fertilizer, is rapidly hydrolysed to $(NH_4)_2CO_3$, which in turn hydrolyses to NH_4OH, CO_2 and H_2O (equation 10.6). The soil pH around a band of dissolving urea may therefore rise above 9 at which

---

*Chlorosis followed by leaf death, usually extending from the tip backwards.

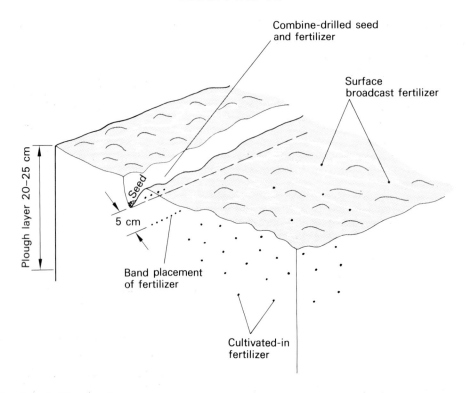

Figure 12.1. Ways of applying fertilizer to soil.

level NH_4OH is unstable and decomposes to NH_3 and H_2O. As the NH_3 volatilizes, the protons released neutralize the OH^- ions from the hydroxide and the pH falls.

A pH rise to 8 and above in the vicinity of an ammonium or urea fertilizer will inhibit *Nitrobacter* organisms; *Nitrosomonas* is not so sensitive and functions well up to pH 9 so that NO_2^- accumulates in the soil. Although oxidation of NH_4^+ to NO_2^- gradually reduces the pH, the high level of NO_2^- may in turn continue to inhibit *Nitrobacter*.

Much of the N from ammonium fertilizers, including urea, that remains in the soil is eventually oxidized to nitrate. Nitrification is an acidifying reaction (equation 10.2) because 2 mol H^+ ion are produced for each mol of NH_4-N that is oxidized. The neutralization of 2 mol H^+ ion would require 1 mol $CaCO_3$ so that the theor-

etical *lime equivalent* of a fertilizer such as $(NH_4)_2SO_4$, where all the N is in ammonium form, is 100 kg $CaCO_3$ per 14 kg N, or 7:1. The theoretical lime equivalent of NH_4NO_3 is 3.5. Thus, regular heavy dressings of ammonium fertilizers will acidify the soil unless liming is practised (section 11.3). One way of countering the acidifying effect of ammonium fertilizers has been to produce mixed fertilizers of NH_4NO_3 and $CaCO_3$, such as 'Nitrochalk' (26 per cent N) and ammonium nitrate limestone (20 per cent N), but these are not as popular as the higher analysis straight N fertilizers or mixed N fertilizers. In practice, the acidity developed is not as great as is theoretically possible because: (a) some NH_3 may be volatilized; (b) NH_4^+ ions may be immobilized by microorganisms or taken up by plants; and (c) NH_4^+ ions may be 'fixed' in the lattice of micaceous clays (section 7.2).

NITRIFICATION AND CONSEQUENT N LOSSES

The NO_3^- produced by nitrification is susceptible to loss by leaching out of the root zone (section 10.2) and by denitrification (section 8.4). *Direct* leaching of fertilizer, that is the removal of soluble N in surface or subsurface runoff into streams, is only likely to occur when heavy rainfall occurs soon after fertilizer application. This has been observed with cereals and N-fertilized grassland in spring in Britain (Figure 12.2).

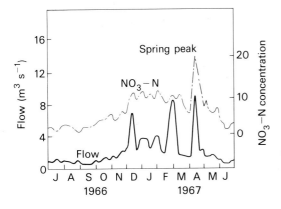

Figure 12.2. Nitrate levels in a river draining fertile agricultural land (after Tomlinson, 1971).

But even after the fertilizer N has entered the soil biomass–organic matter pool, it becomes vulnerable to leaching as the organic N is mineralized and especially when a 'flush' of mineralization follows the rewetting of a dry soil by rain or irrigation (section 10.2). Thus, with the steady increase in the use of N fertilizers in Britain to the current level of 1.6 Mt per year, nitrate concentrations in surface waters and particularly groundwaters have been rising (Figure 12.3). Underground supplies, mainly in aquifers in the Chalk and the Triassic sandstones, make up almost 40 per cent of the potable water used in England and Wales so that maintenance of the quality of this water is vital to public health. Children under 3 months are most at risk from high nitrate in water (> 50 ppm N) because they may develop *methaemoglobinaemia* or 'blue-baby disease', a condition induced by reduction of NO_3^- to NO_2^- in the stomach and the absorption of excess NO_2^-

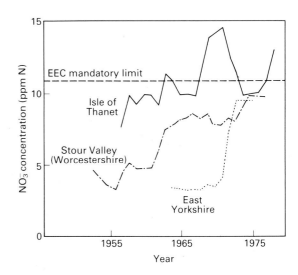

Figure 12.3. Time trends in nitrate concentration in some underground waters in Britain (after Wilkinson and Greene, 1982).

into the blood. A nitrate concentration < 11.3 ppm N is recommended by the World Health Organization for European waters: 11.3–22.6 ppm is considered acceptable and > 22.6 ppm unacceptable. The latest European Community directive specifies a *maximum* acceptable level of 11.3 ppm N and sets the 'guide level' at 5.7 ppm.

Nitrification also leads to accelerated cation leaching (section 10.4), which is another reason why, unless a soil is highly calcareous, regular liming should accompany heavy dressings of NH_4-N fertilizers.

Efficiency of utilization

RATE AND METHOD OF APPLICATION

The recovery of fertilizer N in the crop or animal product depends on the climate, the kind of crop and on the rate, method and timing of the fertilizer application. Potentially the best recovery is achieved when temperature and other factors allow the crop to be grown during the wetter part of the year, as can be done in much of the tropics. For a given climate, N recoveries are usually lowest for plantation crops and short-season arable crops, and highest for permanent grass. Bananas in the Ivory Coast, for example, recovered only 14 per cent of the 430 kg N ha⁻¹ applied, but grass recovered 53 per cent at rates up to 700 kg ha⁻¹. Maize grown at different locations in the USA, Central America and Brazil recovered about 56 per cent of fertilizer N at application rates up to c. 150 kg N ha⁻¹. In Britain, where leaching losses in the wet winter months are appreciable, crop recovery of fertilizer N averages about 50 per cent and recovery by grass fertilized at rates of up to 400 kg N ha⁻¹ y⁻¹ ranges between 50 and 65 per cent. However, when grass is grazed by cattle, recovery of N in the milk and meat is only 5–15 per cent of the input; the other 85–95 per cent is returned to the soil in dung and urine, and, under intensive grazing management, much is lost by a combination of leaching, denitrification and volatilization.

Timing of N applications is crucial for the best utilization by the crop to occur. Spring applications in Britain give better results than those in autumn, especially in areas with > 675 mm annual precipitation (and > 125 mm of excess winter rainfall*). Little nitrate remains in the soil in spring when the excess winter rainfall exceeds 250 mm. During the grass-growing season, N is best applied as a *split dressing*, say 4 or 5 times during the season, or after each cut of hay or silage that is made. In this way, the supply of N is better matched to the growth cycle of the grass and losses by leaching are reduced.

Similarly, in the southern maize belt of the USA it has been found that delaying N application until 3–4 weeks after germination avoids leaching losses from late spring rains, which can be heavy, and provides N at the stage of growth when the plant's demand is highest.

RESIDUAL EFFECTS

A residual effect occurs when sufficient of the fertilizer applied to one crop remains available in the soil to benefit the growth of a succeeding crop. For N fertilizer, residual effects depend on the rate of N applied, the crop, the rain falling between cropping periods and the soil type. Extensive data obtained from

*Defined as the excess of precipitation over evaporation when the soil is at field capacity during the winter months.

field and lysimeter studies in Britain indicate that the nitrate concentration in drainage from arable soils ranges between 5 and 25 ppm N on average, whereas that from fertilized grassland lies between 2 and 10 ppm N (although in the case of intensively managed grazed swards, concentrations may exceed 50 ppm N). It is therefore possible to construct a schedule of probable N leaching losses as shown in Table 12.3. As excess winter rainfall increases, one moves across the table from left to right and diagonally upwards, because nitrate concentration usually falls as the cumulative drainage volume exceeds 100 mm or more.

Table 12.3. Estimated N losses by leaching in relation to excess winter rainfall in Britain.

Mean NO_3-N concentration in drainage water (ppm)	Quantity of N (kg ha⁻¹) leached by an excess winter rainfall (mm) of:				
	100	200	300	400	500
5	5	10	15	20	25
10	10	20	30	40	50
15	15	30	45	60	75
20	20	40	60	80	100

Thus, for an excess winter rainfall of 250–300 mm and allowing for crop uptake, there will be little residual N from a fertilizer application of 100 kg N ha⁻¹ to a preceding crop. The situation can be exacerbated when heavy rain falls on well-structured soils that have been recently fertilized because soil and fertilizer NO_3^- can be washed rapidly down large pores into the drainage water (see Figure 10.6b), even though the soil may be below field capacity.

Residual effects have been observed when deep-rooting cereals, such as winter wheat, follow root crops to which heavy N dressings have been applied (Table 12.4a). In other cases, such as cereal following cereal, part of any residual effect is undoubtedly due to the larger residue of N-rich plant material in the soil, which decomposes to release mineral N. When the preceding crop is a legume, the stimulatory effect of the residues is generally greater than that of even a heavily fertilized grass (Table 12.4b).

Table 12.4. Residual effects of N-fertilized crops and legumes.

(a) Wheat yield (t grain ha^{-1}) following
      potatoes given

	0	and	185 kg N ha^{-1}
	2.84		3.91

(b) Wheat yield (t grain ha^{-1}) following

	N-fertilized ryegrass	clover ley
	4.83	6.12

(After Cooke, 1969.)

12.3 PHOSPHATE FERTILIZERS

Forms of phosphate fertilizer

Phosphate fertilizers may be subdivided into the *orthophosphates*, the condensed orthophosphates or *polyphosphates*, and the insoluble *mineral* and *organic phosphates* (Table 12.5). The natural rock phosphates, consisting of the minerals *apatite* and *fluorapatite* with calcite and other impurities, are the raw material from which other P fertilizers are made by processes that render part or all of the P soluble in water.

ORTHO-P FERTILIZERS

The active constituents are *monocalcium phosphate* (MCP), formula $Ca(H_2PO_4)_2 . H_2O$, and *phosphoric acid*. The former compound is the basis of the *superphosphates* formed by the dissolution of rock phosphate in H_2SO_4 or H_3PO_4; for example:

$$Ca_{10}(PO_4)_6F_2 + 7H_2SO_4 + 3H_2O \rightarrow 3Ca(H_2PO_4)_2 . H_2O$$
$$+ 7CaSO_4 + 2HF \quad (12.1)$$
(single superphosphate)

If rock phosphate is reacted with excess H_2SO_4, wet-process H_3PO_4 is produced, which is then used to dissolve fresh rock phosphate and give higher grades of superphosphate, such as *triple superphosphate* (TSP):

$$Ca_{10}(PO_4)_6F_2 + 14H_3PO_4 + 10H_2O$$
$$\rightarrow 10Ca(H_2PO_4)_2 . H_2O + 2HF \quad (12.2)$$
(triple superphosphate)

Table 12.5. Forms of phosphate fertilizer.

Fertilizer	Composition	P content (%)
Ortho-P		
Phosphoric acid	H_3PO_4	23
Normal or single superphosphate	$Ca(H_2PO_4)_2$; $CaSO_4$	8–10
Concentrated or triple superphosphate	$Ca(H_2PO_4)_2$	19–21
Nitric phosphate	$NH_4H_2PO_4$; $CaHPO_4$ and NH_4NO_3	6–11
Monoammonium phosphate	$NH_4H_2PO_4$	21–26
Diammonium phosphate	$(NH_4)_2HPO_4$	20–23
Poly-P		
Superphosphoric acid	$H_4P_2O_7$ and higher MW polymers	> 33
Ammonium polyphosphate	$(NH_4)_4P_2O_7$ and higher MW polymers	26–27
Insoluble phosphates		
Rock phosphate ore	$Ca_{10}(PO_4)_6(F, OH)_2 . nCaCO_3$ and sesquioxide impurities	6–18
Basic slag	Basic Ca, Mg phosphates, Fe_2O_3 and $CaSiO_3$	3–10
Organic phosphates	Bone meal: guano	5–13

If insufficient H_3PO_4 is used to convert all the rock phosphate to MCP, a partially acidulated phosphate rock (PAPR) fertilizer is produced. Single superphosphate has the advantage of containing 14 per cent S which is of additional value on sulphur-deficient soils.

Roasting rock phosphate ore in an electric furnace with quartz and coke reduces the phosphate minerals to elemental P, which when burnt in oxygen forms the oxide, P_2O_5. This oxide combines with water to form phosphoric acid according to the reaction:

$$P_2O_5 + 3H_2O \rightarrow 2H_3PO_4 \quad (12.3)$$

Phosphoric acid normally contains 52–54 per cent P_2O_5 (23–24 per cent P). Partial neutralization of the acid with NH_3 gives the multinutrient fertilizers, MAP and DAP, discussed earlier (section 12.1). Alternatively, ordinary superphosphate can be mixed with aqua or anhydrous NH_3 to form *ammoniated superphosphate* (*c.* 4 per cent N and 7 per cent P); and TSP can be ammoniated to give a product containing 8 per cent N and 14 per cent P. Unlike MAP and DAP, the ammoniated superphosphates contain some S. About half the P present is water soluble, but all of it is soluble in neutral ammonium citrate (see below).

Nitric phosphates are made by dissolving rock phosphate in concentrated HNO_3 according to the reaction:

$$Ca_{10}(PO_4)_6F_2 + 14HNO_3 \rightarrow 3Ca(H_2PO_4)_2$$
$$+ 7Ca(NO_3)_2 + 2HF \uparrow \quad (12.4)$$

$Ca(NO_3)_2$ solidifies from the mixture on cooling and the remaining solution is ammoniated to give the salts $CaHPO_4$, $NH_4H_2PO_4$ and NH_4NO_3. Consequently, about half of the P is water soluble.

This fertilizer was introduced in the 1950s when a world shortage of S for H_2SO_4 manufacture was thought imminent, but now it is popular only in continental Europe.

The relationship between these ortho-P fertilizers is summarized in Figure 12.4. They supply the bulk of the world's P fertilizer.

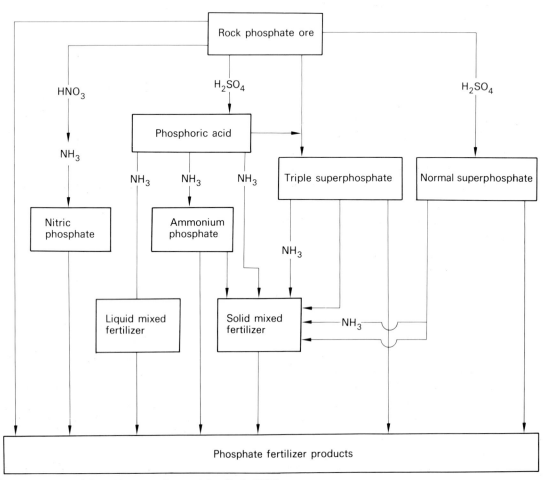

Figure 12.4. Flow diagram of P fertilizer manufacture (after Slack, 1967).

POLY-P FERTILIZERS

Normal H_3PO_4 can be upgraded to *superphosphoric acid* (72 per cent P_2O_5) by reducing the ratio of water to P_2O_5. This acid consists of a mixture of pyrophosphoric acid ($H_4P_2O_7$) and higher molecular weight polymers, as for example:

$$3P_2O_5 + 5H_2O \rightarrow 2H_5P_3O_{10} \qquad (12.5)$$
$$\text{(tripolyphosphoric acid)}$$

Ammoniation of superphosphoric acid produces very soluble fertilizers of high N and P analysis—the *ammonium polyphosphates*. Much less soluble polyphosphate fertilizers, such as the calcium and potassium metaphosphates, have also been produced but are not widely used.

INSOLUBLE PHOSPHATES

Rock phosphates are associated with some igneous and sedimentary rocks. In sedimentary deposits, the crystals of apatite and fluorapatite may be cemented together by calcium carbonate to form the mineral, *francolite*. Crystal size and the magnitude of clay, sesquioxide and silica impurities determine the solubility of the ore in water and acids. *Beneficiation*, the removal of much of the impurity by flotation and washing in water, is the first step in raising the P content of the crushed ore (to 13–18 per cent). Finely ground beneficiated rock phosphate (> 90 per cent passing through a 100-mesh sieve) is used as a fertilizer.

Ground rock phosphate (GRP) costs less per kg of P than soluble P fertilizers, but it is only successful on acid soils for slow-growing grass or tree crops, for which repeated dressings of soluble P fertilizer are uneconomic. Other rock phosphate fertilizers are made by heating finely ground ore to 1200°C and above, a process called *calcination*, or by heating the ore with soda ash and silica (the product is called Rhenania phosphate in Germany). The roasting ignites organic impurities and converts fluorapatite to the more soluble *β-tricalcium phosphate*, $Ca_3(PO_4)_2$.

Sometimes GRP is mixed with *basic slag*, a by-product of steel manufacture from phosphatic iron ores. Basic slag must also be finely ground, and it is best used on acid soils where its liming effect (equivalent to approximately two-thirds its weight of ground limestone) and its micronutrient impurities are beneficial, especially for legumes.

The efficacy of these insoluble fertilizers during the first season after their application is roughly related to their *citrate solubility*, which is measured in 2 per cent citric acid in the UK, M ammonium citrate at pH 7 in the USA, or ammonium citrate at pH 9 on the continent. The availability to plants of several insoluble phosphates compared to superphosphate is illustrated in Figure 12.5.

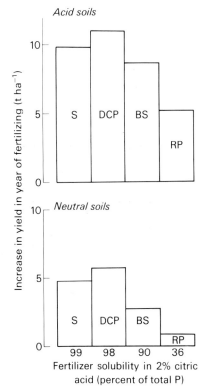

[S = superphosphate; DCP = dicalcium phosphate; BS = basic slag; RP = rock phosphate]

Figure 12.5. Potato yield response in relation to the citrate solubility of P fertilizers (after Mattingly, 1963).

Reactions in the soil

When a granule of MCP, the active constituent of
superphosphate, is placed in dry soil it takes up water
by vapour diffusion, or by osmosis in a moist soil, until a
saturated salt solution forms which flows outwards
from the granule. This solution is very acid (pH ~ 1.5)
and concentrated in P (*c.* 4 M) and Ca (*c.* 1.4 M). It
reacts with the soil minerals, dissolving large amounts
of Ca, Al, Fe and Mn, some of which are subsequently
precipitated as new compounds, the *soil-fertilizer
reaction products*, as the pH slowly rises and solubility
products are exceeded. An MCP granule of 5 mm in
diameter dissolves in 24–36 hours to leave a residue of
dicalcium phosphate dihydrate (DCPD) and anhydrous dicalcium phosphate (DCP) containing about 20

per cent of the original P. The sequence of events is
summarized in Figure 12.6.

The rapidity of this reaction means that the plant
feeds not so much on the fertilizer itself, but on the
fertilizer reaction products, which are metastable and
revert slowly to thermodynamically more stable (but
less soluble) products. Depending on whether the soil
is calcareous or rich in sesquioxides, for example, and
on the presence of other salts, such as NH_4Cl,
$(NH_4)_2SO_4$ or KCl, within the granule, intermediate
products ranging from potassium and ammonium
taranakites $(H_6K_3Al_5(PO_4)_8 . 18H_2O$ and $H_6(NH_4)_3$
$Ak_5(PO_4)_8 . 18H_2O)$, complex calcium–aluminium and
calcium–iron phosphates, to DCPD are formed. The
complex Al and Fe compounds slowly hydrolyse to
amorphous $AlPO_4$ and $FePO_4$, respectively. In acid
soils, DCPD dissolves congruently, but at pH > 6.5 it
hydrolyses to form a less soluble residue, probably

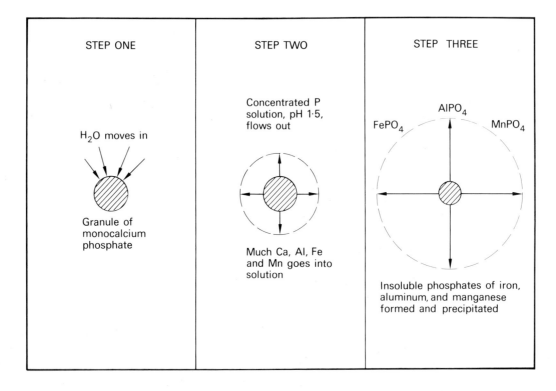

Figure 12.6. Successive stages in the dissolution of an MCP granule in soil (after Tisdale and Nelson, 1975).

octacalcium phosphate (OCP), which in more basic soils may revert to *hydroxyapatite* (HA). Taking account of the interaction of fertilizer type, time and soil, a wide spectrum of P availability confronts the plant, as illustrated in Figure 12.7.

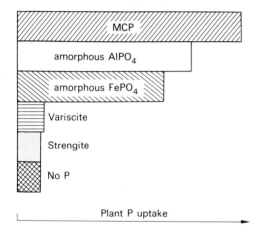

Figure 12.7. Relative availability of different P sources to maize (after Huffman, 1962).

POWDERED VS GRANULAR FORMS

Insoluble fertilizers, such as the basic calcium phosphates, are most effective when used in a finely divided form. Hygroscopic absorption of water does not cause the problem that it does with soluble fertilizers, but because fine powders are difficult to apply, a form of weak aggregation or 'mini-granulating' of the powdered fertilizer is sometimes employed. By contrast, granulation of soluble P fertilizers has the advantage, in addition to ease of handling, of confining the reaction between soil and fertilizer to small volumes around each granule. The rate of phosphate fixation in the soil is therefore diminished.

RESIDUAL EFFECTS

Only 10–20 per cent of fertilizer P may be absorbed by plants during the first year: the remainder is nearly all retained as fertilizer reaction products, which become less soluble with time. A rough guideline is that two-thirds of the water and citrate-soluble P remains after one crop, one-third after two crops, one-sixth after three crops and none after four crops.

Larsen (1971) has suggested that a more precise way of assessing the residual effect is to measure the rate at which the labile pool of soil phosphate (L value) declines after the addition of fertilizer, and to calculate the half-life for the 'decay' of the fertilizer's value to the crop. Half-lives in neutral and alkaline soils vary from 1 to 6 years. Larsen also pointed out that insoluble P fertilizers, in contrast to soluble ones, required a certain time to reach peak effectiveness, as well as having a half-life for immobilization (Figure 12.8).

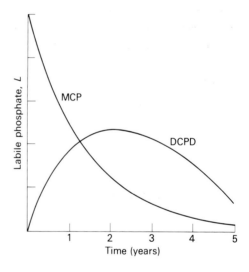

Figure 12.8. Changes in soil labile phosphate after the addition of P fertilizer (after Larsen, 1971).

Eutrophication

CAUSE AND EFFECTS

Eutrophication is a *natural* process whereby surface waters are gradually enriched with organic and inorganic nutrients carried in runoff from the surrounding catchment. Natural eutrophication is evidenced by the formation of fen peats, in both temperate and tropical climates, where base-rich waters collect in drainage depressions. Nevertheless, eutrophication can become a problem when it is *accelerated* by increased loads of sediment and dissolved nutrients in the streams draining urban and rural areas: the higher levels of P, N, Si and organic matter promote the growth of aquatic plants, from minute phytoplankton to rooted higher plants.

Surges of phytoplankton growth are called *algal blooms*. These detract from the amenity of the water and pose problems for water treatment works, because of the fouling of primary filter beds and tainting of the water by essential oils released by the algae. Proliferation of aquatic weeds makes a stream more sluggish, leading to greater sediment deposition and eventual blocking of the channel. Dead vegetation sinks to the bottom where a thick layer of decomposing organic sludge accumulates, consuming large quantities of dissolved oxygen and creating anoxic conditions, which may result in fish death.

Experience indicates that accelerated eutrophication and the appearance of algal blooms are most likely to occur:
(a) in standing bodies of water;
(b) during the high-light, high-temperature conditions of summer;
(c) when the P concentration of the water exceeds *c.* 0.05 ppm and the ratio of dissolved N : P is ~ 20 or less.

SOURCES OF NUTRIENT INPUT
Dissolved and particulate nutrient loads arise from *point* and *non-point* (diffuse) sources. Examples of point sources are effluents from factories, sewage works and intensive livestock units, such as large cattle feedlots and dairies. Non-point sources comprise the drainage from the remaining agricultural and non-agricultural land, grassland, field and forest, which by its very nature is difficult to monitor accurately.

In the UK, the bulk of the N-enrichment of surface waters comes from agricultural land where each year drainage from 1 hectare may carry between 15 and 150 kg N, depending on the land use and fertilizers applied. Conversely, between two-thirds and four-fifths of the phosphate in surface waters appears to come from point sources, especially intensive animal farms, which release much organic P, and sewage works which handle the domestic and industrial detergent load. In contrast to the mobile nitrate anion, orthophosphate ions are strongly adsorbed (section 7.3) and the reaction products formed by P fertilizers in soil are relatively insoluble so that leaching usually contributes < 1 kg P ha^{-1} y^{-1}. Exceptionally losses may rise into the 1–10 kg P ha^{-1} y^{-1} range, when P-rich surface soil is eroded, when organic manure is spread copiously on soil that is either frozen or very wet, or when P fertilizer is liberally applied to very acid peaty soils.

12.4 OTHER FERTILIZERS INCLUDING MICRONUTRIENT FERTILIZERS

Potassium fertilizers
Potassium occurs naturally as KCl (*muriate of potash*) in salt deposits, which include some NaCl, K_2SO_4 and $MgSO_4$. The KCl is separated by flotation or by recrystallization, which depends on the differential response of the solubility of these salts to a change in temperature.

KCl is the most widely used K fertilizer; it is especially favoured in liquid mixed fertilizers because of its high solubility. Other K fertilizers, representing a wide range of K content and water solubility, are listed in Table 12.6. Potassium sulphate is generally made by reacting SO_3 with water and KCl, that is:

$$2KCl + SO_3 + H_2O \rightarrow K_2SO_4 + 2HCl \quad (12.6)$$

It is less hygroscopic than KCl and therefore easier to handle. Despite its higher cost per kg K, it is preferred as a source of K for tobacco, beets, maize and potatoes because of the sensitivity of these crops to high levels of Cl. Potassium nitrate is also favoured for tobacco because it contains no Cl and all the N is present as NO_3^-, the preferred form of N for this crop. Potassium metaphosphate (KMP) and potassium calcium pyrophosphate (KCP) are only slightly soluble in water, but

Table 12.6. Forms of potassium fertilizer.

Fertilizer	Composition	K content (%)
Potassium chloride	KCl	52
Potassium sulphate	K_2SO_4	44
Potassium nitrate	KNO_3	36
Potassium calcium pyrophosphate	$K_2CaP_2O_7$	21–25
Potassium metaphosphate	KPO_3	28–33
Kainit	$KCl.MgSO_4.3H_2O$	10–25

they are citrate-soluble. *Kainit* is an insoluble slow-release K fertilizer.

K balance in cropping

Since the potassium immediately available to plants is held as exchangeable cations, which are not readily leached nor rendered unexchangeable, it is feasible to attempt to balance K removed by crops and in animal products with K applied in fertilizers and released from non-exchangeable sources in the soil.

K RELEASE

Slow release of non-exchangeable K depends on the content of potash feldspars and micaceous clay minerals. Soils with such reserves, even though low in exchangeable K, may supply between 20 and 80 kg $ha^{-1} y^{-1}$ for many years, especially to grasses which are efficient in taking up K at low concentrations; but sandy soils lacking micaceous clays are soon exhausted of K by continuous cropping. On many soils where potassium is adequate initially, regular N and P fertilizing may so stimulate growth that K-deficiency symptoms appear, more often on pasture land where fodder conservation (as hay or silage) is practised.

K UPTAKE BY CROPS

Recovery of K fertilizer during one season varies from 25 to 80 per cent, being highest for grassland, which may remove up to 450 kg K $ha^{-1} y^{-1}$ when highly productive. Clovers and root crops require higher soluble K levels in soil than grasses, so that the ratio of N:K fertilizer applied should be 2:3 compared to

1.5:1 for temperate grasses and 2:1 for tropical grasses. If too much K is used on grazed grassland, luxury uptake of K may induce Mg deficiency in the herbage, which predisposes to *grass tetany* or *hypomagnesaemia* in the stock.

Sulphur fertilizers

Sulphur for fertilizers is produced by roasting iron pyrites, or is recovered from natural gas containing H_2S, or mined as elemental S. For use on its own, it may be made into prills from molten S and clay (up to 90 per cent S). In the soil, S is slowly oxidized to SO_4^{2-} by *Thiobacillus* bacteria according to the reaction:

$$S+3/2O_2+H_2O \rightarrow 2H^+ + SO_4^{2-} \quad (12.7)$$

Sulphur also occurs as $CaSO_4$ in single superphosphate and in agricultural gypsum ($CaSO_4.2H_2O$), which is a byproduct in the manufacture of triple and concentrated superphosphates. $(NH_4)_2SO_4$, containing 24 per cent S, is the most soluble sulphate salt in fertilizers and is used in liquid mixed fertilizers.

In Britain, as in other industrialized countries, amounts of SO_4-S up to 70 kg S ha^{-1} descend each year from the atmosphere in rain or as dry deposition (section 10.3). Because crop requirements average 20–30 kg S ha^{-1} (exceptionally > 50 kg S ha^{-1} for some *Brassica* species), instances of S deficiency in crops and animals in these countries are few, in contrast to areas of the tropics and subtropics where the soils are highly leached and atmospheric inputs of S low (for example northern Australia). However, emissions of S to the atmosphere over Britain have fallen by about 30 per cent since 1970, and are destined to fall further, so that the need for fertilizer S may become more widespread in the future. This is particularly so for grassland that is heavily fertilized with N for which the N:S ratio in the herbage should be kept near to 10:1 for the optimum nutrition of ruminant animals.

Micronutrient fertilizers

The micronutrients Fe, Mn, Zn, Cu, B and Mo may need to be applied individually, or in mixed fertilizers, to correct a soil deficiency. Cobalt may also be required to correct a vitamin B_{12} deficiency in ruminants and for effective N_2 fixation by legumes.

These elements can be added to the soil as sparingly soluble salts or oxides, or as chelates (section 10.5). When used in the latter form, or applied as foliar sprays of soluble salts, such as $ZnSO_4.7H_2O$, they are intended for immediate uptake by the plant. In their less soluble forms they maintain a very low concentration in the soil solution for several years, thus providing a prolonged residual effect. An increasingly popular slow-release form is the *frit* made by fusing the element in a glass that can be crushed and mixed with NPK fertilizers. Table 12.7 gives the normal concentrations of these elements in mixed fertilizers and the maximum rate of application considered safe for most crops.

Table 12.7. Application rates for micronutrients in fertilizer.

Element	Content (%)	Safe maximum (kg ha^{-1})
Mo	0.005	0.2
B	0.025	1.0
Mn	0.15	10
Cu	0.15	10
Zn	0.25	10
Fe	0.25	10

(After Jones, 1982.)

12.5 PESTICIDES IN THE SOIL

Environmental impact

The term 'pesticide' embraces the many natural or synthetic chemical compounds, which have biocidal or biostatic effects: they can be more specifically designated *insecticides, miticides, nematicides, fungicides, herbicides, rodenticides* or *molluscicides* according to the group of organisms they are intended to control. Of these, the herbicides, insecticides and fungicides account for the bulk of agricultural usage, in that order.

The ideal pesticide is one that controls only the target organism and persists long enough to complete only this purpose before degrading into harmless products. It must also be of low toxicity to mammals. In practice, however, the ideal is not always achieved and chemicals have been used in agriculture (and in public health programmes, such as for the eradication of the malaria mosquito) that are 'broad spectrum' in their activity; that is, they kill harmless and beneficial organisms as well as the target organisms. Furthermore, there are pesticides, such as the *organochlorine insecticides*, that are very stable and have the added property of high lipid solubility. They therefore accumulate in the fatty tissues of animals, particularly predators high up in natural food chains. Indiscriminate application of these and other persistent chemicals has led to their residues and breakdown products becoming widely disseminated, and there have been many instances of their detrimental effect on beneficial insects and plants, domestic animals and wildlife.

As a result of public concern over such malpractices, pesticide manufacturers voluntarily, or in response to government legislation, now undertake extensive research into the development and testing of new chemicals before their release as pesticides, as shown in Figure 12.9. These procedures are designed to evaluate not only the efficacy of a chemical for the control of specific pests but also its *environmental impact*; that is, its effect on non-target organisms, particularly mammals, and its persistence in the environment.

The soil plays a key role in determining the fate of a pesticide in the environment. Nematicides, many insecticides and fungicides are deliberately applied to soil to control soil-inhabiting pests. There are also soil-applied herbicides and an increasing number of systemic insecticides and fungicides intended for absorption by roots and transport to the aerial parts of plants. An even wider range of compounds reaches the soil adventitiously because of rainwash from leaves and the incorporation of treated plant residues. Much attention has therefore been paid to improving the efficiency of pesticide application to target organisms, such as through the use of ultra-low volume sprays and granulation or microencapsulation to delay the release of an active constituent.

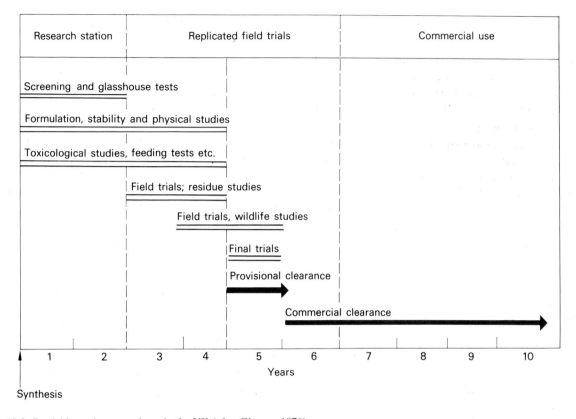

Figure 12.9. Pesticide testing procedures in the UK (after Glasser, 1976).

Pesticide persistence in soil

The main factors governing pesticide stability in soil and the loss of pesticide to surface waters and air may be summarized as follows:
(a) volatilization;
(b) adsorption by soil solids;
(c) chemical and biological decomposition;
(d) transport in the air, liquid or solid phase;
(e) absorption by plants and animals.

The relative importance of each process depends very much on the *properties of the chemical*: its water solubility, volatility, affinity for organic or mineral surfaces; its *formulation* and *mode of application*— whether presented as a spray, dust or in granules, applied to crop surfaces or cultivated into the soil; the *environmental conditions* (temperature and rainfall); the *soil type* and *cropping system.*

VOLATILIZATION

Volatilization can be a major mechanism of pesticide loss from soil, and for very persistent chemicals, such as the organochlorines, the means whereby they can become widely dispersed in the environment. For example, for a relatively volatile compound like *lindane* (Table 12.8), the initial volatilization rate from the surface of a moist soil can be of the order of 0.03 kg ha^{-1} d^{-1}. Incorporating the pesticide into the soil reduces vapour losses substantially. However, pesticide in solution will move to the surface by mass flow as water evaporates (pesticides also diffuse in response to

Table 12.8. Description and key properties of four important organochlorine insecticides.

Common name	Chemical name	Solubility in water (ppm)	Vapour pressure at 30°C (mm Hg)
DDT	1,1-bis(4-chlorophenyl)-2,2,2-trichloroethane	0.001–0.04	7×10^{-7}
Dieldrin (HEOD)	1,2,3,4,10,10-hexa-Chlorocyclopentadiene	0.1–0.25	1×10^{-5}
Lindane (γ-BHC)	γ-1,2,3,4,5,6-hexa-Chlorocyclohexane	7.3–10.0	13×10^{-5}
Toxaphene	Chlorinated camphene containing c. 68% chlorine	0.4	No data

(After Guenzi and others, 1974.)

concentration gradients, but except for the most soluble chemicals the diffusive flux in the soil solution is small when compared with the convective flux). Because water is generally more volatile than the pesticide, the concentration of the latter rises as water evaporates, thereby increasing its vapour pressure and the rate of volatilization. This is known as 'wick action'. However, once the soil surface has become very dry (less than the equivalent of a monolayer of water molecules adhering to the solid surfaces), pesticides become much more strongly adsorbed and their vapour pressure in the soil air is reduced accordingly (Figure 12.10). Retention of a pesticide at the soil surface can hasten the degradation of photo-sensitive compounds.

ADSORPTION

The distribution of a pesticide between solid and liquid phases in soil can be described by an adsorption iso-therm. If the isotherm is linear, or approximately so, the distribution coefficient K_d is given by the equation:

$$Q = K_d C \qquad (12.8)$$

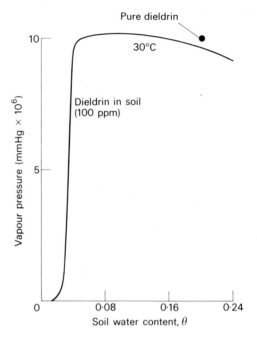

Figure 12.10. Effect of soil water content on the vapour pressure of dieldrin (after Spencer, 1970).

where Q is the amount adsorbed per unit mass of soil and C is the pesticide concentration in solution at equilibrium. The range of K_d values for some common pesticides is given in Table 12.9. For many non-ionized, non-polar molecules, organic matter provides the most important adsorbing surface in the soil. The adsorption characteristics of these chemicals may therefore be expressed by the distribution coefficient, K_{org}, defined as the amount adsorbed per unit mass of organic matter divided by the equilibrium solution concentration. K_d/soil organic matter content, that is K_{org} values, for different pesticides can be estimated with reasonable confidence from their partition coefficient between water and an organic solvent, such as octanol; that is:

$$P^{oct}_{water} = \frac{[\text{Pesticide concentration}]\,\text{octanol}}{[\text{Pesticide concentration}]\,\text{water}} \qquad (12.9)$$

The bipyridyl herbicides, *diquat* and *paraquat*, which have the chemical formula:

and

are very strongly adsorbed by clays because of non-specific (coulombic) and specific adsorption forces (section 7.1). Adsorption renders paraquat inert and also induces a shift in the wavelength of maximum light absorption into the range of sunlight so that the molecule becomes photochemically unstable when adsorbed.

Another group, the triazine herbicides, of which *simazine* is an example:

are very weak bases and in acid media the molecule becomes protonated at one of the amino groups. This enhances its adsorption by clays. Other herbicides, such as the *phenoxyalkanoic acids*, exist as anions at normal soil pH values and hence are not appreciably adsorbed, except on sesquioxides and by hydrogen-bonding to organic matter. The more common representatives of this group are:

2,4-D   2,4-dichlorophenoxyacetic acid
2,4,5-T   2,4,5-trichlorophenoxyacetic acid
MCPA   (4-chloro-o-tolyl)oxyacetic acid

The herbicide *glyphosate* is an interesting example of a chemical that at normal soil pH values probably exists as the zwitterion:

$$^-OCOCH_2\overset{+}{N}H_2CH_2PO_3H^-$$

Because the charged groups are all pH-dependent, the net charge on the molecule changes with soil pH, but it is quite strongly adsorbed over a range of pH.

The combined effects of vapour pressure, water solubility and adsorption affinity can be exemplified by the distribution of several pesticides between solid, liquid and air in soil, calculated by assuming a 50 per cent pore space, of which half is filled with water (Table 12.9). This demonstrates that even for a volatile fumigant, such as ethylene dibromide, the major part of the chemical is retained by the solid phase. This has important consequences for the activity and bio-degradibility of a pesticide, as discussed below.

Table 12.9. Distribution of pesticides in a soil of 50 per cent porosity, half-filled with water.

Compound	K_d (l kg^{-1})	Amount in each phase (%)		
		Air	Liquid	Solid
Ethylene dibromide	0.5	0.7	28.4	70.9
Dimethoate	0.3	1.6×10^{-7}	40	60
Simazine	1.9	1.3×10^{-7}	9.5	90.5
Monuron	2.2	2.0×10^{-7}	8.3	91.7
DDT	5.0×10^4	1.2×10^{-6}	4.0×10^{-4}	100

(After Hartley and Graham-Bryce, 1980.)

CHEMICAL AND BIOLOGICAL
DECOMPOSITION

With the increase in the potency of pesticides, the quantity which is needed for effective pest control has decreased dramatically. For example, at the beginning of this century inorganic pesticides, such as sodium chlorate, lead arsenate and flowers of sulphur, were applied at rates of 10–500 kg ha^{-1}. After World War II, with the advent of organochlorines, organophosphates and selective herbicides, rates fell to 0.5–5 kg ha^{-1}. At present, the new synthetic *pyrethroid* insecticides are effective at rates of 10 g ha^{-1} or less. Thus, the concentration of pesticide in the soil is very small when compared with the other reactive constituents, such as clay, organic matter, etc., so that the disappearance of the chemical usually follows first-order kinetics (cf. equation 10.3).

Some chemicals, such as paraquat (adsorbed state) and the pyrethroids, are degraded photochemically. The natural pyrethroids and the early synthetic compounds decompose within a few hours of exposure to sunlight. On the other hand, the later synthetic compounds, such as *deltamethrin*, are not only photostable but they are also potent at very low concentrations (Table 12.10). The stable synthetic pyrethroids are now being used against many pests previously controlled by DDT because of their potency and minimal environmental impact.

The pyrethroids are also rapidly decomposed by microorganisms in the soil. The same is true of many other pesticides, notable exceptions being DDT and the cyclodienes, such as *dieldrin* and *lindane*. Adsorption of a chemical renders it physiologically

Table 12.10. Activity of some chemical insecticides.

Chemical	Target organism	Median effective dose* (mg/kg)
DDT	Housefly	10
Parathion	Housefly	2
Synergized deltamethrin	Housefly	0.002

*Dose that kills 50 per cent of the population.
(After Graham-Bryce, 1983.)

inactive, although, if it is reversibly adsorbed, its removal from solution by decomposition will cause more to be desorbed and become active. The persistence of the bipyridyls is due to their irreversible adsorption by clays when used at normal rates. But it is also argued that adsorption of a chemical, especially on organic surfaces, may place it in closer proximity to colonies of bacteria or extracellular enzymes, thus hastening its decomposition. Therefore, no sound generalization on the effect of adsorption on the rate of biodegradation of pesticides in soil can be made. Maximum persistence times of the major groups of pesticides, measured as the time for 90 per cent or more of the chemical to disappear from the site of application, are shown in Figure 12.11.

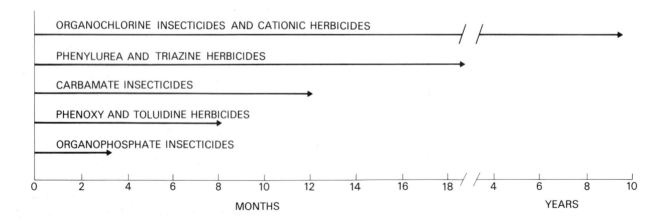

Figure 12.11. Maximum persistence of pesticides in soil under a mild climate (after Stewart *et al.*, 1975).

Biodegradation of the phenoxyalkanoic acid herbicides illustrates a phenomenon also observed with the phenylureas, organophosphates and carbamate insecticides when they are added to a soil not previously exposed to the compound in question. An initial lag phase during which the herbicide concentration remains constant is followed by a period of rapid disappearance (curve A, Figure 12.12). Soil organisms, principally bacteria, which are capable of metabolizing the foreign molecule apparently multiply during the lag phase; these adapted organisms enjoy a competitive advantage for substrate over non-adapted organisms, resulting in their proliferation and the breakdown of the herbicide. The soil is then said to be *enriched* with adapted organisms, a condition persisting for several months in the absence of fresh herbicide. Herbicide reapplied to an enriched soil is detoxified without any lag phase (curve B, Figure 12.12). The enhanced biodegradation of pesticides due to enrichment effects in soil occurs more commonly than was once thought: it is speculated that the rapid adaptation of bacteria may be due in part to the free exchange between organisms of genetic material in plasmids.

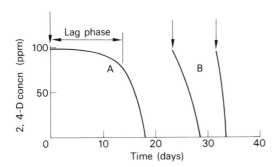

Figure 12.12. Time course of microbial decomposition of 2,4-D in soil (curve A—initial enrichment, curve B—addition of herbicide to enriched soil; arrows indicate time of addition of the herbicide) (after Audus, 1960).

Apart from complete detoxification, many soil organisms can partially degrade pesticides that are analogues of normal substrates, a process called *co-metabolism*. An organism adapted to benzoates, for example, may oxidize a chlorobenzoate herbicide to a catechol derivative, which is more prone to oxidation by the general soil microflora than the original compound.

ABSORPTION AND TRANSPORT OF PESTICIDES
Uptake of pesticides by plant roots depends on their water solubility; absorption through the cuticle of leaves depends more on their lipid solubility. Transport in the liquid phase mainly occurs by mass flow, as discussed previously, and therefore depends on the chemical's water solubility and the rate of water flow. Herbicides, like pichloram, the phenylureas (monuron, fenuron) and the phenoxyalkanoics, can be leached but they are usually biodegraded rapidly in soil. Strongly adsorbed pesticides (DDT, paraquat) can be transported when soil particles are removed during erosion. This can lead to residues of these persistent chemicals building up in sediments. However, there is evidence that DDT is decomposed more rapidly to the less toxic DDE under anaerobic conditions, in the presence of plant residues, so that its disappearance from stream and lake sediments may be accelerated.

12.6 SUMMARY

Substances recognized as fertilizers and pesticides have been used for centuries. But in the short period after World War II there has been a phenomenal increase in the quantity and variety of agricultural chemicals to meet the need for greater production and improved quality of the crops harvested.

The main nutrients supplied are N, P and K either singly in *straight* fertilizers or in various combinations in *mixed* or *multinutrient* fertilizers. In industrialized areas of the world, the S needs of most of the agricultural land are supplied through atmospheric deposition. Most of the simple compounds of N— NH_4NO_3, $NaNO_3$, $Ca(NO_3)_2$, $(NH_4)_2SO_4$ and

$(NH_2)_2CO$—are soluble solids, but N is also used as solutions of NH_4NO_3, urea, or NH_3 in water and as liquefied NH_3 gas. The N of these compounds is immediately available to the plant, but it is vulnerable to leaching in the NO_3^- form. Losses range from 5 to 150 kg N ha^{-1} y^{-1} depending on the rate of fertilizer applied, the land use and the excess winter rainfall. The main K compounds, KCl and K_2SO_4, are also very soluble, although K^+ is retained as an exchangeable cation in most soils. The micronutrients Fe, Mn, Zn, Cu, B and Mo are applied at low concentrations in mixed fertilizers, or individually as *chelates* or soluble salts.

The naturally occurring phosphate minerals are most insoluble and must be treated with strong acids (H_2SO_4, HNO_3 and H_3PO_4) to convert the bulk of the P to a soluble form. Monocalcium phosphate, $Ca(H_2PO_4)_2 \cdot H_2O$, is the active constituent of single and triple *superphosphates*, whereas higher analysis P fertilizers (the *polyphosphates*) are made from superphosphoric acid, formed by reducing the amount of water in which P_2O_5 is dissolved. Soluble P fertilizers react quickly with the soil to form less soluble *fertilizer reaction products* on which plants feed.

Ideally, a *pesticide* should be lethal for a specific organism and remain active only long enough to control that pest. Pesticide chemicals are inactivated in the soil by adsorption on to clays or organic matter, by microbial attack or photochemical decomposition. Of the more stable compounds, the organochlorine insecticides have become widely disseminated in the environment due to volatilization from the soil and their accumulation in animal fatty tissues. More soluble compounds, such as the phenoxyacid and urea herbicides, may be leached, but fortunately most of them are decomposed rapidly by soil microorganisms.

REFERENCES

AUDUS L.J. (1960) Microbiological breakdown of herbicides in soils, in *Herbicides and the Soil* (Eds. E.K. Woodford & G.R. Sagar). Blackwell Scientific Publications, Oxford.

COOKE G.W. (1969) Prediction of nitrogen requirements of arable crops in mainly arable cropping systems, in *Nitrogen and Soil Organic Matter*. MAFF Technical Bulletin No 15, pp. 40–60.
GLASSER R.F. (1976) Pesticides: the legal environment, in *Pesticides and Human Welfare* (Eds. D.L. Gunn & J.G.R. Stevens). Oxford University Press, Oxford.
GRAHAM-BRYCE I.J. (1983) Biting the hand that feeds. *Chemistry and Industry* No. 19, 733–739.
GUENZI W.D. (1974) (Ed.) *Pesticides in Soil and Water*. Soil Science Society of America, Madison.
HARTLEY G.S. & GRAHAM-BRYCE I.J. (1980) *Physical Principles of Pesticide Behaviour*. Academic Press, London.
HUFFMAN E.O. (1962) Reactions of phosphate in soils: recent research by TVA. *Proceedings of the Fertilizer Society, London*, 5–35.
JONES U.S. (1982) *Fertilizers and Soil Fertility*, 2nd Ed. Prentice-Hall, Virginia.
LARSEN S. (1971) Residual phosphate in soils, in *Residual Value of Applied Nutrients*. MAFF Technical Bulletin No. 20, 34–40.
MATTINGLY G.E.G. (1963) The agricultural value of some water and citrate soluble phosphate fertilizers: an account of recent work at Rothamsted and elsewhere. *Proceedings of the Fertilizer Society, London*, 57–94.
SLACK A.V. (1967) *Chemistry and Technology of Fertilizers*. Wiley Interscience, New York.
SPENCER W.F. (1970) Distribution of pesticides between soil, water and air, in *Pesticides in the Soil: Ecology, Degradation and Movement*. Michigan State University, East Lansing.
STEWART B.A., WOOLHISER D.A., WISCHMEIER W.H., CARO J.H. & FRERE M.H. (1970) *Control of water pollution from cropland*, Volume 1. Agricultural Research Service and Environmental Protection Agency, Washington D.C.
TOMLINSON T.E. (1971) Nutrient losses from agricultural land. *Outlook on Agriculture* 6, 272–278.
WILKINSON W.B. & GREENE L.A. (1982) The water industry and the nitrogen cycle. *Philosophical Transactions of the Royal Society, London B* 296, 459–475.

FURTHER READING

AGRICULTURE AND WATER QUALITY (1976) MAFF Technical Bulletin No. 32.
COOKE G.W. (1982) *Fertilizing for Maximum Yield*, 3rd Ed. Granada, London.
FOLLETT R.H., MURPHY L.S. & DONAHUE R.L. (1981) *Fertilizers and Soil Amendments*. Prentice-Hall, New Jersey.
GUNN D.L. & STEVENS J.G.R. (1976) (Eds.) *Pesticides and Human Welfare*. Oxford University Press, Oxford.
HUFFMAN E.O. (1968) The reactions of fertilizer phosphate with soils. *Outlook on Agriculture* 5, 202–207.
ROYAL SOCIETY STUDY GROUP REPORT (1983) *The Nitrogen Cycle of the United Kingdom*. The Royal Society, London.

Chapter 13
Problem Soils

13.1 A BROAD PERSPECTIVE

Nutrient deficiencies, soil acidity, structural instability or a soil's susceptibility to erosion can restrict the growth of natural vegetation or agricultural crops: these problems have been explored in the preceding three chapters. Excess water in the soil can also severely limit the use of land for agriculture.

As discussed in sections 9.2 and 9.5, an excess of precipitation over evaporation for several months each year, impermeable subsurface layers and high groundwater tables, separately or combined, induce *waterlogging* and the attendant problems of:
(1) inadequate aeration for root and soil microbial activity;
(2) poor trafficability of the soil for machinery and animals, with the danger of surface structural collapse and 'poaching'*;
(3) invasion by flood-tolerant weeds and an increase in animal parasites and diseases favoured by wet conditions;
(4) slow warming of soil in spring.
The causal climatic factors are beyond control, but an improvement in soil drainage (section 13.4) does much to mitigate these ill effects.

On the other hand, in areas of high evaporation, serious problems occur when the water table rises to within 2 m of the soil surface and *salinization* occurs due to the upward movement and evaporation of saline groundwater (section 6.6). Such a situation may arise when the steady-state equilibrium of the soil's hydrology is disturbed by natural events, such as earth movements (faulting) or long-term climatic change (which may extend over thousands of years), or by man's intervention through changing the land use or

supplying irrigation (usually a time span of tens of years). Problems associated with hydrological disturbances, and their solutions, fall into two categories:
(a) soils not initially saline with deep water tables, which become saline under irrigation or due to a change in land use;
(b) soils already salinized and of limited cropping potential, but requiring careful management to avoid chemical and physical deterioration under irrigation.

13.2 WATER MANAGEMENT FOR SALINITY CONTROL

Permeation of saline groundwater

The existence of large salty lakes and surrounding salt pans in many inland areas is evidence of the confluence of natural drainage in depressions in which salts accumulate following the evaporation of water. In the southwest of Western Australia, many of the valley bottoms are naturally salinized and support a scrubby heath vegetation tolerant of high salinity. Groundwater levels are kept low under the valley slopes by the deep-rooted indigenous Eucalyptus trees (Figure 13.1a); but following forest clearance for arable farming, the rate of deep percolation increases from about 5 to 10 mm y^{-1} and the water table rises in the lower slopes due to the increased hydraulic gradient. Thus, the area of saline soils creeps gradually upslope (Figure 13.1b).

Soils under irrigation

REGULATING THE WATER SUPPLY
Some soils, notably those of the Nile Valley, have been irrigated for centuries without ill effect. When the Nile

---

*Rutting and damage to the wet soil surface by animals.

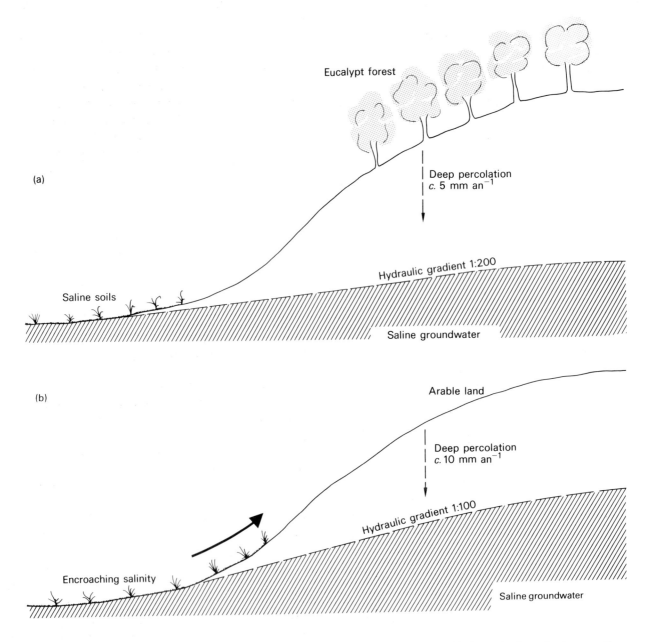

Figure 13.1. **(a)** Natural landscape in south-western Australia in steady-state hydrological equilibrium. **(b)** Same landscape with raised groundwater table after forest clearance (after Holmes, 1971).

flooded in August and September, the water was led onto the land and impounded in basins of 400–16,000 ha for up to 40 days, after which the water receded and crops were planted. The soils were naturally well drained and the annual flooding leached out any accumulated salts. More recently, in order that irrigated crops could be grown throughout the year, dams and barrages have been built to regulate the river's flow. Productivity has increased, but not without cost, because with the application of 1.5–2 m of water annually, most of which is evaporated or transpired by the crop, there has been an inevitable build-up of salts in the soils.

Abundant and regular supplies of water may also encourage farmers to overwater, the excess water contributing to the groundwater, which slowly rises. Another hazard is the seepage loss from canals and ditches that conduct water to the fields: this problem is especially serious in Pakistan where extensive water-logging of the permeable soils of the Indus Valley occurs alongside the irrigation canals, which are long and very wide (Figure 13.2).

Seepage is prevented by lining canals with concrete—the best but most expensive material; asphalt

Figure 13.2. Soils waterlogged by lateral seepage from a large irrigation canal in Pakistan (courtesy of H. van Someren).

or plastic sheeting are cheaper but of limited durability. Water use must be adjusted according to the *soil moisture deficit* (*SMD*). The loss of soil water can be measured directly using a neutron probe (section 6.2). Alternatively, it can be calculated from measured soil water suctions (using tensiometers, or gypsum resistance blocks for suctions > 0.8 bars) and a knowledge of the soil's moisture characteristic curve (section 6.5). More appropriate for large areas is the prediction of *SMD* from evapotranspiration losses (*ET*), calculated from the Penman equation (section 6.6). Thus,

$$ET \text{ (mm day}^{-1}) \times \text{days from last watering} = \text{expected } SMD \text{ per metre depth}$$

Given that available water capacities (*AWC*) normally range from *c.* 80 mm m^{-1} for sandy soils to 200 mm m^{-1} for silty clays and clay loams, an *ET* rate of 6 mm d^{-1} (see Figure 6.14) would remove just over half the available water in 7–17 days, depending on the soil texture, at which point it would be desirable to irrigate again to avoid a check to crop growth.

IRRIGATION METHODS
The method of applying water also influences the likelihood of soil salinization. Near level or uniformly graded land can be irrigated by *flood* or *furrow* methods. *Sprinklers*, *sprays* or *trickle emitters* are suitable for irregular terrain.
(a) With *border check* or *border strip* flooding, water is distributed along bays separated by levees or mounds running parallel to the direction of flow (Figure 13.3). The method is suitable for slopes from 4 to < 1 per cent. Sufficient depth of water (> 50 mm) must be applied at each irrigation to ensure that some water reaches the bottom end of the bay before infiltrating the soil. To avoid overwatering at the input end, the bays should be shorter in permeable soils (< 90 m) than in less permeable clays (up to 270 m). On the lowest gradients (< 1 per cent), construction of levees across the slope at vertical intervals of approximately 5 cm to form rectangular basins gives more uniform water application, and therefore better control of

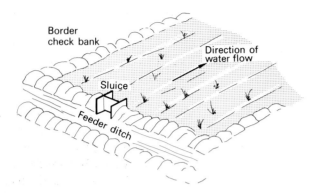

Figure 13.3. Border check flood irrigation.

salinity. However, this method of *basin flooding* is more labour intensive and usually practical only on highly productive orchard and padi rice crops.

(b) *Furrow irrigation* is obviously suited to row crops and can be used on land too steep or uneven to be flooded, provided the furrows are keyed to the contours. But there is the problem of salt redistribution in the soil as water moves by capillarity from the wet furrows to evaporate from the ridges where the plants are growing. The salt concentration at the apex of the ridge may be 5–10 times higher than in the body of the soil, creating an unfavourable environment for seed germination and young seedling growth. Measures to avoid this problem consist of sowing seeds in double rows on broad sloping ridges (Figure 13.4a) or in single rows on the longer slope of asymmetric ridges (Figure 13.4b).

(c) *Spray* or *sprinkler irrigation* allows greater flexibility in the use of terrain and depth of water applied than either flood or furrow methods, but it requires more costly equipment. The water is delivered through upright nozzles fitted to pipes that are fixed or movable. Of the latter, self-propelled water guns and motor-driven 'centre-pivot' spray systems permit more uniform distribution of water and hence greater efficiency of application. This is important for keeping the leaching fraction under irrigation to the minimum necessary for proper control of soil salinity (see below). One drawback of spray irrigation is the

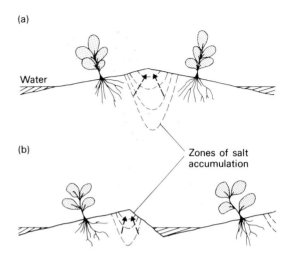

Figure 13.4. Ridge and furrow designs to avoid salinity damage to plants. **(a)** Paired crop rows on broadly sloping ridges. **(b)** Single crop rows on asymmetric ridges.

evaporative losses, especially on windy days, which reduce the efficiency of application. Even with good-quality water, evaporation from the foliage can leave salt deposits and cause leaf scorch.

Trickle or *drip irrigation* is an adaptation of spray irrigation that minimizes evaporative losses and is especially suited to high-value orchard and row crops on soils of low *AWC*. Water is delivered through flow regulating 'emitters', which are placed close to the plants at regular intervals. Salts in the soil are leached to the periphery of the wetted zone, the shape of which depends on the soil's permeability and the rate of water application (Figure 13.5). Thus, the irrigation rate can be adjusted to create a volume of low-salinity soil large enough for most of the roots to function without ill effect. Because the soil water potential in the active root zone remains high, water of higher salinity (> 3 mmho cm^{-1}) that could not be used for intermittent spray or furrow irrigation is often acceptable for trickle irrigation. However, salt accumulation towards the fringe of the wet zone necessitates periodic flushing of the whole soil if salinity problems for subsequent crops are to be avoided.

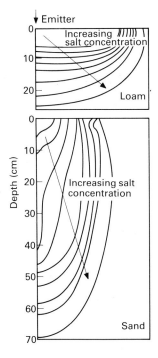

Figure 13.5. The distribution of salt concentration during trickle irrigation into soils of different texture: **(a)** loam—input $4 \, l \, h^{-1}$; **(b)** sand—input $4 \, l \, h^{-1}$ (after Bresler, 1975).

Quality of irrigation water

KINDS OF SALTS AND THEIR CONCENTRATION

The major ions in surface and underground waters used for irrigation are Ca^{2+}, Mg^{2+}, Na^+, Cl^-, SO_4^{2-} and HCO_3^- with small amounts of K^+ and NO_3^-. The total concentration of dissolved salts (*TDS*), sometimes referred to as the '*salinity hazard*', is most conveniently measured by the specific conductance, which is the electrical conductivity (*EC*) of the water, independent of the sample size. Analysis of many surface and well waters has revealed the approximate linear relationship:

$$TDS \, (mg \, l^{-1}) \simeq 640 \, EC \, (mmho \, cm^{-1}) \quad (13.1)$$

The units of *EC*, a reciprocal resistance, are commonly $mmho \, cm^{-1}$ which are equivalent to $dS \, m^{-1}$ in SI units.

Equation 13.1 holds for waters of *EC* up to about $10 \, mmho \, cm^{-1}$.

Experience in the western USA (Richards, 1954) suggests that *EC* class limits of < 0.25, $0.25–0.75$, $0.75–2.25$ and $> 2.25 \, mmho \, cm^{-1}$ at 25°C can be used to define waters of *low, medium, high* and *very high* salinity. The *EC* of most irrigation waters lies between 0.15 and $1.5 \, mmho \, cm^{-1}$ which is roughly equivalent to $1.5–15$ mmol anion $(-)$ or cation $(+)$ charge per litre*. However, this water is concentrated 2–20-fold due to evapotranspiration in the soil, at which level many crops will experience a reduction in yield through osmotic stress. The osmotic potential, ψ_s, of the soil solution can be calculated from the approximate formula:

$$\psi_s \, (bars) = -0.36 \, EC \, (mmho \, cm^{-1}) \quad (13.2)$$

The salinity level of the soil solution has traditionally been measured as the electrical conductivity of the *saturation extract* (EC_e). Sufficient distilled water is mixed with a dry soil sample to make a glistening paste and the saturation extract is then obtained by filtering under vacuum. The EC_e value is 2–3 times less than the true *EC* of the soil solution before dilution. The distinction between *saline* and *non-saline* soils has been drawn at an EC_e of $4 \, mmho \, cm^{-1}$ (section 9.6), but it is now recognized that sensitive crops can be adversely affected at EC_e values as low as $2 \, mmho \, cm^{-1}$ (see Figure 13.8). The *EC* of the soil solution can be measured *in situ* with salinity sensors, which are electrodes imbedded in a porous ceramic. Because of problems with the desaturation of the porous ceramic, however, their use is restricted to soil matric potentials > -2 bars. More recently, bulk soil *EC* (solid and solution phases combined) has been measured using the 'four electrode' technique and 'time domain reflectometry' (TDR). Both methods require the soil water content, θ, the transmission coefficient and the solid phase electrical conductivity to be known before the soil solution *EC* can be calculated. Once calibrated

*Because salt concentrations are usually expressed in milli-equivalents per litre ($me \, l^{-1}$) in the soil salinity literature, these units will be used throughout this chapter.

for a particular soil, however, TDR appears to offer the opportunity to obtain the soil solution salinity and θ from a single measurement of bulk soil EC.

SODIUM HAZARD

Apart from total salinity, the relative proportion of sodium to calcium and magnesium has an important effect on the quality of irrigation water. The Gapon equation (7.11), introduced in section 7.2, shows that the proportion of exchangeable Na^+ to exchangeable Ca^{2+} and Mg^{2+} ions is determined by a selectivity coefficient $k_{Na-Ca,Mg}$ and the ion activity ratio $(Na^+)/(Ca^{2+}+Mg^{2+})^{1/2}$ in the soil solution. In practice, the latter ratio is expressed in terms of concentrations (me l^{-1}) and is called the *sodium adsorption ratio*, that is:

$$SAR = \frac{[Na^+]}{\left[\dfrac{Ca^{2+}+Mg^{2+}}{2}\right]^{1/2}} \qquad (13.3)$$

In this form the SAR is approximately equal to the exchangeable Na percentage (ESP), in me per 100 g soil. As such, it is the preferred index of the *sodium hazard* or *sodicity* of the water. An ESP of 15 (or $SAR \simeq 15$ me l^{-1}) has generally been accepted as the criterion of a *sodic* soil, although the effect of ESP on a soil's physical properties must also be considered in relation to the total salt concentration of the soil solution, as discussed in section 13.3.

To predict the likely ESP to be attained by a soil under irrigation, a knowledge of the SAR of the irrigation water is essential. As the irrigation water is concentrated in the soil by evapotranspiration, assuming insoluble salts are not precipitated, the concentrations of all the ions increases in the same proportion, but the SAR increases disproportionately. For example, for a twofold increase in the ionic concentrations, the SAR increases by $2/\sqrt{2} = 1.4$, and so on. Exchange occurs between cations in solution and on the colloid surfaces until equilibrium is regained at a higher ESP value. Because the *quantity* of exchangeable cations in a given soil volume is usually much greater than the *quantity* of cations in solution,

the changes in ESP are small when compared with the change in SAR of the irrigation water as it evaporates. Nevertheless, continued use of water of high SAR will eventually lead to a comparably high ESP. This factor of relative salt buffering capacities of the solution and solid phases explains why, for a given SAR, waters of high salinity are considered to be of higher sodium hazard than waters of low salinity (Richards, 1954).

In Figure 13.6, the solid line represents the relationship between predicted ESP and the SAR of a number of irrigation waters in western USA. The dashed line shows the predicted relation if the irrigation water were to be concentrated threefold in the soil: the actual ESP values of a number of surface soils fell between the two lines, indicating reasonable agreement between prediction and measurement. However, ESP values deeper in the soil profile are less well predicted, probably because the simple ESP–SAR relation does not take account of (a) precipitation of insoluble carbonates, and (b) the differential effects of ion-pair formation and ionic strength on the activities on the Na^+, Ca^{2+} and Mg^{2+} ions as the water becomes more concentrated through evapotranspiration.

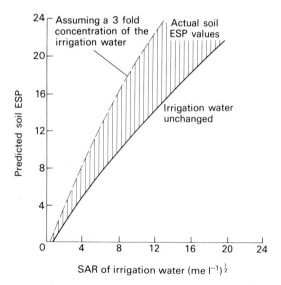

Figure 13.6. Relationship between surface soil ESP and SAR of irrigation water (after Richards, 1954).

Precipitation of $CaCO_3$ can be predicted from the Langelier *saturation index* defined as

$$SI = pH_a - pH_c \qquad (13.4)$$

where pH_a is usually put as 8.3 or 8.4 and pH_c is given by the equation:

$$pH_c = (pK_2 - pK_{SP} + p\gamma_{HCO_3} + p\gamma_{Ca}) + p[Ca] + p[HCO_3] \qquad (13.5)$$

The terms in round brackets are the negative logs of the second dissociation constant of H_2CO_3, the solubility product of $CaCO_3$ and the activity coefficients of HCO_3^- and Ca^{2+} ions, respectively. As the sum of these terms is relatively constant, pH_c will decrease as the concentration of calcium and bicarbonate in solution increases and therefore $CaCO_3$ is more likely to precipitate. This is illustrated in Figure 13.7, which also shows that the smaller the fraction of applied water draining through the soil (the *leaching fraction LF*), the greater is the precipitation of $CaCO_3$. The effects of LF and SI on the SAR_{iw} of the irrigation water and hence on the predicted ESP have been incorporated in

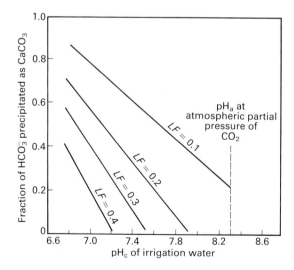

Figure 13.7. Precipitation of insoluble carbonates in relation to pH_c of the irrigation water and LF (after Bower *et al.*, 1968).

two empirical equations:

$$ESP \simeq SAR_u = SAR_{iw}(1+SI) \qquad (13.6)$$

for the upper (u) root zone, and:

$$ESP \simeq SAR_l = k\,SAR_u \qquad (13.7)$$

for the lower (l) root zone. k is a factor that depends on the LF and the rate of mineral weathering (Rhodes, 1968). There is evidence, however, that equations 13.6 and 13.7 overpredict the ESP of many soils under irrigation, and more satisfactory results are obtained with an SAR value that is adjusted for the effects of ionic strength and ion-pair formation (see point (b) above).

Prediction of the soil's ESP is vital to irrigation management because swelling pressures are enhanced as the ESP increases, especially if the total salt concentration of the soil solution is lowered (section 13.3).

SALT BALANCE
Even good quality irrigation water contains some salts. Because crops take up little of this salt (about one-tenth), but transpire nearly all of the water, salts inevitably accumulate in the soil under sustained irrigation. Good irrigation seeks to minimize this salt build-up.

A complete salt balance for an irrigated soil might be written as:

$$S_p + S_{iw} + S_r + S_d + S_f = S_{dw} + S_c + S_{ppt} \qquad (13.8)$$

In this equation, the cumulative contribution of S_p (rainfall and dry deposition), S_r (residual soil salts), S_d (salt released by weathering) and S_f (fertilizer salts) is usually small when compared with S_{iw} (salts in the irrigation water) and tends to be balanced by S_c (crop removal of salt). Historically, the effect of S_{ppt} (precipitation of carbonates and sulphates in the soil) has been discounted and management practices have concentrated on matching S_{iw} to S_{dw} (salt removed in drainage water). This approach leads to the simplified salt balance equation:

$$EC_{iw}d_{iw} = EC_{dw}d_{dw} \qquad (13.9)$$

where the subscripts iw and dw refer to irrigation and drainage water, respectively, d is the volume of water applied per unit area and EC represents the salt concentration. It follows that:

$$\frac{EC_{iw}}{EC_{dw}} = \frac{d_{dw}}{d_{iw}} = \text{Leaching requirement, } LR \quad (13.10)$$

The LR is therefore defined as the fractional depth of water that must pass through the soil to maintain EC_e in the upper two-thirds of the root zone below a specified value. This value is set by the crop's tolerance to salinity, which can be interpolated from diagrams of the kind shown in Figure 13.8. LR values usually range between 0.1 and 0.3.

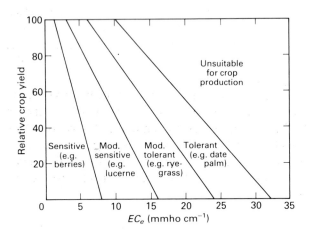

Figure 13.8. Salt tolerance classes for crops based on the EC_e of the saturation extract (after Maas and Hoffman, 1977).

Equation 13.10 can be rewritten by introducing the term d_{cw}, the total depth of irrigation water required to satisfy the crop's consumptive use, that is:

$$EC_{iw}d_{iw} = EC_{dw}(d_{iw} - d_{cw})$$

from which it can be seen that:

$$d_{iw} = \frac{d_{cw}}{1 - LR} \quad (13.11)$$

It is worth noting the distinction between LR and LF, the *actual* fractional depth of the applied water that

passes through the root zone. Although ideally LF should equal LR, because of variability in the application and infiltration of water on a field scale, the average LF is generally $> LR$. Unnecessary leaching of salts into the 'return' flow creates salinity problems for other water users downstream. Modern irrigation management therefore focuses on the dual objective of minimizing saline return flows, while avoiding a build-up of soluble salts in the soil. This can be achieved by improving the uniformity of water application so that LF can be as low as possible (< 0.1) and the precipitation of insoluble salts (S_{ppt}) in the soil is maximized (see Figure 13.7).

SPECIFIC ION EFFECTS

In addition to the osmotic effects of high salt concentrations on plant growth, specific toxicity symptoms arise when individual elements exceed certain levels. For example, fruit trees are likely to show marginal leaf burn and necrotic spots when Na and Cl concentrations exceed 0.2 and 0.5 per cent, respectively, in the dry matter. Li can prove toxic at concentrations > 0.1 mg l^{-1} in the soil solution and irrigation water with > 0.3 mg B l^{-1} is suspect for use on sensitive crops like citrus.

13.3 RECLAMATION OF SALT-AFFECTED SOILS

Permeability changes on leaching

As discussed in section 13.2, the evaporation of water from the soil–plant system leads to a rise in the SAR of the soil solution and eventually to an increase in the soil's ESP. The effect is exacerbated when high-salt irrigation water is used. A saline soil ($EC_e > 4$ mmho cm^{-1}) has been called *saline–sodic* when the ESP value is > 15. Such soils need very careful management because, although the structure remains stable as long as the concentration of the soil solution is high, when soluble salts are depleted by leaching, adverse changes in the structure will occur.

Figure 13.9 illustrates the interactive effect of ESP and salt concentration in the percolating water on a

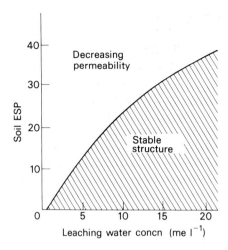

Figure 13.9. Interactive effect of *ESP* and concentration of the leaching water on soil permeability (after Quirk and Schofield, 1955).

soil's permeability (hydraulic conductivity). The threshold concentration below which the hydraulic conductivity falls by 10–15 per cent is 10 ml me l^{-1} at an *ESP* of 20. The fall in hydraulic conductivity is caused by the swelling of clay domains and quasi-crystals (section 7.4), which decreases the macroporosity and may also lead to deflocculation of clay particles. These may then be eluviated and block the conducting pores. Although swelling of flocculated clay is reversible (by increasing the salt concentration), deflocculation and clay translocation are not and they may cause a permanent deterioration in structure (section 9.6). Soils with an *ESP* as low as 5 may be susceptible to deflocculation, especially at the surface, under the influence of rain or high-quality irrigation water (*EC* < 0.1 mmho cm^{-1}). Where $CaCO_3$ is present, its solubility is sufficient to maintain the salt concentration of the soil solution at around 3 me l^{-1}, which should prevent clay deflocculation.

Reclamation of saline, saline–sodic and sodic soils requires leaching with water of a low enough *SAR* to initiate Ca^{2+} exchange for Na^+, but a sufficiently high total salt concentration to preserve the soil's permeability. As the *ESP* is gradually reduced, the salinity

of the leaching water may also be reduced until reclamation is complete.

SOURCES OF CA^{2+} IONS
Calcium is usually supplied as gypsum, which has a solubility of 30 me l^{-1} in distilled water. Gypsum can be spread on the soil or dissolved in the irrigation water, its solubility increasing in more concentrated solutions due to the effect of ionic strength and ion-pair formation in reducing the activity of the Ca^{2+} and SO_4^{2-} ions. Approximately 3.4 t is required to reduce by one unit the *ESP* of a volume of soil 1 ha by 15 cm deep. Calcium carbonate, if present, is normally too insoluble to lower the *SAR* of the leaching water significantly; but in soils containing *Thiobacillus* bacteria the addition of sulphur enhances the solution of $CaCO_3$ due to the acidity generated when S is oxidized to SO_4^{2-} (equation 12.7). Sulphuric acid produced as a waste byproduct of industry has also been used for this purpose.

Metal sulphides in marine sediments behave similarly when first exposed to air (section 11.3), a reaction to the benefit of reclamation in Dutch polders, where the H^+ and SO_4^{2-} ions react with indigenous $CaCO_3$ to produce gypsum sufficient to saturate the soil solution for several years as the reclaimed land is leached by rainwater.

BLENDING OF LOW AND HIGH SALINITY WATERS
The theory of saline-soil reclamation was put to an interesting test with a saline–sodic soil from the Coachella Valley in the USA. The soil, which had an *ESP* of 39 and an EC_e of 4.4 mmho cm^{-1}, was leached with Salton seawater that was progressively diluted with good quality Colorado riverwater. Each water mixture, of which the range in composition is given in Table 13.1, was leached through the soil until the soil was in equilibrium with that mixture. Using four stepwise reductions in the leaching water concentration, the soil permeability did not fall below 0.12 m d^{-1} and the *ESP* was lowered to 5 in only 12 days. However, when the soil was leached with Colorado riverwater

alone, the permeability fell to 0.005 m d^{-1} and reclamation took 120 days.

Table 13.1. Reclamation of a saline–sodic soil with water of stepped-down salt concentration and *SAR*.

Dilution ratio (seawater + riverwater)	Water properties		Soil properties	
	TDS (me l^{-1})	*SAR* (me$^{1/2}$ l$^{-1/2}$)	Hydraulic conductivity (m day^{-1})	*ESP* (%)
Salton Sea	562	57	—	—
1:3	149	27	0.13	28
1:15	45.4	12	0.12	14
1:63	19.6	5	0.11	6
0:1	11.0	2	0.12	5
Colorado River	11.0	2	0.005	5

(After Reeve and Bower, 1960.)

METHOD OF LEACHING

It has been found that intermittent leaching is more efficient than continuous ponding for the reason that, whereas much of the salt in soil is located in the fine intraped pores, percolating water tends to flow preferentially down the larger channels between aggregates (section 6.4). Breaks in the leaching therefore allow time for salts to diffuse into the depleted outer regions of the aggregates, whence they are removed during the next leaching period. Thus, instead of 1 m of water being required to leach 80 per cent of the salt from 1 m of soil under ponding, as little as 30 cm of water applied intermittently can achieve the same result.

Removal of surplus water

It is a corollary of good irrigation management that surplus water, arising out of a leaching requirement or through seepage from a shallow groundwater table, should be disposed of efficiently. In permeable soils, the natural hydraulic properties may be able to cope with the excess water; where this is not the case, soil drainage must be improved by the installation of underground pipes or *tile drains* through which water is rapidly led away (see below).

13.4 SOIL DRAINAGE

Choosing a drainage design

Irrigated soils in dry climates, for which the lowering of the groundwater table and the disposal of water in excess of the crop's needs are prime objectives, require a drainage design different to that of soils of cool humid climates, where the natural permeability may be inadequate to prevent the soil becoming waterlogged for unduly long periods (1 or more days) when precipitation exceeds evaporation. Irrigation schemes are also very capital intensive, producing high returns per hectare, so that the use of more expensive drainage systems may be justified on irrigated, but not on non-irrigated, land. These two broad categories of problem soils therefore warrant discussion individually.

IRRIGATED SOILS

The variables that interact to determine the equilibrium height of the water table in a soil with tile drainage are:

(a) the hydraulic conductivity, K_s, of the saturated soil (section 6.3);

(b) the depth of drains, d, and their height, H, above any impermeable barrier in the soil, that is, a layer with a K_s value ≤ 0.1 of the layer above;

(c) the distance L, between drains;

(d) the mean rate of water percolation, q, through the soil to the groundwater.

The direction of flow to tile drains is predominantly vertical throughout the unsaturated zone. Below the water table, the direction changes from being vertical initially to mainly horizontal and then radial in the saturated zone around the drain, as illustrated in Figure 13.10. The relative extent of these three zones depends on the magnitude of L, H and h, the last being the overall head drop from the water table to drain level. When the rate of percolation, q, to the water table is equal to the drain discharge rate, the system is in steady-state equilibrium and *Hooghoudt's equation* can be used to estimate the required value of L for given values of d, h, H and pipe diameter.

Hooghoudt's approach was to reduce the complex

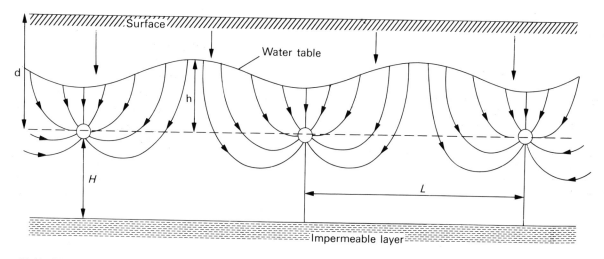

Figure 13.10. Idealized flow of water to tile drains after rain or irrigation.

flow pattern around pipe drains to the equivalent of horizontal flow to vertical trenches, at the same spacing as the pipes, but filled with water to a lesser height H_e above an impermeable layer (Figure 13.11). Because

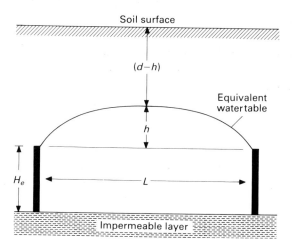

Figure 13.11. Diagram for Hooghoudt's 'equivalent flow' model to drains (after Smedema and Rycroft, 1984).

$H_e < H$, there is less cross-sectional area available for horizontal flow, so the head drop is greater than in the real case. The difference in head loss is just equal to the head loss in radial flow in Figure 13.10, so that the total

head loss in the simulated and real cases is the same. The average thickness through which water flows in the equivalent case is $H_e + h/2$ (Figure 13.11) and the equation giving the discharge rate, q, is:

$$q = 8K_s h(H_e + h/2)/L^2 \qquad (13.12)$$

Rearranging equation 13.12 gives:

$$L = \sqrt{8K_s h(H_e + h/2)/q} \qquad (13.13)$$

The effective height, H_e, above an impermeable layer is a function of H, the true height, L and p, the wetted perimeter of the drain. Equation 13.13 must therefore be solved by successive approximations using trial values of L and H_e, until the calculated value of L is very close to the trial value chosen. The procedure is:

(a) Choose the design criteria—q and the mid-drain water table height $(d-h)$ (Figure 13.11). To prevent salt accumulation in an irrigated soil, $q \simeq 0.001$ m d^{-1} and $(d-h)$ must be > 2 m.

(b) Choose the drain depth, d, consistent with (a).

(c) Establish values for K_s and p.

(d) Knowing H, the true depth to the impermeable layer, find from tables, values of H_e for trial values of L.

Drainage of clay soils presents a somewhat different problem in that the impermeable layer usually lies at or just below drain depth ($H \rightarrow 0$) and deep flow to the drains is inhibited. In this case, equation 13.13 becomes simplified to:

$$L = 2h\sqrt{K_s/q} \qquad (13.14)$$

Often, however, the calculation of a theoretical drain spacing is vitiated by the great spatial variation in K_s so that predictions based on Hooghoudt's equation, or any other model of flow to drains, must be tempered by experience and practical judgement. Nevertheless, the equation demonstrates that drain spacing can be increased if K_s increases or q decreases and also if H increases (provided L is large) or h increases. To increase h, either the water table at the mid-drain point must be allowed to rise or, more usually, drain depth must be increased. There are, however, practical limits to the depth at which drains can be placed, particularly when the permeability of the subsoil is low.

SOILS OF HUMID REGIONS

A widely accepted design criterion for draining wet impermeable soils is to lower the water table to 0.5 m below the surface at the mid-drain point, 24 h after the end of rainfall. This will create a moisture suction of 50 mbars at the surface, corresponding to the *FC* of a well-drained soil. There are two major constraints on the effectiveness of drainage in such soils:

(1) The *pore volume* that is drained at 50 mbars suction, because this determines the maximum air-filled porosity of the drained soil. Its value depends on the moisture characteristic curve of the soil, but if it is less than 10 per cent the improvement in aeration that drainage can achieve will be limited.

(2) The *hydraulic conductivity*, K_s, of the saturated soil which determines the maximum rate of drainage, when the hydraulic head gradient is one. The conductivity of clay soils in Britain ranges from 10 to 100 m day^{-1} in the dry state, when deep fissures are prominent, to 0.1–0.001 m day^{-1} when saturated. Given that the drainage system should be capable of removing 10 mm of surplus water per day, one can calculate that for an impermeable layer close to the drain depth, the theoretical drain spacings for a soil of K_s value between 0.001 and 0.1 m day^{-1} are as small as 0.3–3 m (from equation 13.14). Such close tile-drain spacings are rarely economically feasible.

Underdrainage in practice

DRAIN DEPTH

In order that the water table in quiescent periods does not lie within 50 cm of the surface, tile drains need to be placed more deeply, the actual depth depending on the position of any impermeable layer and the adequacy of the outfall into the collection ditch. The drains are laid with a gentle gradient at a depth between 70 and 120 cm. The deeper the drains the drier the soil in the plough layer at equilibrium, but this condition is not always attained in clay soils in winter due to their low K_s values.

Tile drains are either short lengths of porous clay pipes, usually 75 mm in diameter, laid end-to-end, or continuous plastic pipes of the same diameter, which are

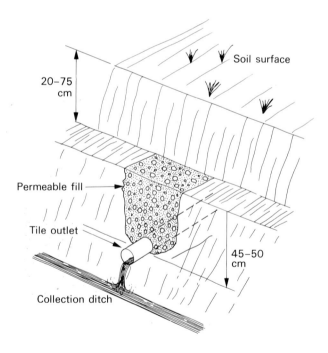

Figure 13.12. Diagram of a tile drain imbedded in permeable fill.

corrugated for strength and perforated to allow water entry. Because water enters the pipe from all sides, the drain's performance is much improved if it is imbedded in a *permeable filling material* (Figure 13.12). Permeable fill, usually consisting of gravel or crushed stone of size range 5–50 mm, is of benefit for three reasons:

(a) it filters out fine soil particles, which may otherwise block the drain;
(b) it reduces the loss in head caused by the restricted number of entry points for water into the drain;
(c) it serves as a connection between the tile drains and any mole drains which are drawn at shallower depths.

If the hydraulic conductivity of the permeable fill is at least ten times the K_s of the surrounding undisturbed soil, the head entry loss of the drain should be close to zero.

DRAIN SPACING

It is clear that laying tile drains at the spacing necessary to drain effectively soils of hydraulic conductivities as low as 0.1–0.001 m day^{-1} is very expensive. The practical alternative is to choose an economic spacing—between 20 and 80 m but commonly 20–40 m—and to try to increase water flow to the drains by secondary treatments, of which *moling* and *subsoiling* are the most important. *Moling* provides cheap drains at the spacing required for efficient drainage in impermeable soils. The depth of the mole drains determines the equilibrium water table position and the spacing governs its rate of rise and fall during and after rainfall. Closer spacings delay the rise and hasten the fall of the water table. *Subsoiling* is intended to improve the hydraulic conductivity of the soil to the point where a drain spacing of 20–40 m provides efficient drainage.

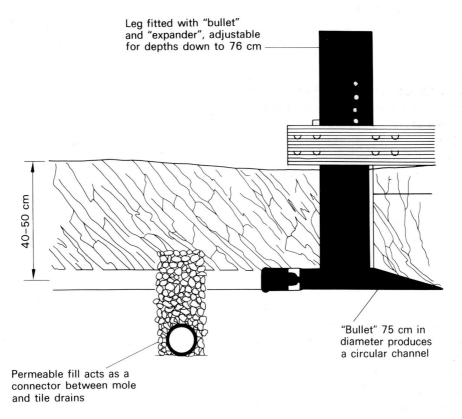

Leg fitted with "bullet" and "expander", adjustable for depths down to 76 cm

40–50 cm

"Bullet" 75 cm in diameter produces a circular channel

Permeable fill acts as a connector between mole and tile drains

Figure 13.13. Diagram of a mole drainer in operation (after MAFF Field Drainage Leaflet No. 11).

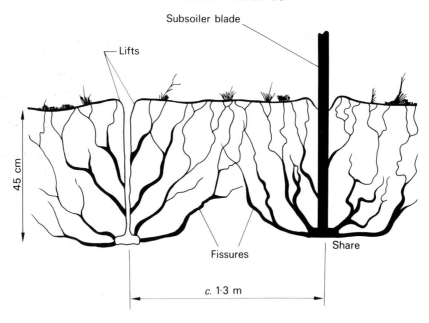

Figure 13.14. Soil fissuring induced by effective subsoiling (after MAFF Field Drainage Leaflet No. 10).

As equation 13.14 shows, however, a 10-fold increase in drain spacing must be complemented by a 100-fold increase in K_s, which is not often achieved because the soil at depth is too moist for effective subsoiling. Nevertheless, subsoiling can be effective where a plough pan markedly reduces the permeability of the soil above the drains.

METHODS OF MOLING AND SUBSOILING
Moles are drawn by the implement illustrated in Figure 13.13. The 'bullet' of 75 mm diameter is pulled through the soil at a depth of 50–60 cm and a spacing of 2–3 m. Behind the bullet, the expander of 100 mm diameter consolidates the walls of the channel and seals the slit left by the vertical blade. Passage of the bullet and blade through the soil should fracture the structure and improve flow to the mole drain. Mole drains should lie at right angles to the tiles, provided their gradient can be kept between 2 and 5 per cent, and pass through the top of the permeable fill (Figure 13.13). Whereas tile drains should last 50 years and more, moles should be

redrawn within 5–10 years for best results.

In subsoiling, a wedge-shaped share is pulled at right angles to the tile drains at 40–50 cm depth, if possible. Ideally, at this depth and at a spacing of about 1.3 m, the structure of all the upper soil profile should be disturbed by fracturing and fissuring (Figure 13.14). The effect is more transient than moling.

The success of moling and subsoiling depends largely on the soil's *consistency* at the time of the operation. Consistency describes the change in the physical condition of the soil with moisture content, and reflects both the cohesive forces between particles (ped shear strength) and the resistance of the soil mass to deformation (bulk shear strength). With reference to the consistency diagram of Figure 13.15, it can be seen that moling is best when the soil is plastic, but nearer to the lower plastic limit, whereas subsoiling is more effective when the soil is friable, nearer to the shrinkage limit. In England and Wales these conditions are generally satisfied if the *SMD* in the whole soil to 1 m depth is at least 50 mm (for moling) and 100 mm (for subsoiling).

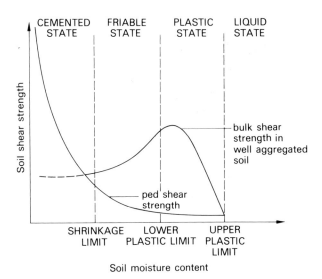

Figure 13.15. Variation in soil shear strength and consistency with moisture content (after MAFF, FDEU Technical Bulletin 75/5).

13.5 SUMMARY

Excess soil water due to a consistently high $P:E$ ratio, an impermeable subsoil or a high groundwater table imposes severe limitations on land use.

In arid areas, groundwater is often saline and irrigation, if not properly managed, tends to raise the water table. Capillary rise of salts and salinization can then occur. Because most of the water added to soil is lost by evapotranspiration, but only about one-tenth of the salts are taken up by the crop, irrigation itself increases soil salinity. Total dissolved salts (TDS) are measured by the *electrical conductivity* (EC), i.e. TDS (mg l^{-1}) $\simeq 640\ EC$ (mmho cm^{-1}). The concentrating effect on the soil solution also causes the *sodium adsorption ratio* (SAR), which is defined by:

$$SAR = \frac{[\text{Na}^+]}{\left[\dfrac{\text{Ca}^{2+}+\text{Mg}^{2+}}{2}\right]^{1/2}}$$

to increase. The equilibrium between exchangeable and solution cations then changes so that adsorbed Ca^{2+} and Mg^{2+} ions are replaced by Na$^+$ and the *exchangeable sodium percentage* (ESP) increases. Such soils require careful management to avoid swelling and loss of permeability, and possibly clay deflocculation and translocation, when leached with good-quality water. The former process is reversible; the latter may lead to an irreversible deterioration of soil structure. The SAR in (me l^{-1})$^{1/2}$ is approximately equal to the predicted ESP and can be used to assess the *sodicity* of irrigation waters. The rate at which the actual ESP increases depends on the total salinity (TDS) of the irrigation water.

The ratio of drainage water to applied irrigation water that will maintain a salt balance in the soil is called the *leaching requirement, LR*. The actual ratio of drainage water to irrigation water is the *leaching fraction, LF*. Usually $LF > LR$ because of unevenness in the application of water, particularly by *flood* or *furrow* methods. *Spray* irrigation is more efficient, but salt deposits left on leaves may cause damage. *Trickle* irrigation is the most efficient and allows the use of water of a higher salinity than normal. In reclaiming soils of high salinity ($EC_e > 4$ mmho cm^{-1}) and high sodicity ($ESP > 15$), water of high EC and low SAR should be used. Drainage is often necessary to remove excess water from irrigated soils.

In cool humid climates, problems of wetness, inadequate aeration, poor trafficability, invasion by flood-tolerant weeds, pests and disease, and low soil temperatures in spring can be mitigated by the installation of *pipe* or *tile drains* to create a suction of at least 50 mbars at the soil surface at the mid-drain point within 24 hours of rain. The efficacy of this treatment will be limited if the drainable pore volume at 50 mbars suction is <10 per cent, or if the saturated hydraulic conductivity, K_s, is <0.1 m day^{-1}. The drain spacing necessary to achieve a certain drain discharge can be calculated from *Hooghoudt's equation*, provided that certain soil characteristics (especially K_s) are known.

If K_s is so low that the tile drains would need to be placed at uneconomically close spacings (1–2 m), the practical solution is to use wider spacings (20–40 m) with secondary treatments such as *moling* and *subsoiling* to improve flow to the tiles. Mole drains are drawn above the tiles at intervals of 2–3 m, preferably when the *SMD* is at least 50 mm in the top metre. Imbedding the tile in 40–50 cm of *permeable fill* promotes peripheral flow into the tile and provides a good connection with the moles above. To be effective, subsoiling must be carried out when the *SMD* is at least 100 mm, but it does not last as long as moling.

REFERENCES

BOWER C.A., OGATA G. & TUCKER J.M. (1968) Sodium hazard of irrigation waters as influenced by leaching fraction and by precipitation or solution of calcium carbonate. *Soil Science* **106**, 24–34.

BRESLER E. (1975) Two-dimensional transport of solutes during non-steady infiltration from a trickle source. *Soil Science Society of America Proceedings* **39**, 604–613.

HOLMES J.W. (1971) Salinity and the hydrologic cycle, in *Salinity and Water Use* (Eds. T. Talsma & J.R. Philip). Macmillan, London.

MAAS E.V. & HOFFMAN G.J. (1977) Crop salt tolerance—current assessment. *Journal of Irrigation and Drainage Division. Proceedings of the American Society of Civil Engineers* **103**, 115–134.

MINISTRY OF AGRICULTURE, FISHERIES AND FOOD (1975) *A design philosophy for heavy soils.* Field Drainage Experimental Unit, Technical Bulletin 75/5.

MINISTRY OF AGRICULTURE, FISHERIES AND FOOD (1980) *Mole drainage.* Field Drainage Leaflet No. 11. MAFF Publications, Middlesex.

MINISTRY OF AGRICULTURE, FISHERIES AND FOOD (1981) *Subsoiling as an aid to drainage.* Field Drainage Leaflet No. 10. MAFF Publications, Northumberland.

QUIRK J.P. & SCHOFIELD R.K. (1955) The effect of electrolyte concentration on soil permeability. *Journal of Soil Science* **6**, 163–178.

REEVE R.C. & BOWER C.A. (1960) Use of high-salt waters as a flocculant and source of divalent cations for reclaiming sodic soils. *Soil Science* **90**, 139–144.

RHOADES J.D. (1968) Mineral-weathering correction for estimating the sodium hazard of irrigation waters. *Soil Science Society of America Proceedings* **32**, 648–652.

RICHARDS L.A. (1954) (Ed.) *Diagnosis and improvement of saline and alkali soils.* United States Department of Agriculture, Handbook No. 60.

SMEDEMA L.K. & RYCROFT D.W. (1983) *Land Drainage.* Batsford Academic, London.

FURTHER READING

ALLISON L.E. (1964) Salinity in relation to irrigation. *Advances in Agronomy* **16**, 139–180.

BRESLER E., McNEAL B.L. & CARTER D.L. (1982) *Saline and sodic soils—Principles, dynamics and modelling.* Springer-Verlag, Berlin.

SHAINBERG I. & OSTER J.D. (1978) Quality of Irrigation Water. International Irrigation Information Centre, Publication No. 2, Bet Dagan.

SHAINBERG I. & SHALHEVET J. (1984) (Eds.) *Soil Salinity under Irrigation—Processes and Management.* Springer-Verlag, Berlin.

TALSMA T. & PHILIP J.R. (1971) (Eds.) *Salinity and Water Use.* Macmillan, London.

THOMASSON A.J. (1975) (Ed.) *Soils and field drainage.* Soil Survey of England and Wales. Technical Monograph No. 7.

VAN SCHILFGAARDE J. (1974) (Ed.) Drainage for Agriculture. *Agronomy* No. 17. American Society of Agronomy, Madison.

Chapter 14
Soil Survey and Classification

14.1 THE NEED TO CLASSIFY SOILS

Soil variability

The point was made in Chapter 1 that the distinctive character of a soil is shaped by a complex interaction of many physical, chemical and biotic forces. This theme was expanded in subsequent chapters on organic matter, structure formation, water movement, reactions at surfaces and so on. However, a study of soil in the field makes it clear that the balance of forces, and the speed with which they act, vary from site to site according to the local expression of the soil-forming factors, as discussed in Chapter 5. Thus, the many *morphological* properties (such as colour, texture, structure, horizon development) and *edaphic*** properties (such as pH, *CEC*, organic C) that characterize a soil display a very wide range in value. Man, in exploiting the soil, also alters the balance of natural processes and so adds to the total pool of *soil variability*.

The existence of soil variability must influence decisions made by potential users at various levels of endeavour:

(a) at the national level, in attempting to reserve for urban, industrial, agricultural and recreational development those areas of land best suited to these purposes;

(b) at the town and regional level, in the allocation of land for residences, parks, waste disposal, roads and 'green belts';

(c) at the farm level, in the choice of which fields to plough, to put down to improved pasture or to keep for rough grazing;

(d) in the laboratory, where the scientist works with small samples of soil, from a few grams to kilograms, assumed to be representative of a much larger volume in the field.

*Properties relevant to plant growth.

The purpose of classification

Faced with the variability within natural populations, our natural response is to try and impose an ordered structure on the fund of knowledge derived from our experience with individuals in the population. This usually involves three steps:

(1) identification of the full range of variability within the population;

(2) creation of groups or *classes* within the population according to the similarity between individuals;

(3) prediction of the likely behaviour of an individual from a knowledge of the characteristics of the class as a whole.

Step (2)—the creation of classes—is called *classification*. Soil classification is more difficult and contentious than the classification of other natural populations because a soil lacks the hereditary characteristics by which individuals within one generation are distinguished, and which are transmitted from one generation to the next. The soil population presents a *continuum* of variation such that arbitrary judgements are inevitable in the creation of classes.

DEFINITION OF A SOIL ENTITY

Soil variability can be crudely partitioned according to the distance over which discernible changes in properties occur. Differences associated with soil faunal and microbial activity and the distribution of pores, for example, achieve maximum expression within 1 m (*short-range variation*). Since as much as half of the total variation in any one property can occur as short-range variation, it is impossible to define rigorously a fundamental *soil unit* or *entity*. Attempts to do so have resulted in the introduction of esoteric terms such as the 'pedon' and 'pedounit'; but whatever the theoretical merits of these terms, the field worker *describes* a soil on the basis of what can be seen of the profile in a pit (usually 1×0.5 m in area, dug to the parent material or to 1.5 m, whichever is the shallower), or exposed in a cutting or quarry face. However, because the excavation of pits is laborious and expensive, the assessment of the full range of soil variability is usually based on the examination of *soil samples* collected by means of augers or coring devices, inserted vertically into the soil (Figure 14.1). Details of sampling techniques used in soil survey are given by Hodgson (1978).

Figure 14.1. A selection of augers and core samplers used in soil survey.

KINDS OF CLASSIFICATION

Another problem confronting the soil classifier is the choice of soil properties on which to separate classes. Many properties can only be evaluated qualitatively—soil colour, structure, type of humus and degree of gleying, for example. Other properties, such as the pore-size distribution and the content of organic matter, clay or calcium carbonate, are difficult to measure. Practical constraints prevent all the soil properties being used; some are selected as the *definitive* properties for class separation, the choice of which depends on the aim of the classification, which is discussed below.

(a) *Special or single-purpose classification.* This kind of classification is made with a specific aim in mind and is based on one or a very few soil properties. The texture classification of Figure 2.3 and the classification of soils for droughtiness and susceptibility to waterlogging (Figure 11.6) are two cases in point. There are many other examples, especially relating to plant growth, such as the *ESP* and salinity (EC_e) status of irrigated soils, and the assessment of 'available' N, P and K levels by soil chemical tests (section 11.2). When classification is based on one or two properties, classes may be separated by subdividing the range of variation in the individual property at the minima in the frequency distribution (Figure 14.2a), or according to the user's requirements (Figure 14.2b).

(b) *General-purpose classification.* Since this kind of classification is intended for many uses, foreseen and unforeseen, the classes should be defined on as many properties as possible. Provided that the properties chosen are *relevant* to the utilization of the soil, the classification will be useful; if not, the classification may be academically elegant, but of little practical value. But where the important soil properties are those requiring tedious and expensive laboratory analysis, it is expedient to choose for the separation of classes certain *diagnostic* properties, not necessarily relevant to soil use but easily assessed in the field, on the assumption (or knowledge) that they are correlated

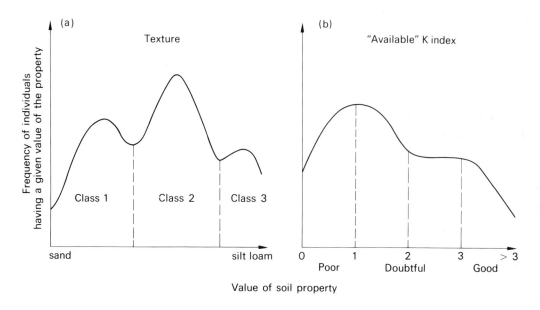

Figure 14.2. (a) and **(b)** Subdivision of the variation in a soil property to define classes for a special-purpose survey.

with the important properties. For example, the colour of the A horizon may be correlated with organic matter content, and mottling of the subsoil indicative of a low hydraulic conductivity. Where such natural grouping or 'clustering' of soil properties exists, the classes created should be more meaningful to the general-purpose user than those defined by arbitrary criteria.

QUALITY OF CLASSIFICATIONS

Implicit in the decision to classify soils is the assumption that the potential user will be able to make better generalizations about the properties of soils in a heterogeneous landscape. In other words, better prediction of soil-dependent behaviour such as regional drainage or crop productivity can be made using the class properties, rather than measurements made on the soil at a few individual sites. For this to be true, a substantial proportion of the variability in the properties of the whole population of soils must occur *between* the classes rather than remain *within* the classes. The accepted mathematical measure of the variability in a population is the *variance, σ^2*. This term, and other statistical terms used in the

quantitative analysis of the results of soil survey and classification, are defined by Webster (1977) to whom the reader should refer for a more detailed treatment.

It is found that, for a given categorical level of classification (see section 14.3), the classes created usually have a similar range of variation so that a pooled *within-class* variance (σ_w^2) can be calculated by taking the weighted average of the class variances, where the weighting factors are the number of degrees of freedom associated with each class variance. The effect of classification is then measured by the index:

$$\tau = 1 - \sigma_w^2/\sigma_t^2 \qquad (14.1)$$

where σ_t^2 is the total variance in the soil population. Clearly, when $\sigma_w^2 \simeq \sigma_t^2$, the spread of values within a class is comparable with the total spread of values, and the proportion of the total variance accounted for by the classification (τ) approaches 0—the classification is therefore of little value. It is easiest to calculate τ for a special purpose classification based on measurements of only one soil property. Typically, about half the variance in a physical or mechanical property of the soil, but only about one-tenth of the variance in many chemical properties, can be attributed to differences between classes. The within-class variance can be decreased by increasing the number in the sample from which the mean and variance are calculated, but it is uncommon for σ_w^2 to be reduced much below 0.25 of the total variance unless a great deal of effort is expended. If the effort, and hence the cost, of classification becomes too great, the alternative of measuring the soil properties directly, when required, may be better than attempting prediction.

14.2   A GUIDE TO SOIL SURVEY

Defining the aims of survey
Sampling to assess the full range of variability, and describing the soil at known locations within a target area, are the first stages of *soil survey*. This information is required so that decisions on land use, and the management best suited to different facets of land use, can be made. The effort put into survey and the procedures adopted are determined by the size of the area in relation to the human and financial resources available, the intensity of the proposed land use and the

exclusiveness of the classes to be created. The output of survey is usually a classification and a *soil map*, which shows the distribution of the different soil classes at a scale commensurate with the density of sampling on the ground. Soil variability influences the density of sampling required to achieve a certain degree of precision in the classification (measured by τ in equation 14.1). Nevertheless, variability occurring within blocks of 1–100 m^2 (i.e. < 0.01 ha) cannot be *mapped* at an acceptable scale (1:1,000 or smaller), so that this restriction sets an upper practical limit to the density of sampling.

The relation between the type and purpose of the survey and the map scale is summarized in Figure 14.3. The main types of survey are as follows.
(1) *Reconnaissance* and *semi-detailed* surveys, ranging in scale from 1:1,000,000 to 1:25,000, have the same broad objectives: to provide a resource inventory for large areas (provinces, states, countries). The preliminary work relies heavily on air-photograph interpretation (API) to delineate differences in vegetation, land form and management that may be the outward expression of soil differences. Once boundaries have been drawn on the air photographs, surveyors go into

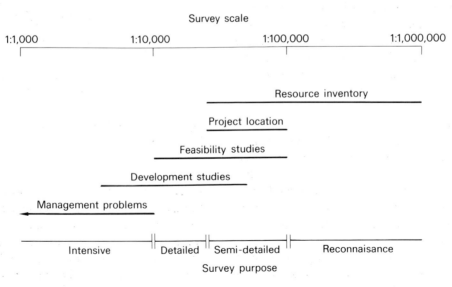

Figure 14.3. Scale of soil survey in relation to purpose (after Young, 1973).

the field to check their significance, relocating them if necessary, and to describe the main features of the soils and environment within each landscape unit. Areas suitable for various forms of land use can be located and guidance obtained as to the most suitable type of survey to be performed at the next stage.

(2) *Detailed* and *intensive* surveys range in scale from 1:25,000 to 1:1,000 and are carried out for one of three reasons:

(a) to assess the technical and economic *feasibility* of a proposed project in a given area;

(b) to assess the preliminary work necessary for the *development* of a specified project;

(c) within a developed area, to provide data for the solution of *management* problems.

The last-mentioned surveys are usually special-purpose surveys, for example the mapping of salt-affected soils in an irrigation area or soils subject to erosion in an area of continuous cereal cropping.

Survey procedures

Any soil survey should begin with a reconnaissance and general fact-finding exercise in the area. The use of air photographs has been mentioned and, in addition, the surveyor needs to study the geology, climate and vegetation, as well as to consult local land users and agricultural historians. When the preliminaries are complete, the surveyor goes into the field to sample the soil and create a classification that will provide the basis for his mapping. The procedure adopted is determined primarily by the scale of mapping.

GRID SURVEY

This procedure is preferred for special-purpose surveys that are mapped at large scales (> 1:25,000). The sample sites are located on a grid pattern, and the density of sampling should be adjusted according to the area surveyed so that the number of observations per cm² of the final map is independent of the scale. Commonly, 4–9 observations per ha are recommended, which at a scale of 1:10,000 produces 4–9 observations per cm² of the map. Boundaries are drawn to join points of equal value of a specific

property, or where general-purpose classes are defined, between points at which dissimilar soils occur (Figure 14.4). Grid survey is expensive, the cost increasing roughly in proportion to the square of the map scale.

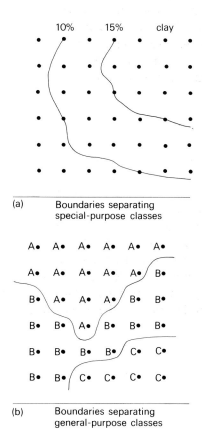

Figure 14.4. (a) and (b) Soil boundaries drawn from grid surveys (after Beckett, 1976).

FREE SURVEY

In free survey, the surveyor chooses sampling points on the assumption that changes in surface features (soil colour, relief, vegetation or land use) are indicative of soil differences. The density of sampling can be varied as the surveyor concentrates on confirming the inferred boundaries and checking the uniformity of the soil within each boundary (Fig. 14.5a). At small scales,

(a)

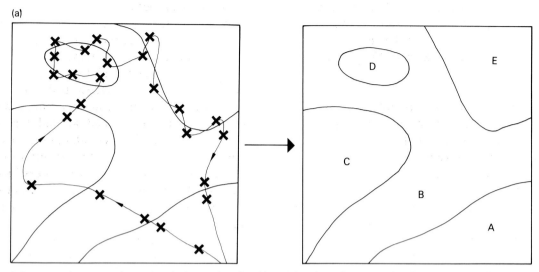

Soil surveyor's traverse (——►) and observation sites (✖) giving the soil map on the right at a scale of
1:50,000. Boundaries of map units A and C were inferred from external features; those of units
D and E were located by sampling

(b)

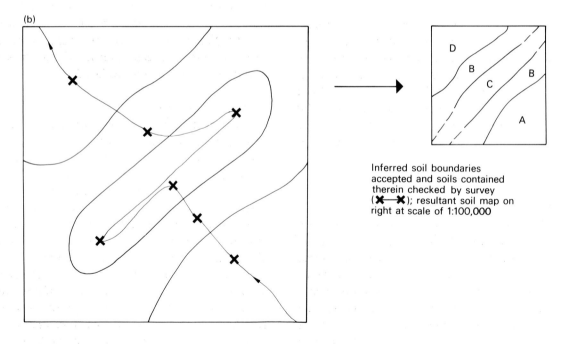

Inferred soil boundaries
accepted and soils contained
therein checked by survey
(✖—✖); resultant soil map on
right at scale of 1:100,000

Figure 14.5. (a) and **(b)** Sampling pattern and soil boundaries drawn from free survey (after Beckett, 1976).

often the inferred boundaries must be accepted and the limited field effort directed to discovering the soils lying within the boundaries and the elucidation of any recurrent patterns, such as catenas (Figure 14.5b).

CLASSIFICATION

The surveyor's prime aim is to be able to place any soil in the area in one class. With special-purpose classifications, the class limits are rigidly defined according to the values of one or two properties: for example, one classification of soils for irrigation purposes sets the following class limits:

saline soils $EC_e > 4$ mmho cm^{-1}, $ESP < 15$
saline–sodic soils $EC_e > 4$ mmho cm^{-1}, $ESP > 15$
sodic soils $EC_e < 4$ mmho cm^{-1}, $ESP > 15$

For general purposes, however, each class is usually built around a central concept, covering a limited range in the values of several soil properties, and a certain amount of overlap in the values of any one property is permissible at the class limits.

The list of classes is the classification. Data for individual profiles are subsumed into a definitive description of each class, and one or more properties may be chosen as *diagnostic* or *key* properties that enable an unknown soil to be quickly allocated to a class. Each class is given a locality name and referred to as a *soil series*. Soils of the same series, which differ only in the texture of the A horizon, are separated as *types*.

Survey results and land evaluation

SOIL MAPPING

Because all the soil profiles of any one series need not be contiguous, the distribution of soil series in the landscape is frequently intricate and necessitates some simplification for mapping. This is achieved by means of a *mapping unit*, consisting of one or more classes, which is the smallest area of the map that can be delineated by a single boundary at the scale used. At large map scales, the objective is to use *simple* mapping units, each representing a narrowly defined class (series), and to attain a purity of 80 per cent or more within that unit. But as the map scale decreases (the

threshold lying between 1:30,000 and 1:75,000), or even at larger scales where a complex mosaic of classes occurs in a small area, *compound* mapping units must be used. The legend to the map lists the classes in each unit, their proportions and if possible their pattern of occurrence. At the smallest scales ($< 1:500,000$) complex units appear as *soil associations*, in which the classes are recognized and located first and then grouped for ease of presentation, or as *land systems*, which are first distinguished by their surface expression (e.g. by API) and the included classes identified by subsequent sampling. The 'nested' arrangement of mapping units according to map scale is illustrated in Figure 14.6.

In addition to the legend, most soil maps provide a *memoir*, which contains a detailed description of each soil series, with general comments about the geology and vegetation of the area and the management of soils grouped by mapping units. If the original data from the survey are recorded in a computer-compatible format, a data store can be created which is a valuable supplement to the memoir. In response to specific requests by users, selected data can be retrieved from the store and computer-printed maps produced.

LAND EVALUATION

Some workers argue that soil survey should go further than just classification and mapping to provide estimates of crop yields for identified soils at specified management levels (e.g. traditional, improved and advanced) and estimates of the productivity of alternative systems of land use (as net return per ha or *per capita*). This requires the collaboration of soil surveyors, agronomists and economists in analysing the results of field trials on experimental stations and records from farms on known soil types. Except for very intensive agriculture, provision of estimates on a series-by-series basis is unrealistic; for most purposes soil series can be grouped into *soil management units*, which are defined by criteria that can be pedological (e.g. parent material) or agricultural (e.g. average slope, soil depth). Soil survey and classification are only steps in the assessment of *land use capability*,

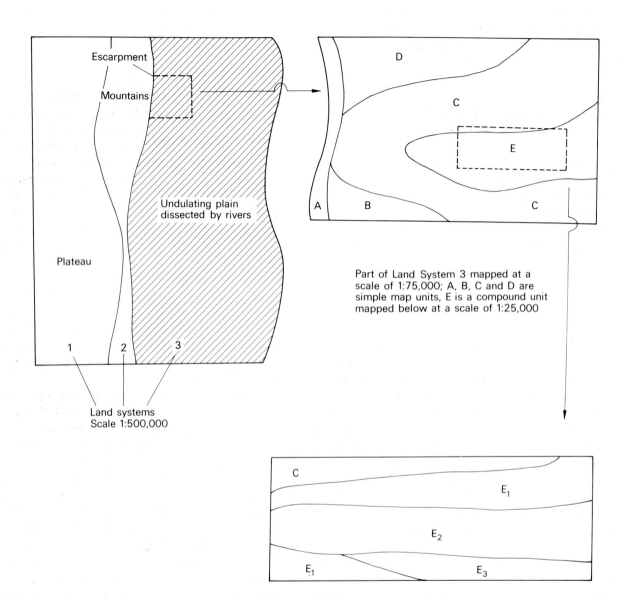

Part of Land System 3 mapped at a
scale of 1:75,000; A, B, C and D are
simple map units, E is a compound unit
mapped below at a scale of 1:25,000

Figure 14.6. 'Nested' arrangement of soil mapping units.

which must also take note of the effects of climate, altitude and aspect, slope and regional drainage on the agricultural productivity of land. This topic is discussed by McRae and Burnham (1981).

14.3 AN OVERVIEW OF SOIL CLASSIFICATION

Diverse approaches to classification

When carried out as an adjunct to soil survey, classification provides the framework for making local generalizations about soil, entailing the recognition of soil series. The upwards expansion of a classification to allow progressively vaguer generalizations about the soil at higher levels or *categories* has preoccupied many individual soil scientists and corporate bodies concerned with soil survey, resulting in national or even global classifications of a hierarchical structure*. Although such classifications are normally developed from the bottom upwards by the aggregation of small homogeneous classes to form more inclusive groups, identification of an unknown soil always proceeds stepwise from the highest category downwards.

Opinions differ widely on the criteria to be used in establishing the different categories and for separating classes within categories. Traditionally, the soil taxonomist may choose definitive properties on the basis of:

(a) the inferred *genesis* of the soil, on the assumption that the soil is in equilibrium with its environment and reflects the influence of prevailing soil-forming factors; or

(b) the soil's *morphology*, being the outward expression of the processes which shaped the soil.

The success of either approach in producing a general-purpose classification depends on the taxonomist's skill and experience in choosing definitive properties that are correlated with a large number of other soil properties. The non-traditionalist may prefer a *numerical method* in which soil profiles are objec-

*A structure in which individuals are collected in small groups, the small groups belonging to larger groups, and so on.

tively sorted into classes by similarity-grouping techniques (Webster, 1977). The profiles are characterized in terms of those soil properties that are observable or measurable, that is, no inferred genetic processes. Some properties, such as water content at the time of sampling, are not used because they are transient and influenced by external factors not germane to the objective of the classification. They may need to be specified, however, because of their effect on other definitive properties, such as the effect of water content on soil colour and consistency. Generally, each property is given equal weighting when soil profiles are compared, although some taxonomists consider that the value of a property in a surface horizon or layer should be given greater weight than that property's value in the subsoil. The classes so created contain maximum information but frequently do not correspond to any natural soil groupings in the field and so they are difficult to map. Apart from the problem of mapping, numerical classifications are unstable in the sense that the inclusion of newly described profiles in the similarity-grouping process may completely alter the classification.

In the USA, where the most effort has been devoted to detailed soil survey (> 10,000 series defined), from the 1950s onwards a new classification has been evolved which, although broadly 'traditionalist', places the primary emphasis on soil properties that can be observed or measured and preferably those which are *quantitative*, such as percentage clay, rather than *qualitative*, such as colour or grade of structure. It aspires, as far as possible, to take no cognizance of the imperfectly understood relationships between pedogenesis and the soil-forming factors in establishing classes. Somewhat paradoxically, however, one of its underlying premises is that 'the properties selected should be those that either affect soil genesis or result from soil genesis' (Soil Survey Staff, 1960). In order to make the new classification as comprehensive as possible, there was extensive consultation with pedologists in other countries as the classification advanced through several 'approximations', culminating in the publication of *Soil Taxonomy* in 1975. The outcome of this and other

Table 14.1. Structures of a selection of national and supra-national soil classifications.

Classification	Categories (the number of actual or potential classes in each category is shown in parentheses)
USDA Comprehensive System, Soil Taxonomy (Soil Survey Staff, 1975)	Orders (10) → Suborders (47) → Great Groups (185) → Subgroups (970) → Families (4,500) → Series (> 10,000)
World Soil Classification (FAO–UNESCO, 1974)	'World Classes' (26) → Soil Units (106)
CCTA Soil Map of Africa Classification (D'Hoore, 1964)	Main Groups (16) → Subgroups (25) → Soil Types (63)
Soil Classification for England and Wales (Avery, 1973)	Major Groups (10) → Groups (43) → Subgroups (109) → Series
Russian Soil Classification (Gerasimov and Glazovskaya, 1960)	Genetic Types → Subtypes → Species → Subspecies
Factual Key for Australian Soils (Northcote, 1960–1971)	Divisions (3) → Subdivisions (11) → Sections (54) → Classes (271) → Principal Profile Forms (855)
Canadian System (Canada Department of Agriculture, 1974)	Orders (8) → Great Groups (23) → Subgroups (165) → Families (*c.* 900) → Series (*c.* 3,000)
Classification of the Australian Handbook of Soils (Stace *et al.*, 1968)	Great Soil Groups (43) → (Families) → Series
Binomial System for South Africa (MacVicar *et al.*, 1977)	Forms (41) → Series (504)

(After Butler, 1980.)

attempts at national or supra-national classification of soil are summarized in Table 14.1.

CONCEPTUAL CLASSES

No matter how divergent current classifications are in principle, language and format, two main levels of generalization stand out: a lower level encompassing the surveyor's series, which are narrow classes in harmony with the landscape, and a higher level of broad classes with somewhat overlapping boundaries, each of which embodies a central, usually pedogenetic, concept. The latter *conceptual classes* are valuable in teaching soil science because they each convey a measure of knowledge about the soil, distilled from the experience of scientists in many different lands. For this reason, the USDA classification of Baldwin, Kellogg and Thorp (1938), revised by Thorpe and Smith (1949), was presented in Table 5.3: the suborder descriptions and Great Soil Group names denote the influence of one or more of the soil-forming factors on pedogenesis. The same can be said of *Soil Taxonomy* at the highest categorical level—the *order*. With the exception of the Entisols, which have no distinctive horizons, and the Histosols (organic soils), each order has a broad climatic connotation. A comparison between the ten new orders of *Soil Taxonomy* and the Great Soil Groups of the classification of Baldwin *et al.*, is made in Table 14.2.

Below the order level, however, the classes in *Soil Taxonomy* are defined progressively more narrowly according to the measured soil properties and the microclimate in which the soil forms. For example, the *suborders* are separated on the basis of differentiae associated with hydromorphism, vegetation (as it influences podzolization, lessivage), parent material, time and intensity of weathering (producing ferrallitization) and so on.

Separation of classes in the next inferior category—the *Great Group*—is intended to divide the soil continuum into more-or-less equally spaced groupings according to the variation in soil genesis. Much use is made of the concept of *diagnostic horizons*, which are

Table 14.2. Soil orders of *Soil Taxonomy* and some of the included Great Soil Groups from the Baldwin *et al.* classification.

Order	Brief description	Included Great Soil Groups
Entisols	Recently formed soils, no diagnostic horizons	Azonal soils and some Low Humic Gley soils
Vertisols	With swell-shrink clays, high base status	Grumusols
Inceptisols	Slightly developed soils without contrasting horizons	Acid brown soils, some Brown Forest soils. Low Humic Gley and Humic Gley soils
Aridosols	Soils of arid regions	Sierozem, Solonchak, some Brown and Reddish Brown soils and Solonetz
Mollisols	Soils with mull humus	Chestnut, Chernozem, Prairie, Rendzinas, some Brown Forest, Solonetz and Humic Gley soils
Spodosols	Soils with iron and humus B horizons	Podzols, Brown Podzolic soils and Groundwater Podzols
Alfisols	Soils with a clay B horizon and > 35% base saturation	Gray-brown Podzolic, Gray Wooded soils, Degraded Chernozem and associated Planosols
Ultisols	Soils with a clay B horizon and < 35% base saturation	Red-yellow Podzolic, Reddish-brown Lateritic soils and associated Planosols
Oxisols	Sesquioxide-rich, highly weathered soils	Lateritic soils, Latosols
Histosols	Organic hydromorphic soils	Bog soils (Peats)

(After Butler, 1980; Duchaufour, 1982; and Soil Survey Staff, 1960.)

connotative of an underlying pedogenic process (just as the symbol Ea connotes a bleached sandy horizon from which sesquioxides have been eluviated). However, a diagnostic horizon need not coincide with a pedological horizon, such as Ah, Ea or Bt: it may include parts of more than one pedological horizon or soil layer. For example, a *mollic epipedon*—one of the six diagnostic horizons recognized at the surface—may include all of the Ah horizon and the upper part of the Bt horizon, if organic matter has been uniformly incorporated by faunal activity to a considerable depth. More weight in the classification is given to *subsurface* horizons, of which there are ten, because they are less prone to disturbance through ploughing or truncation through erosion. Thus, the *Great Groups* are separated on the presence or absence of diagnostic horizons and their arrangement in the profile.

Towards a common perception of classification

Although *Soil Taxonomy* has now been widely deployed outside the USA, it has been criticized on conceptual grounds (Ragg and Clayden, 1973) and, more importantly, proficiency in the use of the ponderous and unfamiliar terminology at any level below order and suborder is notoriously difficult to achieve for the non-taxonomist. For the latter reason it is not of great value for teaching general soil science, but it has inspired the development of a supra-national classification—the *World Soil Classification*—which conveys useful generalizations about the genesis of soils in relation to the interactive effects of the main soil-forming factors. This classification arose as the list of classes in the legend to the FAO–Unesco Soil Map of the World (1974). The classification is a mono-categorical one comprising 106 classes or 'soil units'. Like *Soil Taxonomy*, it makes class separations on the basis of *diagnostic horizons*, which are connotative of the underlying pedogenesis. Some of the descriptive terminology, in simplified form, has been adopted from *Soil Taxonomy*, but many of the traditional Great Soil Group names have been retained, as well as new names coined which do not suffer in translation nor do they have different meanings in different countries. The units are mapped as soil associations, designated by the dominant soil unit, with three *textural* classes (coarse, medium and fine) and three *slope* classes superimposed (level to gently undulating, rolling to hilly, and steeply dissected to mountainous).

For the sake of logical presentation and generalization at a higher level, the soil units have been grouped 'on the basis of generally accepted principles of soil formation' to form the 26 'World Classes', which are set out in Table 14.3. Although formulated on inferred pedogenic processes, such as gleying, salinization and lessivage which have a bearing on soil use,

Table 14.3. Higher classes from the World Soil Map with some included classes from national classifications.

Fluvisols	(L. *fluvius*, river Fr. *sol*, soil) connotative of flood plains and alluvial deposits; incl. Alluvial soils (Aust.), Sols tropicaux récents (Fr.), Fluvents (USA).
Gleysols	(Russian word for mucky soil mass) connotative of excess water; incl. Haplaquepts, Humaquepts (USA).
Regosols	(Gr. *rhegos*, blanket) implying a loose mantle overlying a hard rock core; incl. Skeletal soils (Aust.), Orthents (USA).
Lithosols	(Gr. *lithos*, stone) connotative of very shallow soils over rock.
Arenosols	(L. *arena*, sand) suggesting immature coarse-textured soils; incl. Psamments (USA).
Rendzinas	(Po. *rzedzic*, noise) connotative of a plough scraping through a stony limestone soil; incl. Rendolls (USA).
Rankers	(Aus. *Rank*, steep slope) hence shallow soils but on siliceous parent material.
Andosols	(Jap. *An*, dark and *Do*, soil) material rich in volcanic glass commonly having a dark surface horizon; incl. Andepts (USA).
Vertisols	(L. *verto*, turn) connotative of churning surface soil; incl. Black Earths (Aust.).
Solonchaks	(R. *sol*, salt) connotative of salinity; incl. Salorthids (USA).
Solonetz	Salt-affected; incl. Natrustalfs, Natrixeralfs (USA).
Yermosols	(L. *eremus*, desolate) hence wide open, dry spaces; incl. Desert loams (Aust.), Typic Aridisols (USA).
Xerosols	(Gr. *xerox*, dry) desert environments; incl. Mollic Aridisols (USA).
Kastanozems	(L. *castaneo*, chestnut; R. *zemlja*, earth) soils rich in organic matter and brown or chestnut in colour; incl. Ustolls (USA).
Chernozems	(R. *chern*, black) soils rich in organic matter and black in colour; incl. Haploborolls, Vermiborolls (USA).
Phaeozems	(Gr. *phaios*, dusky) soils rich in organic matter with a dark A horizon; incl. Hapludolls (USA), Degraded Chernozems (R.).
Greyzems	(English *grey*, the colour) soils rich in organic matter and grey in colour; incl. Argiborolls, Aquolls (USA).
Cambisols	(L. *cambiare*, to change) suggesting changes in colour, structure and consistence due to weathering; incl. Chocolate soils (Aust.), Eutrochrepts, Ustochrepts (USA).
Luvisols	(L. *luvi*, washed) connotative of lessivage; incl. many podzolic soils, Red Brown Earths (Aust.), Hapludalfs, Haploxeralfs (USA).
Podzoluvisols	Intermediate in development between *Luvisols* and *Podzols*; incl. Glossudalfs (USA).
Podzols	(R. *pod*, under; *zola*, ash) connotative of soils with strongly bleached subsurface horizons; incl. Orthods, Ferrods (USA).
Planosols	(L. *planus*, flat) related to the flat or depressed topography with poor drainage in which these soils form; incl. Soloths, Solods (Aust., R.), Albaqualfs (USA).
Acrisols	(L. *acris*, very acid) hence low base saturation; incl. Hapludults (USA); Red-yellow podzolics (Aust.).
Nitosols	(L. *nitidus*, shiny) connotative of bright shiny ped faces; incl. Krasnozems (Aust.), Tropudalfs and Paleudalfs (USA).
Ferralsols	(L. *ferrum*, iron and *aluminium*) soils high in sesquioxides; incl. Sols ferralitques (Fr.), Oxisols (USA).
Histosols	(Gr. *histos*, tissue) soils rich in fresh or partly decomposed organic matter; incl. Moor peats (Aust.), Sols hydromorphes organiques (Fr.), Histosols (USA).

Abbreviations used: L. Latin, Gr. Greek, Po. Polish, Aus. Austrian, Jap. Japanese, R. Russian, Fr. French, Aust. Australian, incl. including.

the World Classes have been compiled so broadly and the total variation within each class is so large that their value in the prediction of land use is limited. Their merit is that they constitute a gallery of conceptual classes which convey meaning to a worldwide clientele.

14.4 SUMMARY

Soil classification is necessary for the orderly presentation to soil users of the accumulated knowledge of soil behaviour. Classification involves grouping soils into classes according to the degree of similarity between individual soil profiles. Such grouping can be made using statistical techniques, but the results of such a *numerical classification* are often difficult to present in a soil map. For a classification to be useful for predictive purposes, the variation remaining *within* classes should be substantially less than the variation *between* class means for a range of important soil properties. Classification is made more difficult and the results contentious because there is no definable soil entity; the soil mantle varies continuously so that subdivision of the variation in one or more properties to create classes is usually arbitrary.

Soil survey starts with the assessment of the range of

variability and description of the soil in the field. This is followed by the creation of a *list of classes* (the classification), which allows any soil in the area to be placed in one class, and ends with the presentation of results, usually as a *soil map*. Depending on the map scale, which may vary from 1:1,000,000 for reconnaissance surveys to 1:1,000 for intensive surveys of management problems, surveying proceeds on a grid (for large scales) or free survey pattern (for small scales). Interpretation of air photographs is helpful at small scales in locating soil boundaries that coincide with landscape features.

Classes are defined in terms of one or two soil properties relevant to a specific purpose (*single-purpose classification*), or on many properties that may be relevant for various soil uses (*general-purpose classification*). *Diagnostic* or *key* properties, which are not necessarily relevant to usage but are easily assessed in the field, are often chosen to define classes on the assumption they are correlated with more important properties.

The classes recognized by the surveyor (*soil series*) are usually aggregated to form progressively less homogeneous (more inclusive) classes, which facilitate national and international exchange of soil knowledge. Because of the wide choice of criteria on which to separate classes at any one level and to establish the different levels or *categories* of generalization, many classifications have been proposed. The Comprehensive System of the USDA (*Soil Taxonomy*) is a classification purporting to use only those soil properties that can be observed or measured, preferably quantitatively, in establishing criteria for class separation. The advantage of this approach for taxonomic purposes, particularly at the lower, more exclusive categorical levels, is that it avoids reliance on the imperfectly understood relationships between pedogenesis and the soil-forming factors. However, non-taxonomists tend to seek broader generalizations about soil differences, which are embodied in the higher level *conceptual classes*. For this purpose *Soil Taxonomy's* orders and *suborders* may be used, or the *World Soil Classification*, which comprises 106 soil

units grouped into 26 World Classes on the basis of generally accepted concepts of soil formation.

REFERENCES

BALDWIN H., KELLOGG C.W. & THORP J. (1938) Soil classification in *Soils and Man*. Yearbook of Agriculture, Washington DC.
BECKETT P.H.T. (1976) Soil survey. *Agricultural Progress* **51**, 33–49.
BUTLER B.E. (1980) *Soil Classification for Soil Survey*. Clarendon Press, Oxford.
DUCHAUFOUR P. (1982) *Pedology, Pedogenesis and Classification*. (Translated by T.R. Paton). Allen & Unwin, London.
FAO–UNESCO (1974) *Soil map of the World 1:5,000,000* Volume 1. Legend, Paris.
HODGSON J.M. (1978) *Soil Sampling and Soil Description*. Clarendon Press, Oxford.
MCRAE S.G. & BURNHAM P. (1981) *Land Evaluation*. Clarendon Press, Oxford.
RAGG J.M. & CLAYDEN B. (1973) *The classification of some British soils according to the Comprehensive System of the United States*. Soil Survey of England and Wales, Technical Monograph No. 3. Harpenden.
SOIL SURVEY STAFF (1960) *Soil classification, a comprehensive system, 7th approximation*. Washington, D.C.
SOIL SURVEY STAFF (1975) *Soil Taxonomy. A Basic System of Soil Classification for Making and Interpreting Soil Surveys*. United States Department of Agriculture Handbook No. 436, Washington DC.
THORP J. & SMITH G.D. (1949) Higher categories of soil classification. *Soil Science* **67**, 117–126.
WEBSTER R. (1977) *Quantitative and Numerical Methods in Soil Classification and Survey*. Oxford University Press, Oxford.
YOUNG A. (1973) Soil survey procedures in land development planning. *Geographical Journal* **139**, 53–64.

FURTHER READING

AVERY B.W. (1973) Soil classification in the Soil Survey of England and Wales. *Journal of Soil Science* **24**, 324–338.
BECKETT P.H.T. & WEBSTER R. (1971) Soil variability—a review. *Soils and Fertilizers* **34**, 1–15.
MULCAHY M.J. & HUMPHRIES A.W. (1967) Soil classification, soil surveys and land use. *Soils and Fertilizers* **30**, 1–8.
NORTHCOTE K.H. (1971) *A Factual Key to the Recognition of Australian Soils*. 3rd Ed. Rellim, South Australia.
VAN DER EYK J.J., MACVICAR C.N. & DE VILLIERS J.M. (1969) *Soils of the Tugela Basin—a study in subtropical Africa*. Natal Town and Regional Planning Reports, Volume 15. Pietermaritzburg.
YOUNG A. (1976) *Tropical Soils and Soil Survey*. Cambridge University Press, Cambridge.

Index

Numbers in *italic* refer to figures; numbers in **bold** refer to tables